OCEAN CIRCULATION

THE OCEANOGRAPHY COURSE TEAM

Authors
Joan Brown
Angela Colling
Dave Park
John Phillips
Dave Rothery
John Wright

Editor
Gerry Bearman

Design and Illustration
Sue Dobson
Ray Munns
Ros Porter
Jane Sheppard

This Volume forms part of an Open University course. For general availability of all the Volumes in the Oceanography Series, please contact your regular supplier, or in case of difficulty the appropriate Pergamon office.

Further information on Open University courses may be obtained from: The Admissions Office, The Open University, P.O. Box 48, Walton Hall, Milton Keynes, MK7 6AA.

Cover illustration: Satellite photograph showing distribution of phytoplankton pigments in the North Atlantic off the US coast in the region of the Gulf Stream and the Labrador Current. *(NASA and O. Brown and R. Evans, University of Miami.)*

OCEAN CIRCULATION

PREPARED BY AN OPEN UNIVERSITY COURSE TEAM

PERGAMON PRESS
OXFORD · NEW YORK · BEIJING · FRANKFURT · SÃO PAULO · SYDNEY
TOKYO · TORONTO

in association with

THE OPEN UNIVERSITY
WALTON HALL, MILTON KEYNES, MK7 6AA, ENGLAND

U.K.	Pergamon Press plc, Headington Hill Hall, Oxford OX3 0BW, England
U.S.A.	Pergamon Press, Inc., Maxwell House, Fairview Park, Elmsford, New York 10523, U.S.A.
PEOPLE'S REPUBLIC OF CHINA	Pergamon Press, Room 4037, Qianmen Hotel, Beijing, People's Republic of China
FEDERAL REPUBLIC OF GERMANY	Pergamon Press GmbH, Hammersweg 6, D-6242 Kronberg, Federal Republic of Germany
BRAZIL	Pergamon Editora Ltda, Rua Eça de Queiros, 346, CEP 04011, Paraiso, São Paulo, Brazil
AUSTRALIA	Pergamon Press Australia Pty Ltd., P.O. Box 544, Potts Point, N.S.W. 2011, Australia
JAPAN	Pergamon Press, 5th Floor, Matsuoka Central Building, 1-7-1 Nishishinjuku, Shinjuku-ku, Tokyo 160, Japan
CANADA	Pergamon Press Canada Ltd., Suite No. 271, 253 College Street, Toronto, Ontario, Canada M5T 1R5

Copyright © 1989 The Open University.

All Rights Reserved. No part of this publication may be reproduced, stored in a retrieval system or transmitted in any form or by any means: electronic, electrostatic, magnetic tape, mechanical, photocopying, recording or otherwise, without written permission from the copyright holders.

First edition 1989

Library of Congress Cataloging in Publication Data
Ocean circulation.
1. Ocean circulation. I. Open University. Oceanography Course Team.
GC228.5.025   1989   551.47—dc 19   89-3819

British Library Cataloguing in Publication Data
Ocean circulation
1. Oceans. Currents
I. Open University. *Oceanography Course Team*
551.47'01
ISBN 0-08-036370-9 Hardcover
ISBN 0-08-036369-5 Flexicover

Jointly published by the Open University, Walton Hall, Milton Keynes, MK7 6AA and Pergamon Press plc, Headington Hill Hall, Oxford OX3 0BW.

Designed by the Graphic Design Group of The Open University.

Printed in Great Britain by BPCC Wheatons Ltd., Exeter

CONTENTS

ABOUT THIS VOLUME — 4

CHAPTER 1 — INTRODUCTION

1.1 THE RADIATION BALANCE OF THE EARTH–ATMOSPHERE SYSTEM — 9
1.2 SUMMARY OF CHAPTER 1 — 11

CHAPTER 2 — THE ATMOSPHERE AND THE OCEAN

2.1 THE GLOBAL WIND SYSTEM — 13
2.2 POLEWARD TRANSPORT OF HEAT BY THE ATMOSPHERE — 17
   2.2.1 Large-scale circulation in mid-latitudes — 17
   2.2.2 Vertical convection in the atmosphere — 21
2.3 ATMOSPHERE–OCEAN INTERACTION — 24
   2.3.1 Easterly waves and tropical cyclones — 24
2.4 SUMMARY OF CHAPTER 2 — 29

CHAPTER 3 — OCEAN CURRENTS

3.1 THE ACTION OF WIND ON SURFACE WATERS — 33
   3.1.1 Frictional coupling within the ocean — 33
   3.1.2 Ekman motion — 36
3.2 INERTIA CURRENTS — 38
3.3 GEOSTROPHIC CURRENTS — 40
   3.3.1 Pressure gradients in the ocean — 40
   3.3.2 Barotropic and baroclinic conditions — 43
   3.3.3 Determination of geostrophic current velocities — 48
   3.3.4 Pressure, density and dynamic topography — 54
3.4 DIVERGENCES AND CONVERGENCES — 59
3.5 THE ENERGY OF THE OCEAN: SCALES OF MOTION — 63
   3.5.1 Kinetic energy spectra — 63
   3.5.2 Mesoscale eddies — 65
3.6 SUMMARY OF CHAPTER 3 — 69

CHAPTER 4 — THE NORTH ATLANTIC GYRE: OBSERVATIONS AND THEORIES

4.1 THE GULF STREAM — 73
   4.1.1 Early observations and theories — 73
4.2 THE SUBTROPICAL GYRES — 79
   4.2.1 Vorticity — 79
   4.2.2 Why is there a Gulf Stream? — 84
   4.2.3 The equations of motion — 93

4.3	**MODERN OBSERVATIONS OF THE NORTH ATLANTIC GYRE**	96
4.3.1	The Gulf Stream system	98
4.3.2	Geostrophic flow in the Gulf Stream	98
4.3.3	Direct current measurements	101
4.3.4	Mapping the Gulf Stream using water characteristics	106
4.3.5	Gulf Stream 'rings'	108
4.3.6	Other methods of current measurement	112
4.4	**COASTAL UPWELLING IN EASTERN BOUNDARY CURRENTS**	115
4.5	**SUMMARY OF CHAPTER 4**	119

CHAPTER 5 — OTHER MAJOR CURRENT SYSTEMS

5.1	**EQUATORIAL CURRENT SYSTEMS**	122
5.1.1	The Equatorial Undercurrent	125
5.1.2	Upwelling in low latitudes	131
5.2	**MONSOONAL CIRCULATION**	134
5.2.1	Monsoon winds over the Indian Ocean	134
5.2.2	The current system of the Indian Ocean	135
5.3	**THE ROLE OF LONG WAVES IN OCEAN CIRCULATION**	138
5.3.1	Oceanic wave guides and Kelvin waves	140
5.3.2	Rossby waves	143
5.4	**EL NIÑO**	145
5.5	**CIRCULATION IN HIGH LATITUDES**	148
5.5.1	The Arctic Sea	148
5.5.2	The Southern Ocean	151
5.6	**SUMMARY OF CHAPTER 5**	155

CHAPTER 6 — GLOBAL FLUXES AND THE DEEP CIRCULATION

6.1	**THE OCEANIC HEAT BUDGET**	159
6.1.1	Solar radiation	159
6.1.2	The heat-budget equation	163
6.2	**CONSERVATION OF SALT**	168
6.2.1	Practical application of the principles of conservation and continuity	169
6.3	**OCEANIC WATER MASSES**	172
6.3.1	Upper and intermediate water masses	174
6.3.2	Deep and bottom water masses	178
6.4	**OCEANIC MIXING AND TEMPERATURE–SALINITY DIAGRAMS**	185
6.4.1	Mixing in the oceans	185
6.4.2	Temperature–salinity diagrams	187
6.5	**TRACERS**	195
6.6	**GLOBAL FLUXES OF HEAT AND FRESHWATER**	199
6.6.1	Postscript: Why is no Deep Water formed in the Pacific?	203
6.7	**SUMMARY OF CHAPTER 6**	204

SUGGESTED FURTHER READING	209
ANSWERS AND COMMENTS TO QUESTIONS	210
ACKNOWLEDGEMENTS	233
INDEX	235

ABOUT THIS VOLUME

This is one of a Series of Volumes on Oceanography. It is designed so that it can be read on its own, like any other textbook, or studied as part of S330 *Oceanography*, a third level course for Open University students. The science of oceanography as a whole is multidisciplinary. However, different aspects fall naturally within the scope of one or other of the major 'traditional' disciplines. Thus, you will get the most out of this Volume if you have some previous experience of studying physics. Other Volumes in this Series lie more within the fields of geology, chemistry or biology (and their associated sub-branches).

Chapter 1 establishes the essential causes of the circulation patterns of air and water in the atmosphere and the oceans. These are first, the need for heat to be redistributed over the surface of the Earth, and secondly that the Earth over which air and water move is both spherical and rotating.

Chapter 2 describes the global wind system, and shows how the atmospheric circulation transports heat from the Equator to the poles. It emphasizes that the oceans and atmosphere are continuously interacting, and illustrates this through the spectacular example of tropical cyclones.

Chapter 3 explains the basic principles that determine the pattern of the wind-driven circulation. The theoretical basis is established with the minimum use of mathematical equations, which are explained and applied in context to enable you to understand their relevance. The Chapter ends with a discussion of meso-scale eddies—the 'weather' of the ocean. These eddies, which were only discovered in the 1970s, are believed to contain almost all the energy of the ocean circulation.

Chapter 4 uses the example of the North Atlantic subtropical gyre—and of the Gulf Stream in particular—to explore how ideas about ocean circulation have evolved from the sixteenth century up to the present day. Some of the theoretical models devised by oceanographers to help them understand why subtropical gyres have the form they do are described, with a minimum amount of mathematics. The equations of motion, that form the basis of such models, are introduced in a qualitative way. Practical ways of studying the ocean circulation are also discussed, from the deployment of different types of current meters to the use of satellite images like the one on the cover of this Volume.

Chapter 5 considers the major current systems outside the subtropical gyres, in both low and high latitudes. The characteristics of equatorial current systems are described and explained in terms of the direct and indirect effects of the wind—the prevailing Trade Winds in the Atlantic and Pacific, and the seasonally reversing monsoonal winds in the Indian Ocean. The importance of the Equator as a wave guide which 'channels' disturbances of the upper ocean is also considered briefly. The Chapter ends with a discussion of the Antarctic Circumpolar Current which flows eastwards around the globe, carrying up to 140 million cubic metres of water per second.

The last Chapter takes up the theme of the first, and considers the transport of heat and water through the ocean–atmosphere system. It begins with a discussion of processes occurring at the air–sea interface (because it is here that water masses acquire their characteristics) and ends with a fascinating example of how atmospheric conditions may influence the formation of deep water masses.

Finally, you will find questions designed to help you to develop arguments and/or test your own understanding as you read, with answers provided at the back of this Volume. Important technical terms are printed in **bold** type where they are first introduced or defined.

CHAPTER 1   INTRODUCTION

The waters of the ocean are continually moving. This motion ranges from powerful currents like the Gulf Stream, down to small swirls and eddies (Figure 1.1)

Figure 1.1 These spiral eddies in the central Mediterranean Sea were photographed from the Space Shuttle *Challenger*. Measuring some 12–15 km across, they are only one example of the wide range of gyral motions occurring in the oceans.

What drives all this motion?

The short answer is: energy from the Sun, and the rotation of the Earth.

The most obvious way in which the Sun drives the oceanic circulation is through the circulation of the atmosphere—that is, winds. Energy is transferred from winds to the upper layers of the ocean through frictional coupling between the ocean and the atmosphere at the sea-surface.

The Sun also drives ocean circulation by causing variations in the temperature and salinity of seawater which in turn control its *density*. Changes in temperature are caused by fluxes of heat across the air–sea boundary; changes in salinity are brought about by the addition or removal of freshwater, mainly through evaporation and precipitation, but also, in polar regions, by the freezing and melting of ice. All of these processes are linked directly or indirectly to the effect of solar radiation.

If surface water becomes denser than the underlying water, the situation is unstable and the denser surface water sinks. The vertical, density-driven circulation that results from cooling and/or increase in salinity—i.e. changes in the content of heat and/or salt—is known as **thermohaline circulation**. The large-scale thermohaline circulation of the ocean will be discussed in Chapter 6.

How does the rotation of the Earth contribute to ocean circulation patterns?

Except for a relatively thin layer close to the solid Earth, frictional coupling between moving water and the Earth is weak, and the same is true for air masses. In the extreme case of a projectile moving above the surface of the Earth, the frictional coupling is effectively zero. Consider, for instance, a missile fired northwards from a rocket launcher positioned on the Equator (Figure 1.2(a)). When it leaves the launcher, the missile is moving eastwards at the same velocity as the Earth's surface as well as moving northwards at its firing velocity. As the missile travels north, the Earth is turning eastwards beneath it. Initially, because it has the same eastwards velocity as the surface of the Earth, the missile appears to travel in a straight line. However, the eastwards velocity at the surface of the Earth is greatest at the Equator and decreases towards the poles, so as the missile travels progressively northwards, the eastwards velocity of the Earth beneath it becomes less and less. As a result, *in relation to the Earth*, the missile is moving not only northwards but also *eastwards*, at a progressively greater rate (Figure 1.2(b)). This apparent deflection of objects that are moving over the surface of the Earth without being frictionally bound to it—be they missiles, parcels of water or parcels of air—is explained in terms of an apparent force known as the **Coriolis force**.

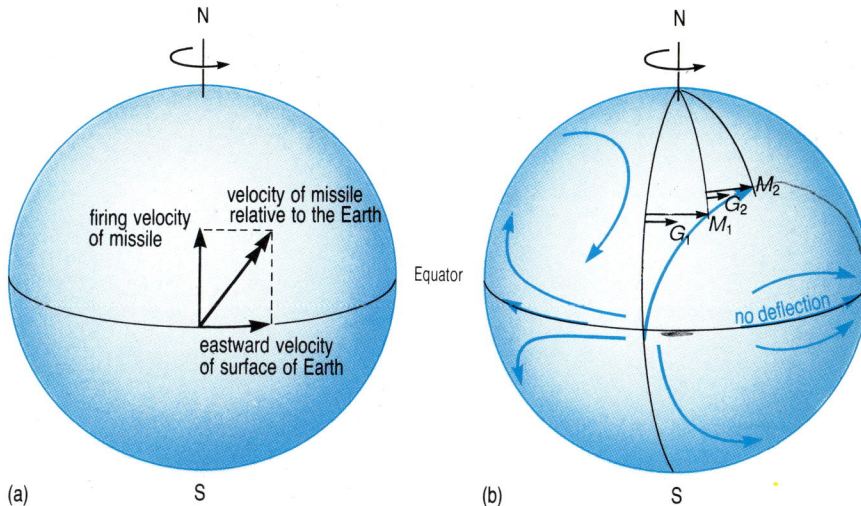

Figure 1.2 (a) A missile launched from the Equator has not only its northward firing velocity but also the same eastward velocity as the surface of the Earth at the Equator. The resultant velocity of the missile is therefore a combination of these two, as shown by the double arrow.

(b) The path taken by the missile in relation to the surface of the Earth. In time interval T_1, the missile has moved eastwards to M_1 and the Earth to G_1; in time interval T_2, the missile has moved to M_2 and the Earth to G_2. Note that the apparent deflection attributed to the Coriolis force (the difference between M_1 and G_1 and M_2 and G_2) increases with increasing latitude. The other blue curves show likely paths for missiles or any other bodies moving over the surface of the Earth without being strongly bound to it by friction.

Figure 1.3 Diagram of a hypothetical cylindrical Earth, for use with Question 1.1.

QUESTION 1.1 Bearing in mind what you have just read, especially in connection with Figure 1.2, what can you say about the Coriolis force for a hypothetical *cylindrical* Earth rotating about its axis, as shown in Figure 1.3?

The example given above, of a missile fired northwards from the Equator, was chosen because of its simplicity. In fact, the rotation of the (spherical) Earth about its axis causes deflection of currents, winds, and projectiles, *irrespective of their initial direction* (Figure 1.2(b)). The reasons for this will be discussed more fully in Chapter 4, but for now you need only be aware of the following important points:

1   The magnitude of the Coriolis force increases from zero at the Equator to a maximum at the poles.

2   The Coriolis force acts at right angles to the direction of motion, so as to cause deflection to the *right* in the Northern Hemisphere and to the *left* in the Southern Hemisphere.

How these factors affect the direction of current flow in the oceans, and of winds in the atmosphere, is illustrated by the blue curves in Figure 1.2(b).

When missile trajectories are determined, the effect of the Coriolis force is included in the calculations, but the allowance that has to be made for it is fairly small. This is because a missile travels at high speed and the amount that the Earth has 'turned beneath' it during its relatively short period of travel is small. Winds and ocean currents, on the other hand, are relatively slow moving, and so are significantly affected by the Coriolis force. Consider, for example, a current flowing with a speed of $0.5\,\mathrm{m\,s^{-1}}$ (about 1 knot) at about 45° of latitude. Water in the current will travel about 1800 metres in an hour, and during that hour the Coriolis force will have deflected it about 300m from its original path (assuming that no other forces are acting to oppose it).

Deflection by the Coriolis force is sometimes said to be *cum sole* (pronounced 'cum so-lay'), or 'with the Sun'. This is because of the direction in which the Sun appears to move across the sky—towards the right in the Northern Hemisphere and towards the left in the Southern Hemisphere.

The Coriolis force thus has the visible effect of deflecting ocean currents. It must also be considered in any study of ocean circulation for another, but less obvious, reason. Although it is not a real force in the fixed framework of space, it *is* real enough from the point of view of anything moving in relation to the Earth. This means that, like any other force, it must always be balanced by an equal and opposite force if acceleration is not to occur. For this reason, we can study, and make predictions about, currents in which horizontal forces resulting from pressure gradients are balanced by the Coriolis force. These are known as geostrophic currents and will be discussed in Chapter 3.

Because solar heating, directly and indirectly, is a fundamental cause of atmospheric and oceanic circulation, the second half of this introductory Chapter will be devoted to the radiation balance of the planet Earth.

1.1 THE RADIATION BALANCE OF THE EARTH–ATMOSPHERE SYSTEM

The solid red curve on Figure 1.4(a) shows the average daily amount of solar radiation reaching the Earth and atmosphere, as a function of latitude. The average daily amount of incoming radiation decreases from Equator to poles because low latitudes receive relatively large amounts of radiation all year, while at high latitudes the increasingly oblique angle of the Sun's rays, combined with the long periods of winter darkness, result in the average amounts of radiation received being low.

However, the Earth not only receives short-wave radiation from the Sun, it also *re*-emits radiation, of a longer wavelength. Little of this long-wave radiation is radiated directly into space; most of it is absorbed by the atmosphere, particularly by carbon dioxide, water vapour and cloud droplets. Thus, the atmosphere is heated from beneath and itself re-emits long-wave radiation into space. This generally occurs from the top of the cloud cover where temperatures are surprisingly similar at all latitudes. The intensity of the radiation emitted into space therefore does not vary greatly with latitude; this can be seen from the dashed curve on Figure 1.4(a).

The variation with latitude of the net amount of radiation energy supplied to the Earth–atmosphere system—i.e. the difference between the solid curve and the dashed curve in Figure 1.4(a)—is shown in Figure 1.4(b). There is clearly a net gain of radiation energy at low latitudes and a net loss at high latitudes.

QUESTION 1.2 (a) At what latitude does the radiation balance change from a net surplus to a net loss?

(b) The radiation budget for the Earth–atmosphere system as a whole must balance, i.e. the system cannot be continually gaining more radiation energy than it loses (or vice versa). How does Figure 1.4(b) illustrate this balance?

Despite the positive radiation balance at low latitudes, and the negative one at high latitudes, there is no evidence that low-latitude regions are steadily warming or that high-latitude regions are steadily cooling. There must therefore be a transfer of heat energy between low and high latitudes, and this is brought about by means of wind systems in the atmosphere and current systems in the ocean.

There has been much debate about the relative importance of the atmosphere and ocean in the polewards transport of heat, but it is believed that the ocean contributes more in the tropics and the atmosphere contributes more at higher latitudes (Figure 1.5).

QUESTION 1.3 (a) How does the total polewards transport of heat in the ocean compare with that in the atmosphere, according to Figure 1.5?

(b) What do you think is the significance of the negative values shown on Figure 1.5?

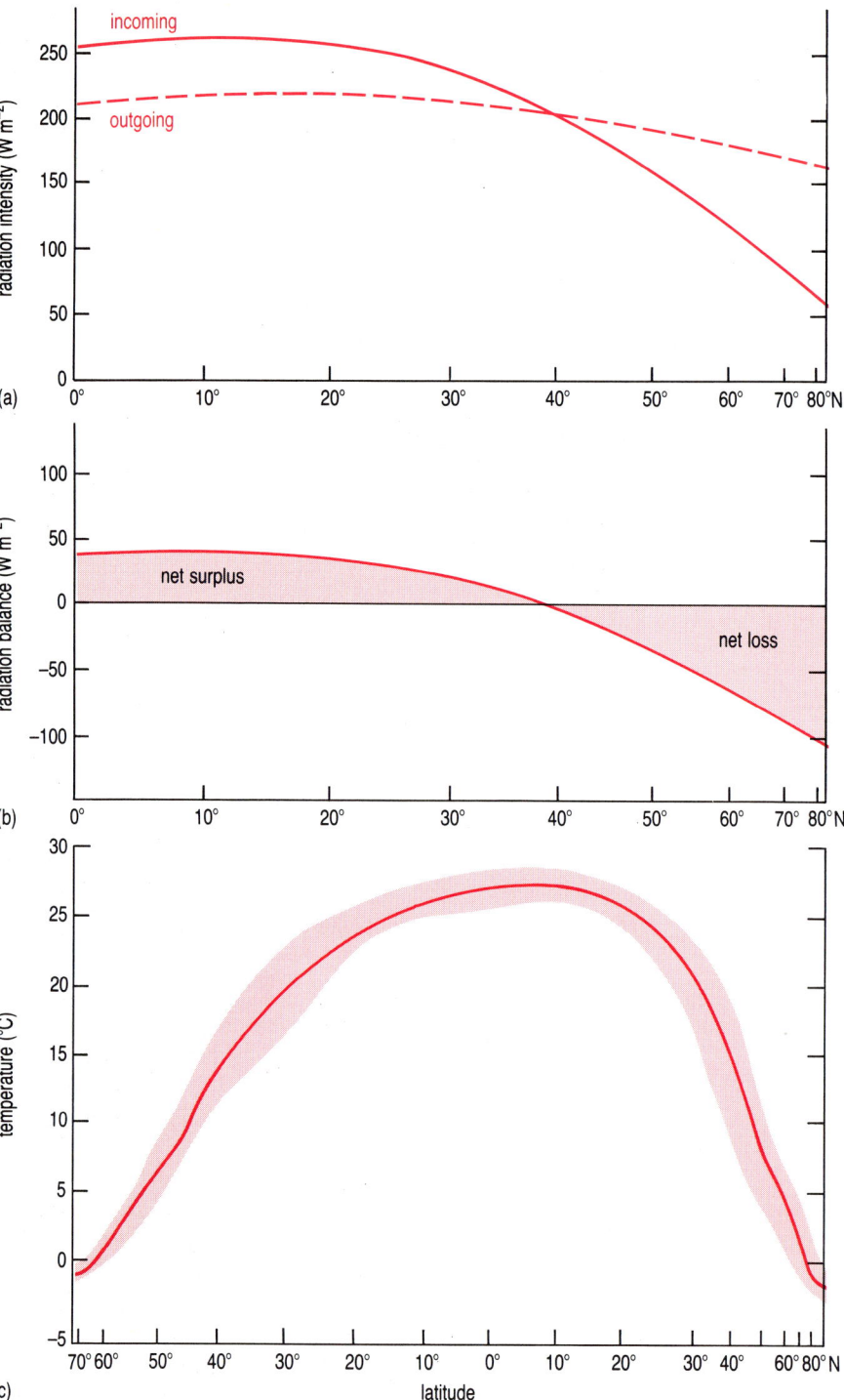

Figure 1.4 (a) Incoming solar radiation (red solid line) and the radiation lost to space (red dashed line) in relation to latitude, scaled according to the Earth's surface area. This diagram is based on data for the Northern Hemisphere only.

(b) The radiation balance for the Earth–atmosphere system (i.e. the difference between the solid and dashed curves in (a)).

(c) The mean annual temperature of surface waters at different latitudes; at a given latitude, there will be surface waters whose mean temperatures are higher or lower than shown by the curve. Mean annual ranges are represented by the thickness of the envelope. (This curve has been included for comparison with that in (a); however, as will be discussed later, sea-surface temperatures largely are determined by the *radiation balance at the sea-surface* rather than that at the top of the atmosphere.)

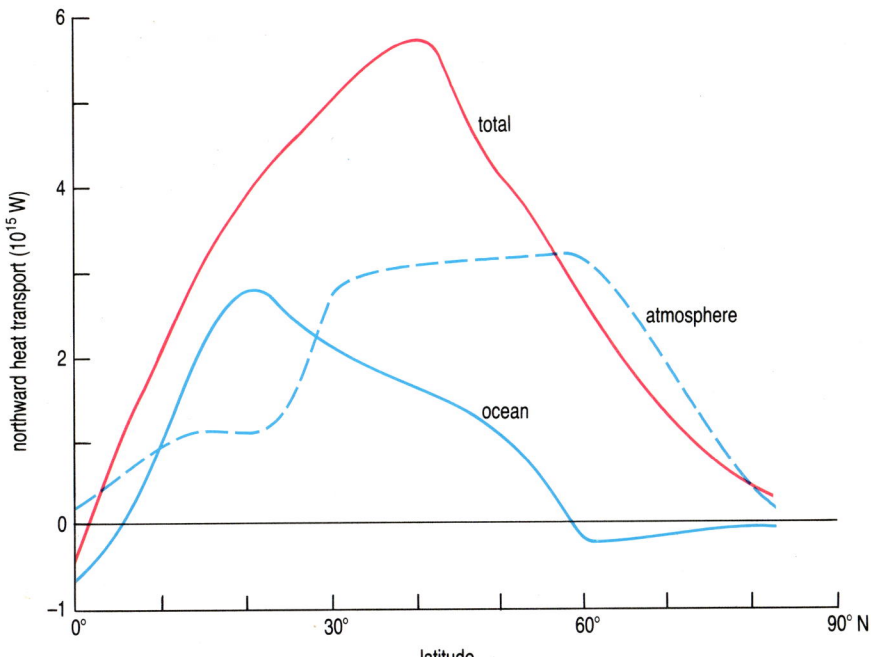

Figure 1.5 Estimates of the contributions to poleward heat transport by the ocean (solid blue curve) and the atmosphere (dashed blue curve), for the Northern Hemisphere. The curve for total heat transfer (red) is the sum of the two blue curves. The estimates were made by considering the difference between the radiation budget at the top of the atmosphere and the heat fluxes thought to be carried by the atmosphere.

While attempting Question 1.3, you may have been surprised to see that there are significant transports of heat across the Equator. The oceans are responsible for a *southward* transport of heat across the Equator; this is greater than the *northward* transport in the atmosphere, and so there is a *net* transport of heat from the Northern to the Southern Hemisphere.

Wind systems redistribute heat partly by the **advection** of warm air masses into cooler regions (and vice versa), and partly by the transfer of latent heat which is taken up when water is converted into water vapour, and released when the water vapour condenses in a cooler environment. The tropical storms known as cyclones or hurricanes are thought to transport significant amounts of heat away from the tropical oceans in the form of latent heat. The generation of cyclones and their role in transporting heat will be described in Chapter 2.

1.2 SUMMARY OF CHAPTER 1

1 Circulation in both the oceans and the atmosphere is driven by energy from the Sun and the Earth's rotation.

2 The radiation balance of the Earth–atmosphere system is positive at low latitudes and negative at high latitudes. Heat is redistributed from low to high latitudes by means of wind systems in the atmosphere and current systems in the ocean. There are two principal components of the ocean circulation: wind-driven surface currents and the density-driven (thermohaline) deep circulation.

3 Air and water masses moving over the surface of the Earth are only weakly bound to it by friction and so are subject to the Coriolis force. The Coriolis force acts at right angles to the direction of motion, so as to deflect winds and currents to the right in the Northern Hemisphere and to

the left in the Southern Hemisphere; the deflections are significant because winds and currents travel relatively slowly. The Coriolis force is zero at the Equator and increases to a maximum at the poles.

Now try the following questions to consolidate your understanding of this Chapter.

QUESTION 1.4 (a) A missile is fired southwards from the Equator. Explain what will happen to the direction of its path in relation to the Earth.

(b) In which direction would a current be deflected by the Coriolis force if it flowed initially (i) eastwards at 45° N, (ii) westwards on the Equator?

QUESTION 1.5 The *average* amount of solar radiation received at the Equator is large because of the long daylength there. True or false?

QUESTION 1.6 In Figure 1.4(a) and (b), the horizontal axis is scaled according to the surface area of the Earth in different latitude bands. Why do you think the horizontal axis of Figure 1.4(c), showing the mean annual temperature of surface waters, is more compressed at northern than southern high latitudes?

CHAPTER 2   THE ATMOSPHERE AND THE OCEAN

Anyone who has seen pictures of the Earth from space, like that in Figure 2.1, will have been struck by how much of our planet is ocean, and will have wondered about the swirling cloud patterns. In fact, the atmosphere and the ocean form one system and, if either is to be understood properly, must be considered together. What occurs in one affects the other, and the two are linked by complex feedback loops.

Figure 2.1   The Earth as seen from a geostationary satellite positioned over the Equator. The height of the satellite (about 35 800 km) is such that almost half of the Earth's surface may be seen at once. The outermost part of the image is extremely foreshortened, as can be seen from the apparent position of the British Isles. Colours have been constructed using digital image-processing to simulate natural colours.

The underlying theme of this Chapter is the redistribution of heat by, and within, the atmosphere. We first consider the large-scale circulation of the atmosphere, and then move on to consider smaller-scale phenomena that characterize the moist atmosphere over the oceans.

2.1   THE GLOBAL WIND SYSTEM

Figure 2.2(a) shows what the global wind system would be if the Earth were completely covered with water. As you will see later, the existence of large land masses significantly disturbs this theoretical pattern.

In the lower atmosphere, pressure is low along the Equator, and air converges here and rises. At about 30°N and S air descends, and there is high atmospheric pressure at the Earth's surface. There is therefore a

Figure 2.2 (a) Wind system for a hypothetical water-covered Earth, showing major winds and zones of low and high pressure. Vertical air movements and circulation cells are shown in exaggerated profile on the left of the diagram; characteristic surface conditions are given on the right of the diagram. The two north–south cells on either side of the Equator make up the Hadley circulation.

(b) Section through the atmosphere, from polar regions to the Equator, showing the general circulation and regions of tropical cloud formation. Note that the return flow in the atmosphere takes place in the upper part of the troposphere (the part of the atmosphere in which temperature decreases with distance above the Earth); the tropopause is the top of the troposphere. The Intertropical Convergence Zone is the zone along which the wind systems of the Northern and Southern Hemispheres meet. (This and other details are discussed further in the text.)

(c) The spiral circulation patterns of which the Trade Winds form the surface expression, seen from above.

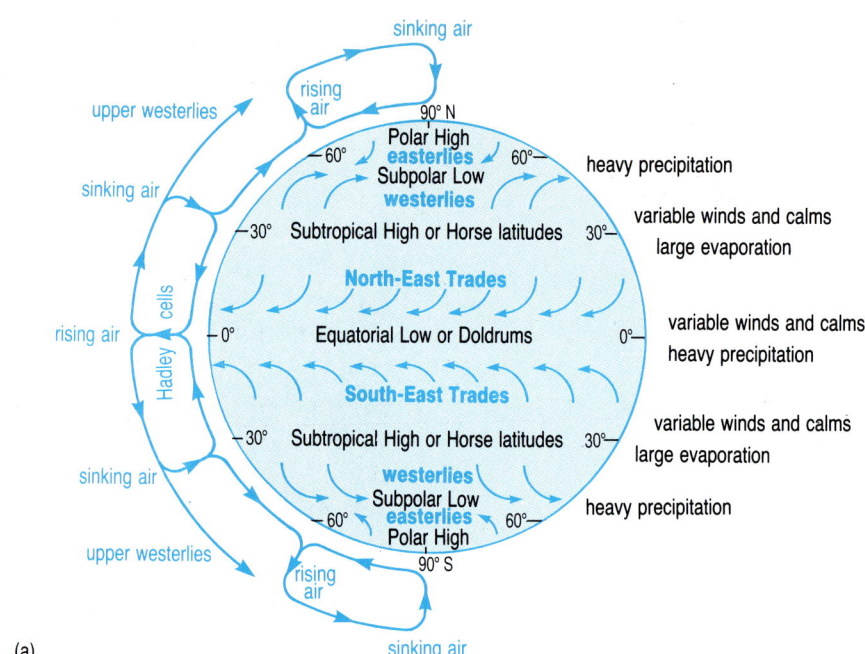

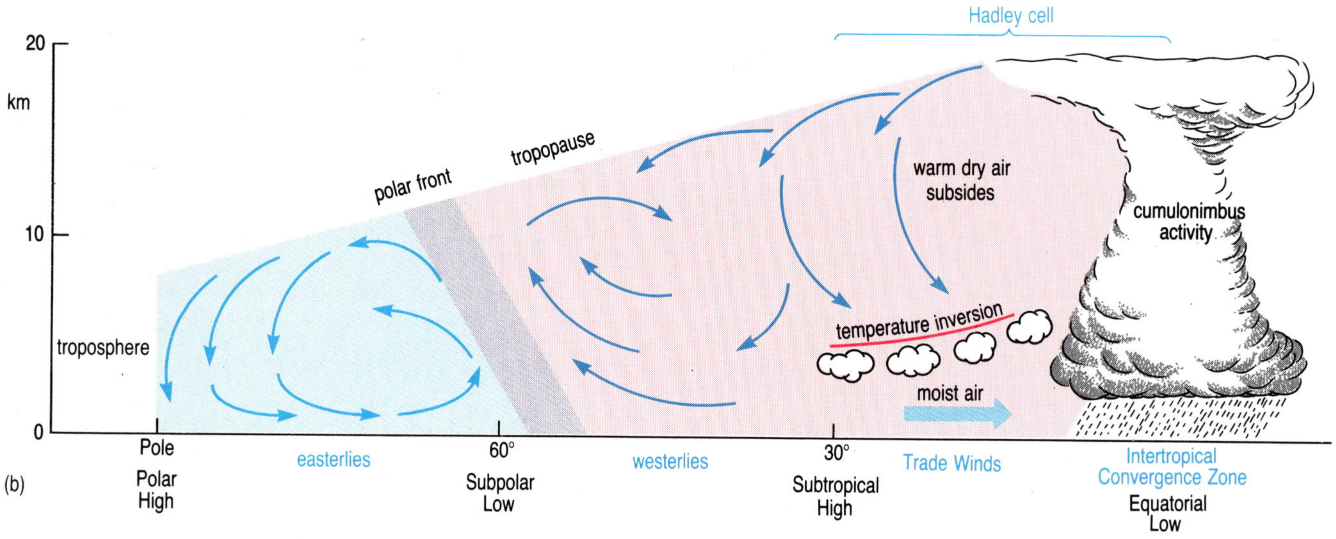

pressure gradient from the subtropical highs towards the equatorial low (Figure 2.2(a)) and, as winds blow from areas of high pressure to areas of low pressure, equatorward winds result. These are the Trade Winds.

As Figure 2.2(a) shows, the Trade Winds blow from the north-east and the south-east, and do *not* blow directly from the north and south. Why is this?

The answer, of course, is because of the Coriolis force. Away from the Equator, the Coriolis force acts to deflect winds and currents to the right in the Northern Hemisphere and to the left in the Southern Hemisphere.

Note that the Trade Winds are named the South-East and North-East Trades because they come *from* the south-east and the north-east. However, you should bear in mind that although winds are always described in terms of where they are blowing *from*, currents are described in terms of where they are flowing *towards*. Thus, a southerly current flows *towards* the south and a southerly wind blows *from* the south. To avoid confusion, we have generally used south*ward* rather than south*erly* (for example) when describing current direction.

The Trade Winds form part of the atmospheric circulation known as the Hadley circulation, or **Hadley cells**, which can be seen on Figure 2.2(a) and, in more detail, on Figure 2.2(b) and (c). Strictly speaking, the term 'Hadley cell' refers only to the north–south component of the circulation (as shown on the left-hand side of Figure 2.2(a) and in Figure 2.2(b)). Because the flow is deflected by the Coriolis force, in three dimensions the circulation follows an approximately spiral pattern, as shown schematically in Figure 2.2(c).

Figure 2.3 shows the prevailing winds at the Earth's surface in (a) July and (b) January.

How closely do the actual winds over the Earth (Figure 2.3) correspond with the hypothetical wind system shown in Figure 2.2(a)?

In general, not that closely, although the actual and hypothetical winds *are* very similar over large areas of ocean, away from the land.

If you compare Figure 2.3(a) and (b) you will see that the greatest seasonal change occurs in the region of the Eurasian land mass. During the northern winter, the direction of prevailing winds is outwards from the Eurasian land mass; by the summer, the winds have reversed and are generally blowing in towards the land mass. This is because continental masses cool down and heat up faster than the oceans (their thermal capacity is lower than that of the oceans) and so in winter they are colder than the oceans, and in summer they are warmer. Thus, in winter the air above the Eurasian land mass is cooled, becomes denser and sinks, so that a high pressure area develops. Winds blow out *from* this to regions of lower pressure. In the summer, the situation is reversed: air over the Eurasian land mass heats up and becomes less dense, so that there is a region of low pressure which winds blow *towards*. The oceanic regions most affected by these seasonal changes are the Indian Ocean and the western tropical Pacific, where the seasonally reversing winds are known as the monsoons.

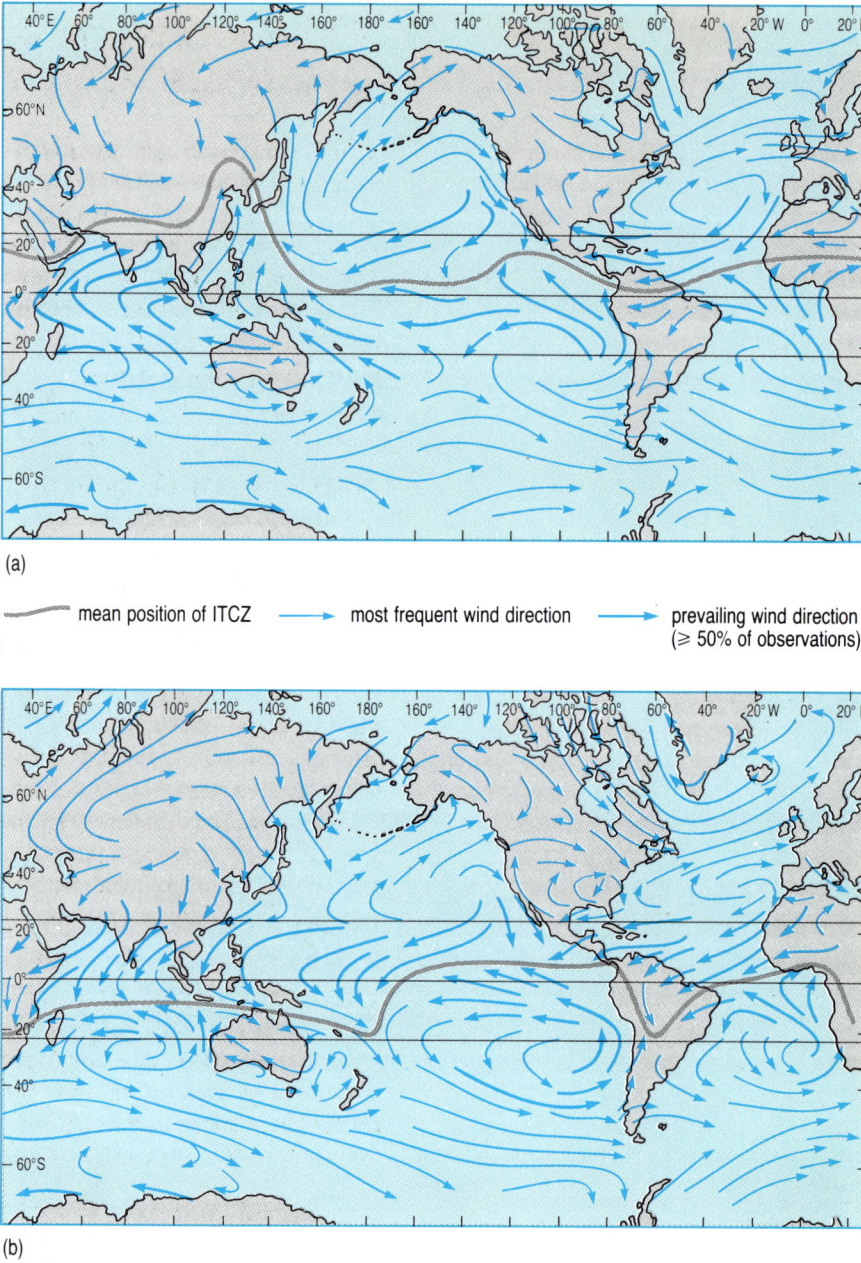

Figure 2.3 The prevailing winds at the Earth's surface, and the position of the Intertropical Convergence Zone in (a) July (northern summer/southern winter) and (b) January (southern summer/northern winter).

The distribution of ocean and continent also influences the position of the zone along which the wind systems of the two hemispheres converge. This zone of convergence—known as the **Intertropical Convergence Zone** or **ITCZ**—is generally associated with the zone of highest surface temperature. Because the continental masses heat up faster than the ocean in summer and cool faster in winter, the ITCZ tends to be distorted southwards over land in the southern summer and northwards over land in the northern summer (Figure 2.3).

2.2 POLEWARD TRANSPORT OF HEAT BY THE ATMOSPHERE

Heat is transported to polar regions by the atmosphere both directly and indirectly. If you look at Figure 2.2(a) and (b) you will see that motion in the upper troposphere is generally polewards. Air moving equatorwards over the surface of the Earth takes up heat from the oceans and continents so that when, after rising at low pressure regions such as the Equator, it moves polewards, heat is also transported polewards. Thus, any mechanism that transfers heat from the surface of the Earth to the atmosphere also contributes to the poleward transport of heat. The most spectacular example of heat transfer from ocean to atmosphere is the generation of tropical cyclones, which will be discussed in Section 2.3.1.

The Hadley cells, of which the Trade Winds are the surface expressions, may be seen as simple convection cells, in the upper limbs of which heat is transported polewards. The situation at higher latitudes, to which we now turn, is not so straightforward.

2.2.1 LARGE-SCALE CIRCULATION IN MID-LATITUDES

It may have occurred to you that whereas the polar high pressure regions and the equatorial low pressure zone (Figure 2.2 (a)) may be seen as direct results of the uneven heating of the Earth's surface, the subtropical zones of high pressure and the subpolar lows shown on Figure 2.2(a) and (b) cannot be explained in this way. However, like the Hadley cells, the low and high pressure centres characteristic of mid-latitudes are a manifestation of the need for heat to be moved polewards, to compensate for the radiation imbalance between low and high latitudes (Figure 1.4).

If, in the long term, no given latitude zone is to heat up, heat must be transported polewards at *all* latitudes. One model of the atmospheric circulation that could be proposed to achieve this would be a simple wind system in which surface winds blew from the polar high to the equatorial low, and the warmed air rose at the Equator and returned to the poles at the top of the troposphere to complete the convection cell (see Figure 2.4).

However, when we take into account the fact that the Earth is rotating, complications arise. Because of the rotation of the Earth and the resulting Coriolis force, the equatorward winds would be deflected to the right in the Northern Hemisphere and to the left in the Southern Hemisphere, and hence would acquire an easterly component in both hemispheres—as we have seen, the Trade Winds blow towards the Equator from the south-east and north-east. But the Coriolis force increases with latitude from zero at the Equator to a maximum at the poles. At low latitudes, therefore, winds are deflected relatively little and form Hadley cells, while at higher latitudes the degree of deflection is much greater and atmospheric vortices tend to form. These are the depressions and anticyclones familiar to those who live in temperate regions. Their predominantly near-horizontal, or slantwise, circulatory patterns contrast with the near-vertical circulatory cells of low latitudes.

Before proceeding further, we should briefly summarize the conventions used in describing atmospheric vortices.

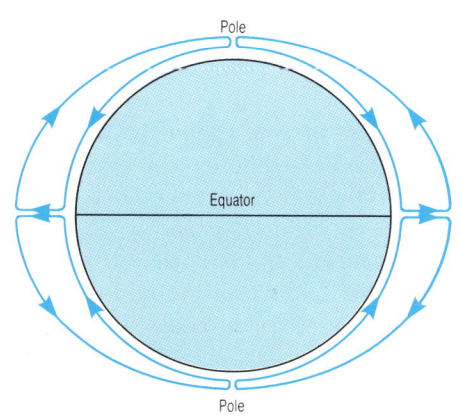

Figure 2.4 Simple atmospheric convection system for a hypothetical non-rotating Earth.

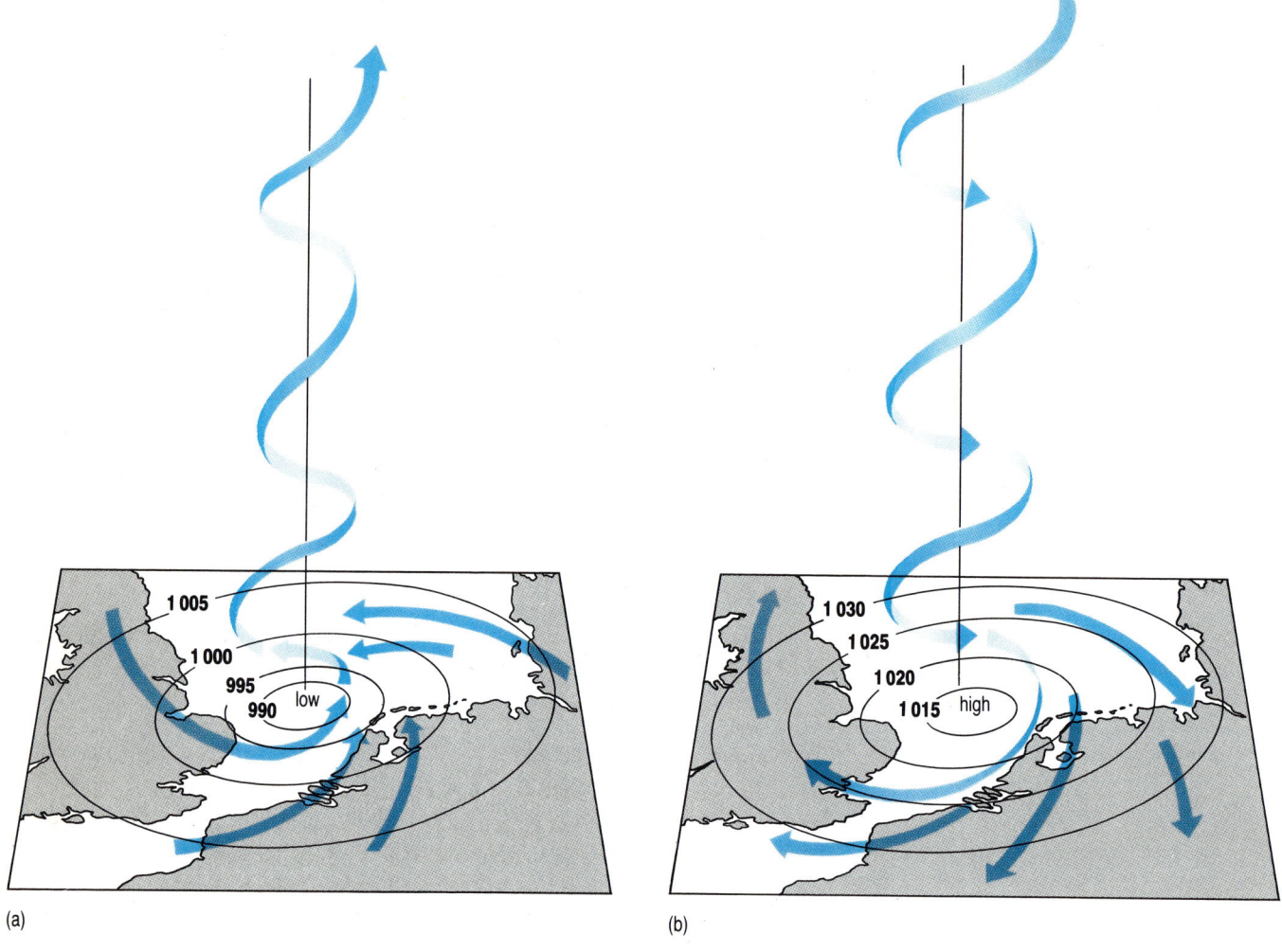

Figure 2.5 Diagrams to illustrate (a) how air spirals inwards and upwards in low pressure areas (cyclones) and (b) downwards and outwards in high pressure areas (anticyclones). These diagrams show the situation in the Northern Hemisphere. The contours are **isobars** (lines joining points of equal atmospheric pressure) and the numbers give typical pressures at ground level, in millibars.

Circulations around *low* pressure centres, whether in the Northern Hemisphere (in which case they are anticlockwise) or in the Southern Hemisphere (in which case they are clockwise) are known as **cyclones** (or lows, or depressions). The way in which air spirals in and up in cyclonic circulations is shown schematically in Figure 2.5(a).

Circulations around *high* pressure centres (clockwise in the Northern Hemisphere, anticlockwise in the Southern Hemisphere) are known as **anticyclones** (Figure 2.5(b)).

How can mid-latitude cyclones and anticyclones contribute to the poleward transport of heat?

Moving air masses mix with adjacent air masses and heat is exchanged between them. For example, air moving northwards in a Northern

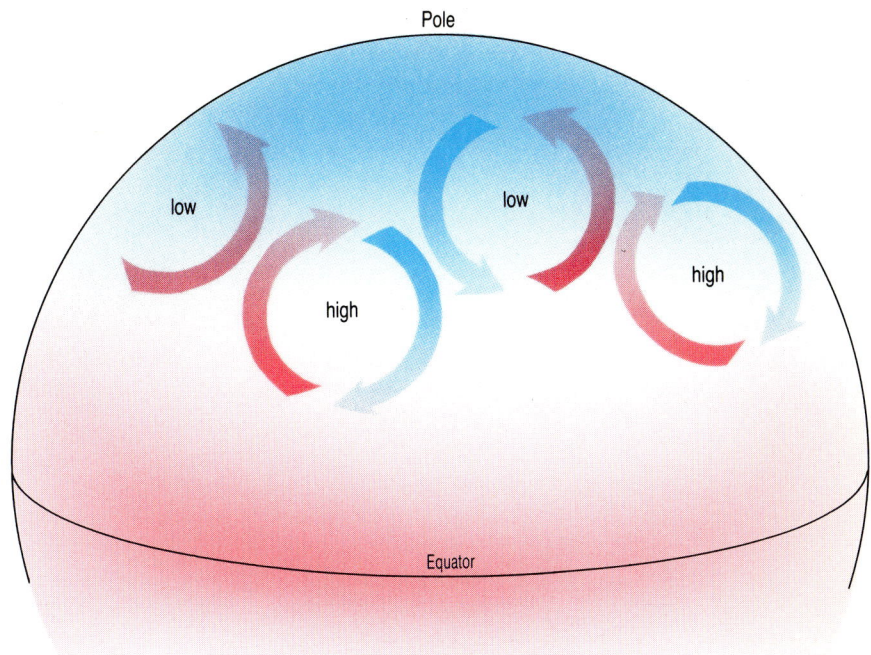

Figure 2.6 Highly schematic diagram to show the poleward transport of heat in the atmosphere by means of mid-latitude vortices (cyclones and anticyclones).

Hemisphere cyclone or anticyclone will be transporting relatively warm air polewards, while the air that returns equatorwards has been cooled. This may be likened to the stirring of bath water to encourage the effect of hot water from the tap to reach the far end of the bath, and is shown schematically in Figure 2.6. How it works in practice is perhaps easiest to see in the case of mid-latitude cyclones or depressions, which correspond to the zone labelled 'Subpolar Low' on Figure 2.2(a). Between the warm westerlies and the polar easterlies, there is a more or less permanent boundary region known as the polar front (Figure 2.2(b)). Undulations in the polar front may develop into depressions, and Figure 2.7(a) shows how formation of these depressions enables heat to be transported polewards. The paths taken by mid-latitude depressions and anticyclones

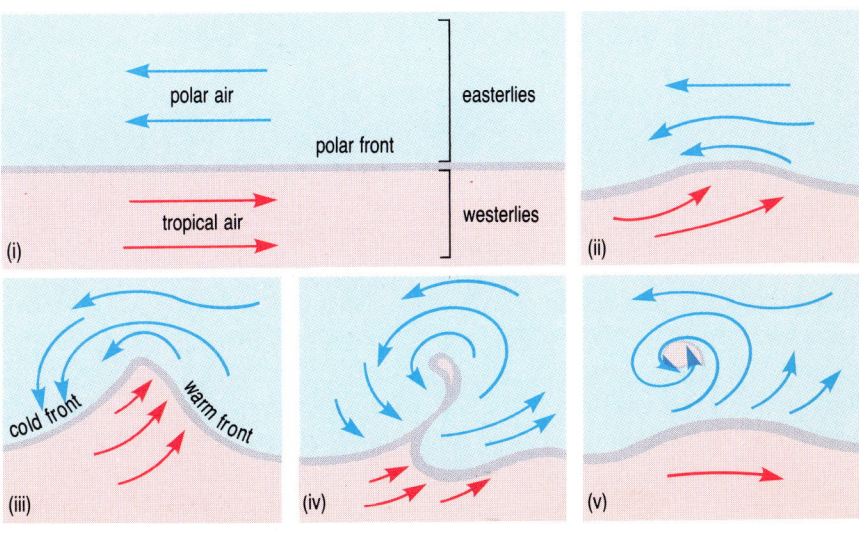

Figure 2.7 (a) Stages in the development of a mid-latitude cyclone (depression) in the Northern Hemisphere, showing how it contributes to the poleward transport of heat. Note that the frontal boundary slopes (warm air overlying cold air) as shown in Figure 2.2(b), and may be anything from 2–3 to 50 km across. That part of the front where cold air is advancing is known as a cold front, and that part where warm air is advancing, as a warm front.

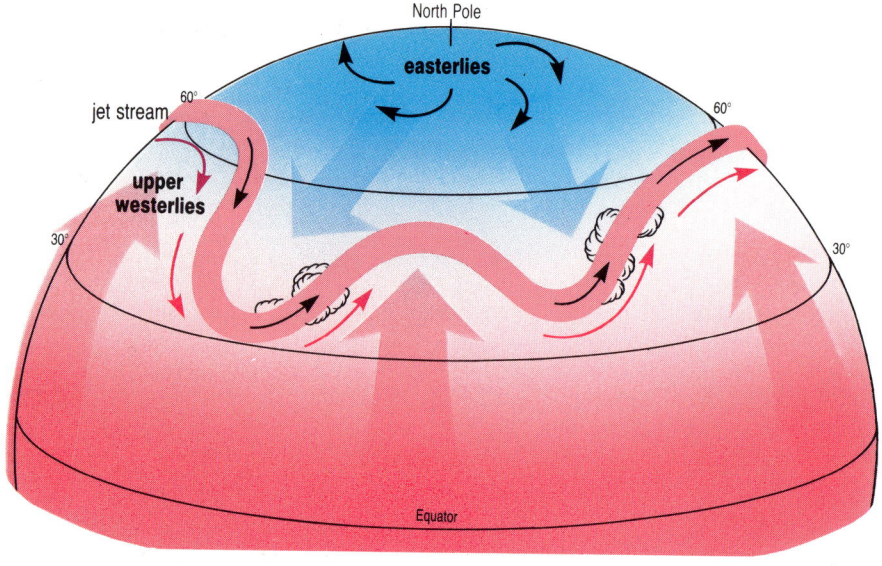

(b)

(c)

Figure 2.7 (b) Schematic diagram showing typical undulations in the northern jet stream, its relation to the flow of warm air (red) and cold air (blue), and the paths taken by depressions. Note that the jet stream flows near the tropopause at heights of 8–15 km. For clarity, we have omitted the subtropical high pressure regions in mid-latitudes and the Hadley circulation in low latitudes.

(c) Satellite photomosaic of the Southern Hemisphere showing cloud cover for one day in April 1983 (i.e. during the southern winter). Cyclonic storms develop in association with the poleward-trending sections of the jet stream and so its undulating path is shown up by the cloud pattern.

are determined by the path taken by the jet stream, the high-level, fast air current that flows around the Earth along the polar limit of the westerlies. The jet stream is characterized by large undulations, typically three to six in number, as shown in Figure 2.7(b).

We now move on to consider how heat is redistributed *within* the atmosphere, by means of predominantly vertical motions.

2.2.2 VERTICAL CONVECTION IN THE ATMOSPHERE

Processes occurring at the air–sea interface are greatly affected by the degree of turbulent convection that can occur in the atmosphere above the sea-surface. This in turn is dependent on the degree of *stability* of the air, i.e. on the extent to which, once displaced upwards, it tends to continue rising.

Two ways in which density may vary with height in the atmosphere (or any other fluid) are illustrated in Figure 2.8. Situation (a), in which density increases with height, is unstable, and upper air will tend to sink and lower air to rise. Situation (b), in which density decreases with height, is stable: a parcel of air (at, say, position O) that is displaced upwards will be denser than its surroundings and will sink back to its original position.

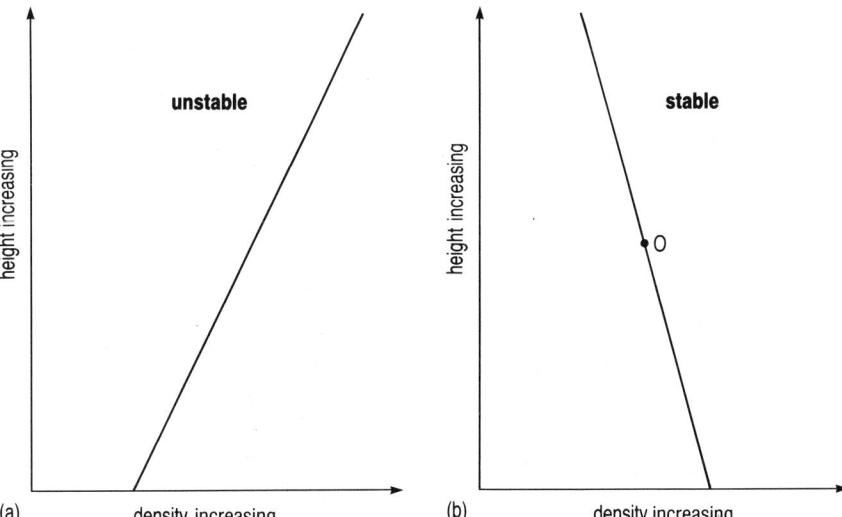

Figure 2.8 Possible variations of air density with height in the atmosphere, leading to (a) unstable and (b) stable conditions.

The density of air depends on its pressure and its temperature. It also depends on the amount of water vapour it contains—water vapour is less dense than air—but for most practical purposes water vapour content has a negligible effect on density. Thus, the variation of density with height in a column of air is determined by the variation in *temperature* with height.

However, the situation is complicated by two factors. The first is that air, like all fluids, is compressible. When a fluid is compressed, the internal energy it possesses by virtue of the motions of its constituent atoms, *and which determines its temperature*, is increased. Conversely, when a fluid expands, its internal energy decreases. Thus, a fluid heats up when compressed (a well-known example of this is the air in a bicycle pump), and cools when it expands. If these changes in temperature occur without gain or loss of heat from or to the surroundings, they are described as **adiabatic**.

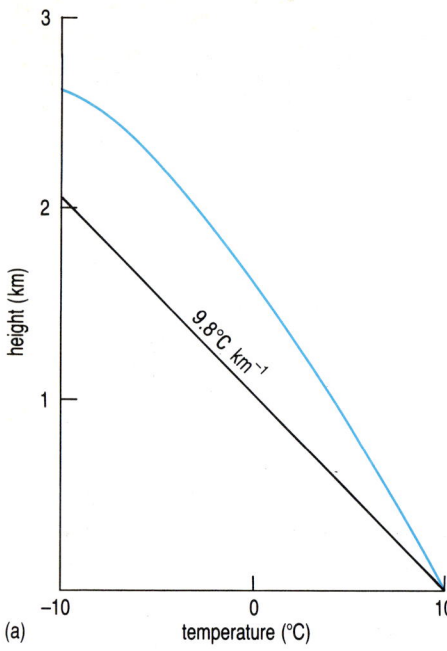

Figure 2.9 (a) Graph to illustrate that the rate at which dry air cools adiabatically (black line) is greater than the rate at which saturated air cools adiabatically (blue curve). (In the example shown, the air in question has a temperature of 10°C at ground level; in practice, it is atmospheric pressure rather than height as such that is important.)

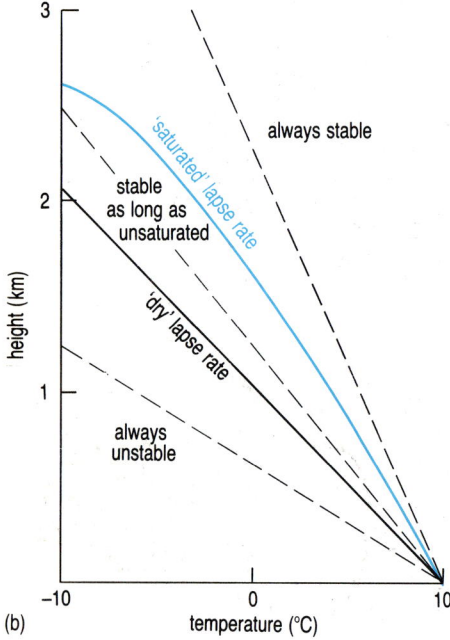

Figure 2.9 (b) The implications of the 'dry' and 'saturated' adiabatic lapse rates for columns of air with various temperature–height profiles.

When air rises, it is subjected to decreasing atmospheric pressure and so expands and becomes less dense. At the same time, it is experiencing an adiabatic decrease in temperature, which tends to increase its density. Air above the sea-surface may move upwards in random, turbulent eddying motions; whether it continues to rise depends on the relative sizes of these two effects. If the adiabatic decrease in temperature of a rising parcel of air is *less* than the decrease of temperature with height in the atmosphere, the rising parcel of air will be warmer than the surrounding air and will continue to rise. The situation will therefore be unstable and conducive to upward convection of air. If, on the other hand, the adiabatic cooling of the rising parcel of air is sufficient to reduce its temperature to a value below that of the surrounding air, it will sink back to its original level and the situation will be stable.

The other complicating factor is the effect of the water vapour in the air, not because of its lower density but because of its latent heat content. The rate at which rising *dry* air cools adiabatically is a constant 9.8°C per km (see the black curve on Figure 2.9(a); over the oceans in particular this 'dry' adiabatic lapse rate is of limited relevance. If rising air is saturated with water vapour—or becomes saturated as a result of adiabatic cooling—continued rise and associated adiabatic cooling results in the condensation of water vapour onto atmospheric nuclei, such as salt or dust particles, to form water droplets. Condensation releases latent heat of evaporation, which partly offsets the adiabatic cooling, so the rate at which air containing water vapour cools on rising is less than the rate for dry air.

This reduced adiabatic lapse rate for saturated air varies with temperature (see the blue curve in Figure 2.9(a)), because a small decrease in temperature at high temperatures results in more condensation than a similar decrease at low temperatures.

QUESTION 2.1 (a) The temperature of a column of dry air decreases from 10°C at ground level to −10°C at a height of 2.5 km. Use Figure 2.9(a) to explain why a parcel of air displaced upwards will *not* continue to rise, so that the situation is stable.

(b) What happens to the stability of this air if it becomes saturated as a result of increase in water vapour content or through adiabatic cooling?

The implications of the 'dry' and 'saturated' lapse rates for the stability of various atmospheric temperature profiles are summarized in Figure 2.9(b).

Over most of the oceans, particularly in winter, the variation of temperature with height in the atmosphere, and the water content of the air, are such that conditions are unstable, air rises, and convection occurs. This is further promoted by turbulence resulting from strong winds blowing over the sea-surface; when turbulence is a more effective cause of upward movement of air than the buoyancy forces causing instability, convection is said to be *forced*. The cumulus clouds that are characteristic of oceanic regions within the Trade Wind belts are a result of such forced convection (Figure 2.10(a)).

As illustrated in Figure 2.2(b), upward development of the Trade Wind cumulus clouds is inhibited by the subsidence of warm air from above. This leads to an *increase* of temperature with height, or a temperature

Figure 2.10 (a) Cumulus clouds over the ocean in the Trade Wind belt.

(b) Cumulonimbus clouds and, at lower levels, cumulus clouds, in the Intertropical Convergence Zone over the Java Sea. Like cumulus clouds, cumulonimbus clouds form where moist air rises and cools, so that the water vapour it contains condenses to form droplets. However, cumulonimbus clouds generally extend to much greater heights than cumulus clouds (see (a)) and their upper parts consist of ice crystals.

inversion (Figure 2.11). Rising air encountering a temperature inversion is no longer warmer than its surroundings and ceases its ascent. The warmer air therefore acts as a 'ceiling' so far as upward convection is concerned.

Extremely vigorous upward movement of moist air occurs along the Intertropical Convergence Zone (Figure 2.2(b)). This gives rise to towering cumulonimbus clouds (Figure 2.10(b)), which enable the ITCZ to be easily seen on images obtained via satellites (Figure 2.12). Convection in the ITCZ extends much higher than that associated with cumulus formation, and is the principal way in which heat is distributed throughout the troposphere in low latitudes.

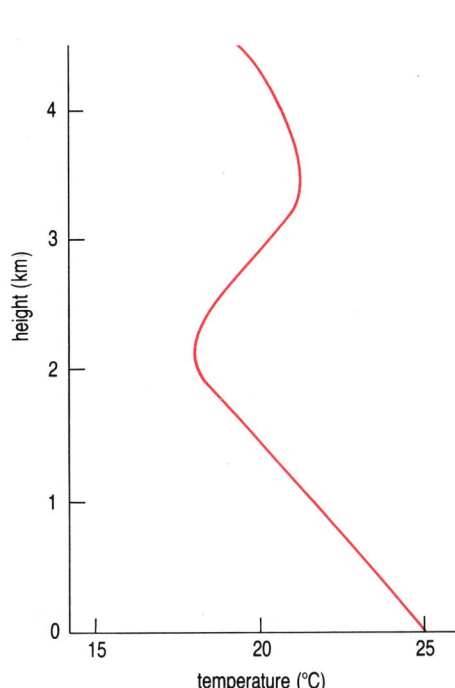
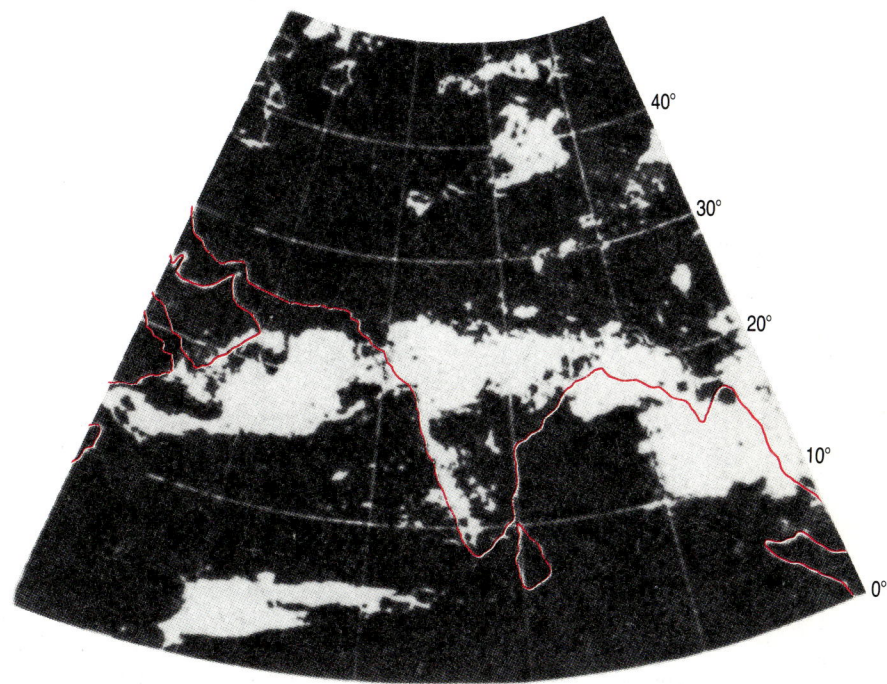

Figure 2.11 Variation of temperature with height in the Trade Wind zone, at about 5° of latitude, showing the temperature inversion.

Figure 2.12 The position of the ITCZ over India and the Indian Ocean, as indicated by cloud cover. This satellite image was taken in July 1973.

2.3 ATMOSPHERE–OCEAN INTERACTION

The discussion in the previous Section illustrated how the ocean influences the atmosphere by affecting its moisture content and hence its stability. This is just one aspect of the complex interaction between the atmosphere and the ocean.

Another aspect of this interaction is the way in which the distribution of sea-surface temperature influences atmospheric circulation. For example, the intensity of the Hadley circulation is influenced by sea-surface temperature; and, as mentioned earlier, the position of the ITCZ generally corresponds to the zone of highest sea-surface temperature.

Why might this be?

It is common for the surface of the sea to be warmer than the overlying air; the higher the temperature of the sea-surface, the more heat may be transferred from the upper ocean to the lower atmosphere. Warmer air is less dense and rises, causing a low pressure region, towards which winds blow. Thus, for example, a region of exceptionally high surface temperature in the region of the Equator could lead to an increase in the intensity of the Trade Winds and the Hadley circulation. The position of the ITCZ will also be related to the low pressure zones associated with high sea-surface temperature; in addition, the warmer the sea-surface, the more buoyancy is supplied to the lower atmosphere and the more vigorous the vertical convection that will result.

In the next Section, you will see a striking example of the influence of sea-surface temperature on the atmosphere.

2.3.1 EASTERLY WAVES AND TROPICAL CYCLONES

A large amount of heat is transported away from low latitudes by strong tropical cyclones—also known as hurricanes and typhoons. Tropical cyclones only develop over oceans and so it is difficult to study the atmospheric conditions associated with their formation. It is known, however, that they are triggered by small low pressure centres, such as may occur in small vortices associated with the ITCZ. Cyclones may also be triggered by linear low pressure areas that form at right angles to the direction of the Trade Winds and travel with them (Figure 2.13). These linear low pressure regions produce wave-like disturbances in the isobaric patterns; because they move with the easterly Trade Winds they are known as easterly waves.

Easterly waves are most common in the western parts of the large ocean basins, between about 5° and 20°N. They occur most frequently during the late summer, and this is thought to be connected with the fact that the Trade Wind temperature inversion (Figures 2.2 (b) and 2.11) is weakest at that time. The temperature of the air in the lower Trade Winds is largely determined by the sea-surface temperature. This is at its highest during late summer, and so that is when the temperature increase across the temperature inversion is least.

Although only a small proportion of easterly waves give rise to cyclones, they are important because they bring large amounts of rainfall to areas

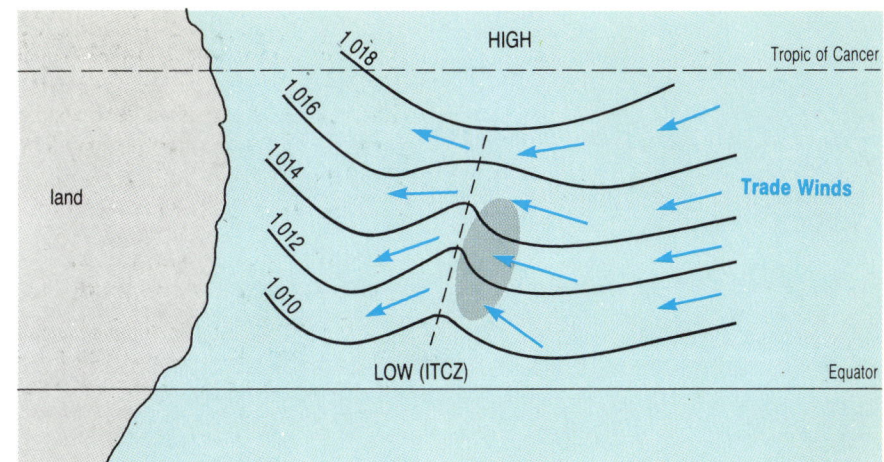

Figure 2.13 Schematic diagram of an easterly wave in the Northern Hemisphere. The black lines are isobars and the values are atmospheric pressure at ground level in millibars. The blue arrows show the wind direction. Ahead of the wave's 'axis', the Trade Wind inversion is strengthened and lowered, leading to particularly fine weather; behind it, the inversion is temporarily destroyed, causing rainfall. The main rainfall area is shown in greyish-blue.

that remain generally dry as long as the Trade Winds are unperturbed (see Figure 2.13).

Once formed, the tropical cyclone is characterized by almost circular isobars closely packed around a centre of very low pressure (typically about 950 mbar). Large pressure gradients near the centre of the cyclone cause air to spiral rapidly in towards the low pressure region (anticlockwise in the Northern Hemisphere, clockwise in the Southern Hemisphere), and wind speeds commonly reach 100–200 km hour^{-1}. The core or 'eye' of the cyclone is an area of light winds and little cloud, but around it is a region where there is violent upward convection of warm humid air (see Figure 2.14(a) and (b)).

(a)

Figure 2.14 (a) Satellite image of Hurricane *Allen* passing from the Atlantic into the Caribbean. The image was obtained using the visible part of the spectrum, and the outlines of the land masses have been superimposed for clarity.

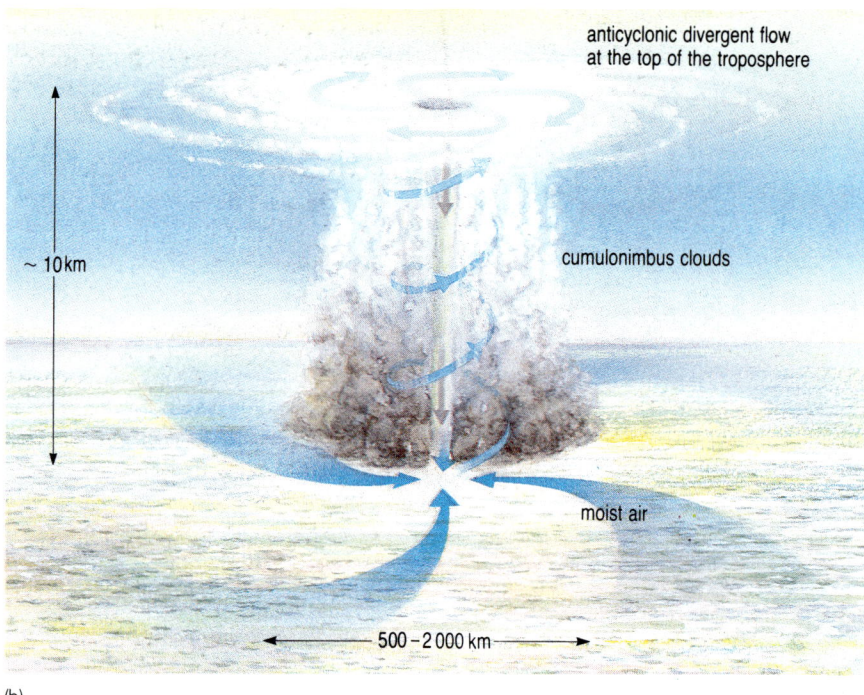

(b)

Figure 2.14 (b) Schematic diagram of a tropical cyclone showing air movements and areas of cloud formation and heavy precipitation. The cumulonimbus clouds are arranged in bands which form a spiral pattern around the core region (or eye) of the storm. Subsidence of air, and adiabatic warming, occurs in this core region, which is a region of light winds and little cloud.

The energy that drives the cyclone comes from the release of latent heat as the water vapour in the rising air condenses into clouds and rain; the resultant warming of the air around the central region of the cyclone causes it to become less dense and to rise yet more, intensifying divergent anticylonic flow of air in the upper troposphere that is necessary for the cyclone to be maintained (Figure 2.14 (b)). Given their source of energy, it is not surprising that tropical cyclones occur only over relatively large areas of ocean where the surface water temperature is high.

In practice, the critical sea-surface temperature needed to generate the increased vertical convection which leads to extensive cumulonimbus cloud development and rain and/or cyclone formation is about 27–29°C. Why should this value be critical? One factor seems to be that the higher the temperature of air the more moisture it can hold, and the greater the upward transfer of latent heat that can occur. Given the positive feedback of the system, a rise in temperature from 27 to 29°C has a much greater effect on the overlying atmosphere than a rise from, say, 19 to 21°C. The full answer to this question is, however, as yet unknown.

QUESTION 2.2 (a) With the help of Figure 2.3 and bearing in mind what you have been reading about the conditions that favour the initiation and development of easterly waves and cyclones, can you explain why they occur more often in the Northern Hemisphere than the Southern Hemisphere?

(b) Can you also explain why cyclones do not develop within about 5° of the Equator?

Tropical cyclones also occur more often in the western than the eastern parts of the Atlantic and Pacific oceans. This is because sea-surface temperatures are higher there, for reasons that will become clear in later Chapters.

The violent winds of tropical cyclones generate very large waves on the sea-surface. These waves travel outwards from the central region and as the cyclone progresses the sea becomes very confused. The region where the winds are blowing in the *same* direction as the cyclone is travelling is particularly dangerous because here the waves have effectively been blowing over a greater distance (i.e. they have a greater **fetch**). In addition, ships in this region may be blown into the cyclone's path.

Cyclones also affect the deeper structure of the ocean over which they pass. Near the centre of the storm the action of the wind causes the surface waters to diverge so that deeper, cooler water upwells to replace it. Thus, not only are cyclones *affected* by sea-surface temperature, but they also *modify* it, so that their tracks are marked out by surface water with anomalously low temperatures, perhaps as much as 5°C below that of the surrounding water. Figure 2.14(c) shows in more detail the changes in the water column that have been observed to accompany tropical cyclones.

Characteristic tracks followed by cyclones as they move away from their sites of generation are shown in Figure 2.15. Note that they nearly always move polewards, which is why they form such a powerful mechanism for the transport of heat to higher latitudes. If they move over land areas they begin to die away, as the energy conversion system needed to drive them can no longer operate. Their decay over land may be hastened by increased surface friction and the resulting increased variation of wind velocity with height (vertical **wind shear**) which inhibits the maintenance of atmospheric vortices. The lower temperatures of land masses (especially at night) may also play a part in their decay. The average lifespan of a tropical cyclone is about a week.

There is a great variation in the frequency of tropical cyclones. Over the past few decades there seems to have been an increase in their

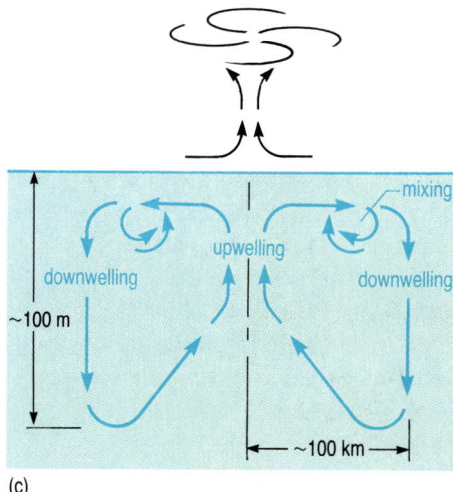

Figure 2.14 (c) The effect of a cyclone on the surface ocean. Near the centre of the storm, surface water diverges and upwelling of deeper, cooler water occurs; some distance from the centre, the surface water that has travelled outwards from the centre converges with the surrounding ocean surface water and downwelling, or sinking, occurs. In between the zones of upwelling and downwelling, the displaced surface water mixes with cooler subsurface water. (Note that the way in which cyclonic winds lead to upwelling of subsurface water will be explained in Chapter 3.)

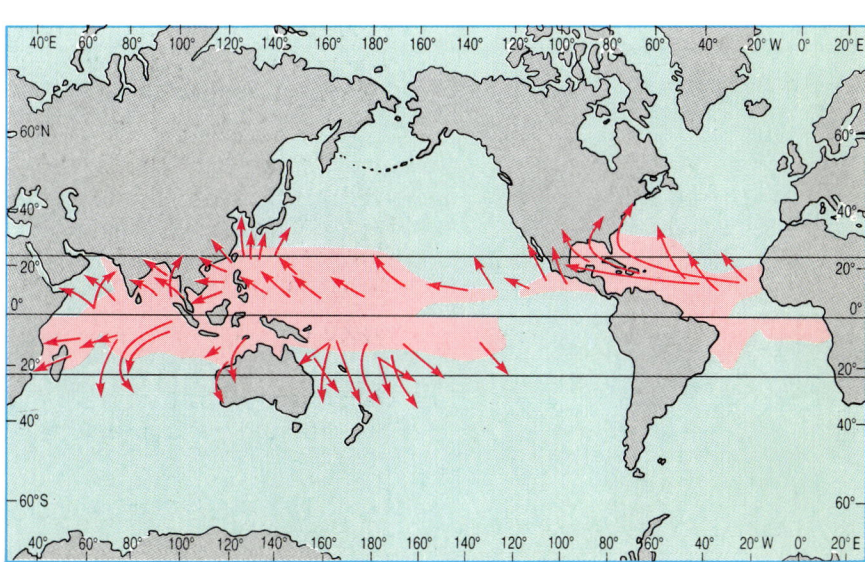

Figure 2.15 Characteristic tracks followed by tropical cyclones. Initially they move westwards, but then generally curve polewards and eastwards around the subtropical high pressure centres. The red shaded regions are those areas where the sea-surface temperature exceeds 27°C in summer (mean values for September and March, in the Northern and Southern Hemispheres, respectively).

Table 2.1 Mean number of cyclones occurring over a ten-year period.

North Pacific (western part)	208
North Atlantic, Caribbean	85
Bay of Bengal	75
South-west Indian Ocean	41
North-east of Australia	31
North-west of Australia	23
Arabian Sea	19
North Pacific (eastern part)	10

occurrence, but it is not clear whether this is a genuine increase or part of a cyclical variation. The mean frequency in various regions is given in Table 2.1.

Cyclones have a great impact on life in the tropics. They are largely responsible for late summer or autumn rainfall maxima in many tropical areas. The strong winds and large waves associated with them may cause severe damage to natural environments such as reefs; indeed, these catastrophic events play a major role in determining the patterns of distribution of species and life forms within such communities. They may lead to great loss of human life, particularly where there are large populations living near to sea-level, on the flood plains of major rivers or on islands. The low atmospheric pressures associated with cyclones may cause a rise in the sea-level, quite apart from the high seas caused by the winds, and widespread flooding of low-lying areas can result. The cyclone disaster which devastated the delta region of the Bay of Bengal in May 1985 was largely the result of such a **storm surge**, further amplified by the occurrence of high tides.

Water spouts

Water spouts are similar to cyclones in that they are also associated with cyclonic air movements, vigorous atmospheric convection, and cumulus and cumulonimbus cloud formation (see Figure 2.16). They are funnel-shaped vortices of air with very low pressures at their centres, so that air and water spiral rapidly inwards and upwards. The funnels extend from the sea-surface to the 'parent' clouds that travel with them; they whip up a certain amount of spray from the sea-surface but are visible mainly because the reduction of pressure within them leads to adiabatic expansion and cooling which causes atmospheric water vapour to condense.

Figure 2.16 A water spout near Lower Malecumbe Key, Florida.

Water spouts are a much smaller-scale phenomenon than cyclones. They range from a few metres to a few hundred metres in diameter, and they rarely last more than fifteen minutes. They occur most often in the spring and early summer. Unlike cyclones they are not confined to the tropics, although they occur most frequently there, usually in the spring and early summer. They are particularly common over the Bay of Bengal and the Gulf of Mexico, and also occur frequently in the Mediterranean.

2.4 SUMMARY OF CHAPTER 2

1   The global wind system acts to redistribute heat between low and high latitudes.

2   Winds blow from regions of high pressure to regions of low pressure. Because of the differing thermal capacities of continental masses and oceans, wind patterns are greatly influenced by the distribution of land and sea.

3   Winds are deflected by the Coriolis force, to an extent that increases with increasing latitude. In mid-latitudes, the predominant wind systems are cyclones, which blow around low pressure centres or depressions, and anticyclones, which blow around high pressure centres. At low latitudes, the atmospheric circulation consists essentially of the spiral Hadley cells, of which the Trade Winds form the lowermost limb. The Intertropical Convergence Zone is the region where the wind systems of the two hemispheres meet; it is generally associated with the zone of maximum sea-surface temperature.

4   Heat is transported polewards in the atmosphere as a result of warm air moving into cooler latitudes. It is also transported as *latent* heat: heat used to convert water to water vapour is released when the water vapour condenses (e.g. in cloud formation). Over the tropical oceans, turbulent convection of the overlying air transports large amounts of heat from the sea-surface high into the atmosphere, leading to the formation of cumulus and, especially, cumulonimbus clouds. An extreme expression of this convection is the generation of tropical cyclones.

Now try the following questions to consolidate your understanding of this Chapter.

QUESTION 2.3   Which of statements (a)–(f) concerned with the global wind system are true and which are false?

(a)   Regions with high surface atmospheric pressure are regions where air is rising.

(b)   Heavy precipitation is characteristic of regions where air is sinking.

(c)   In both seasons of the year, the atmospheric circulation over the North Atlantic is predominantly anticyclonic.

(d)   In both seasons of the year, the atmospheric circulation over Eurasia is predominantly cyclonic.

(e)   The Trade Wind systems of the two hemispheres meet along a zone of high pressure.

(f)   In polar regions, the tropopause is about 12–15 km above the Earth.

QUESTION 2.4 (a) Draw a plan-view sketch of a tropical cyclone in the Northern Hemisphere, showing isobars and wind direction. Indicate on your diagram the regions that are particularly dangerous for navigation.

(b) What is the 'fuel' that drives a tropical cyclone?

QUESTION 2.5 Figure 2.17 is a satellite image of the Earth made at the same time as the image in Figure 2.1 but using the electromagnetic wavelengths to which water vapour is opaque. The warmest areas are black and the coldest white. Thus, dark areas represent regions in which the upper troposphere is dry so that radiation from lower, warmer layers is able to reach the satellite; white areas correspond to radiation from the upper troposphere, which is cold. By comparing this image with that in Figure 2.1, use information from this Chapter to identify the following features:

(a) cumulonimbus cloud formation in association with the ITCZ;

(b) the northern and southern polar fronts;

(c) the subtropical high pressure regions.

Figure 2.17 Satellite image of the Earth, constructed using the 'water vapour' channel (wavelength 6 000 nm); the colour is artificial. This image and that in Figure 2.1 show the same view of the Earth and are for 26 March 1982.

QUESTION 2.6 In Chapter 5, we will be discussing how disturbances may be transmitted from one part of the ocean to another by means of large-scale wave motions. From your reading of Chapter 2, give *two* examples of *atmospheric* disturbances that propagate as waves.

CHAPTER 3   OCEAN CURRENTS

In Chapter 2, we saw how the atmospheric circulation transports heat from low to high latitudes. Figure 3.1 shows how the same is true in the oceans, where surface currents warmed in low latitudes carry heat polewards, while currents cooled at high latitudes flow equatorwards.

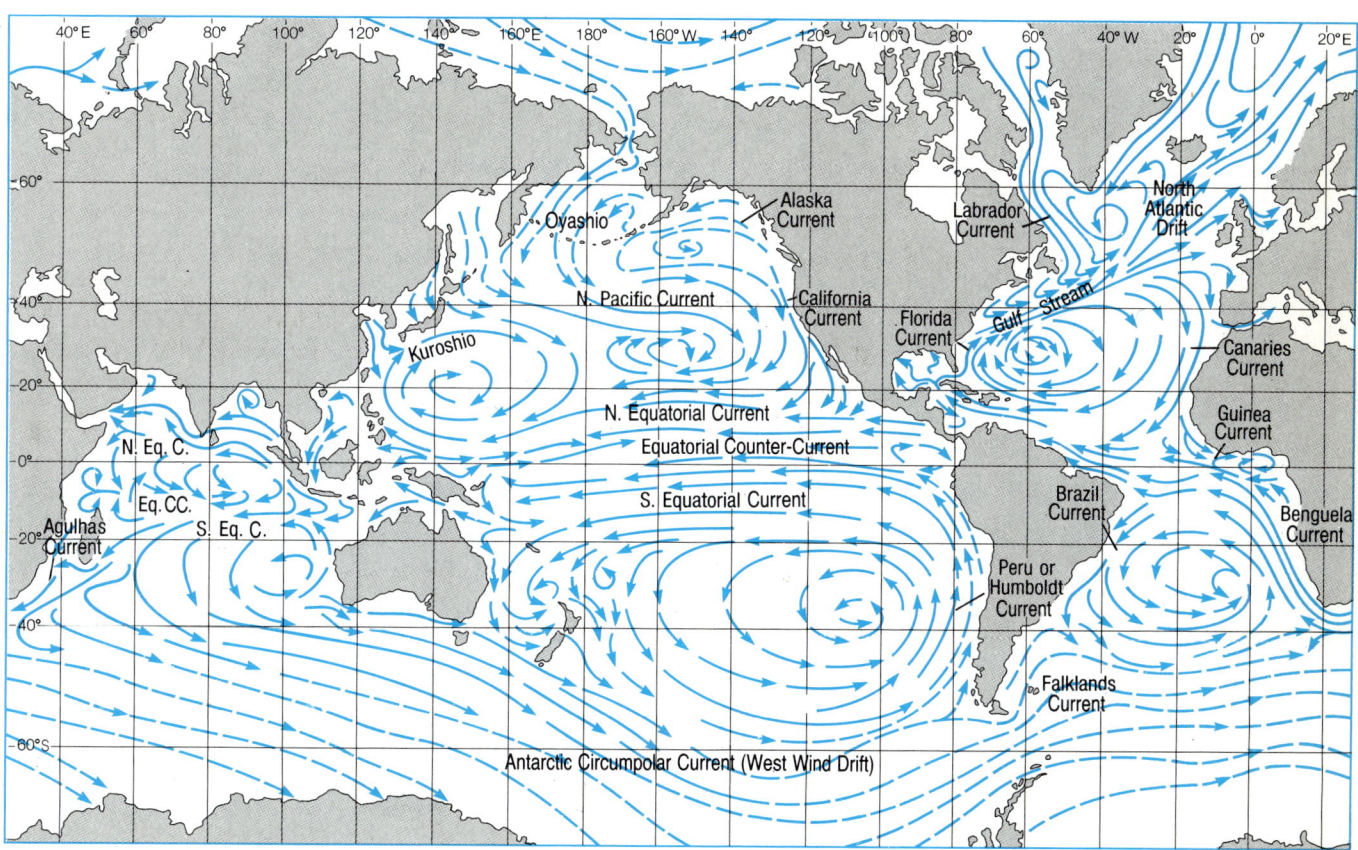

Figure 3.1 The global surface current system. Cool currents are shown by dashed arrows; warm currents are shown by solid arrows. The map shows average conditions for winter months in the Northern Hemisphere; there are local differences in the summer, particularly in regions affected by monsoonal circulations.

QUESTION 3.1 (a) Which are the main such cool surface currents in (i) the North and South Pacific, and (ii) the North and South Atlantic?
(b) What do these currents have in common?

So far as the transport of heat is concerned, the 'warm' and 'cool' surface currents shown in Figure 3.1 are only part of the story. In certain polar regions, water that has been subjected to extreme cooling sinks and flows equatorwards in the thermohaline circulation (Chapter 1). In order to know the net poleward heat transport in the oceans at any latitude, we would need to know the direction and speed of flow of water, and its temperature, at all depths. In fact, the three-dimensional current structure of the oceans is complex and very poorly understood.

If you compare Figures 2.3 and 3.1 you will see immediately that the surface wind field and the surface current system have a general similarity. The most obvious difference results from the fact that the flow

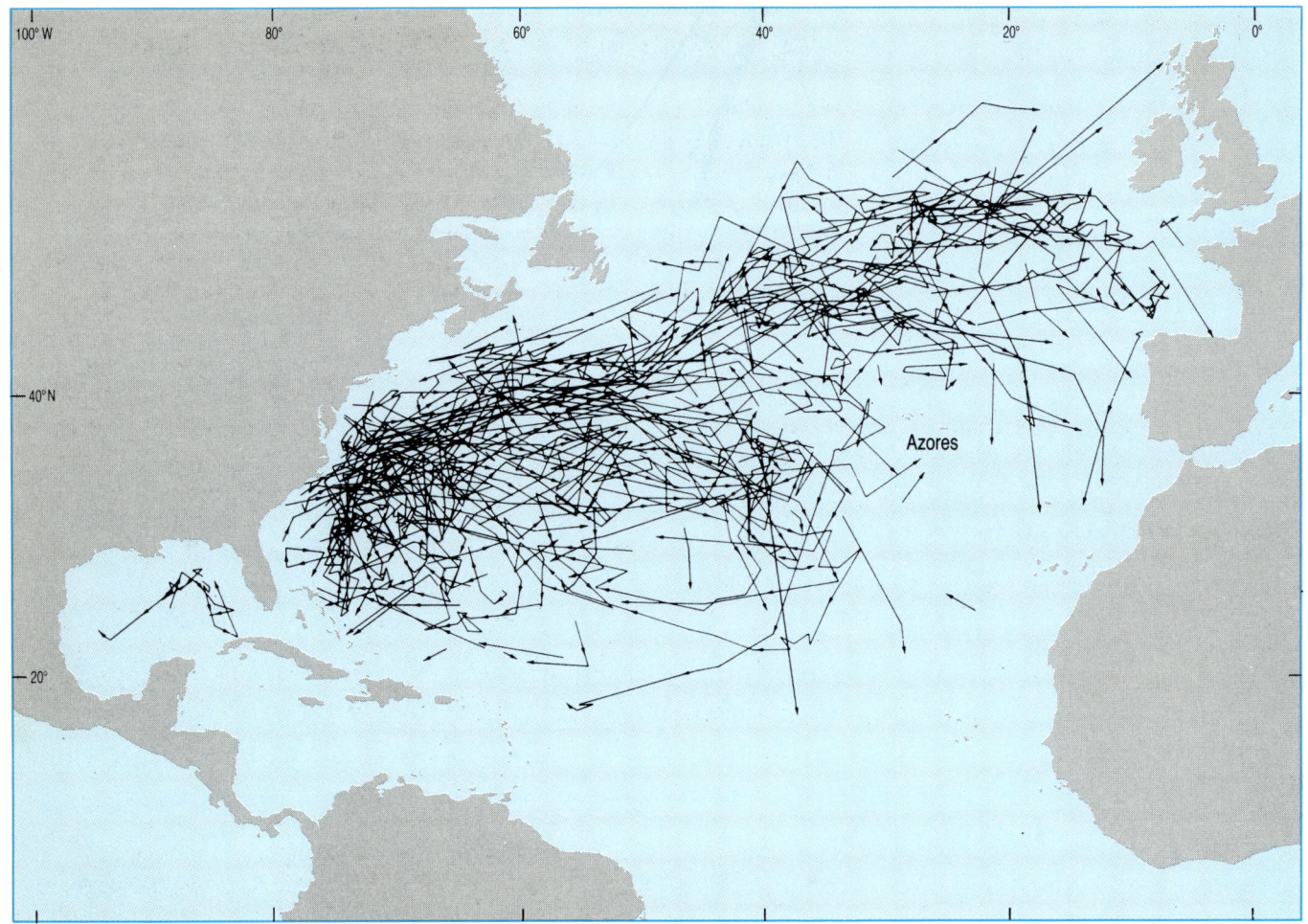

Figure 3.2 The paths of drifting derelict sailing vessels (and a few drifting buoys) over the period 1883–1902. This chart was produced using data from the monthly Pilot Charts of the US Navy Hydrographic Office. The paths are extremely convoluted and cross one another, but the general large-scale anticyclonic circulation of the North Atlantic may just be distinguished.

of ocean currents is constrained by coastal boundaries so that the tendency for circular or gyral motion seen in the atmospheric circulation is even more noticeable in the oceans. The way in which ocean currents result from the atmospheric circulation is not, however, so obvious as it might at first appear, and in this Chapter we consider some of the mechanisms involved.

This is perhaps a suitable point to emphasize that maps like Figures 2.3 and 3.1 represent *average* conditions. If you were to observe the wind and current at, for instance, a locality in the region labelled 'Gulf Stream' on Figure 3.1 you could well find that the wind and/or current directions were quite different from those shown by the arrows, perhaps even in the opposite direction. Moreover, currents should not be regarded as river-like. The actual spatial and temporal variations in velocity (speed and direction) are much more complex than could be shown by the most detailed series of current charts. Nevertheless, Figure 3.2 gives some indication of the variability of surface currents in the North Atlantic.

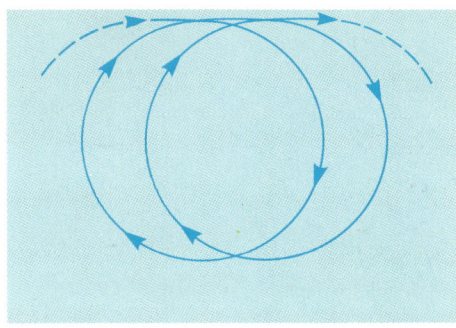

Figure 3.3 In a surface wave, water particles advance slightly further in the crest than they retreat in the trough, so a small net forward motion (known as wave drift) results. In deep water, this motion may be of the order of several millimetres to several centimetres per second.

3.1 THE ACTION OF WIND ON SURFACE WATERS

When wind blows over the ocean, energy is transferred from the wind to the surface layers. Some of this energy is expended in the generation of surface gravity waves—which lead to a small net movement of water in the direction of wave propagation (Figure 3.3)—and some to drive currents. The processes whereby energy is transferred between waves and currents are complex; it is not a simple task to discover, for example, how much of the energy of a breaking wave is dissipated and how much is transferred to the surface current.

Nevertheless, it is still possible to make some general statements and predictions about the action of wind on the sea. The greater the speed of the wind, the greater the frictional force acting on the sea-surface, and the stronger the surface current. The frictional force acting on the sea-surface as a result of the wind blowing over it is known as the **wind stress**. Wind stress, which is usually given the symbol τ (Greek 'tau'), has been found by experiment to be proportional to the square of the wind speed, W. Thus:

$$\tau = cW^2 \tag{3.1}$$

where c depends on the prevailing atmospheric conditions. The more turbulent convection that there is in the atmosphere overlying the sea-surface (Section 2.2.2), the higher the value of c.

How would you expect the value of c to be affected by the wind speed?

The value of c will increase with increasing wind speed, which not only increases the amount of turbulent convection in the overlying atmosphere (Section 2.2.2) but also increases the roughness of the sea-surface.

Because of friction with the sea-surface, wind speed increases with height, and so c also depends critically on the *height* at which the wind speed is measured; this is commonly about 10m, the height of the deck or bridge of a ship. As a rough guide, we can say that a wind with a speed of $10\,\mathrm{ms^{-1}}$ (nearly 20 knots) at 5–10m height, will give rise to wind stress on the sea-surface of the order of $0.2\,\mathrm{Nm^{-2}}$. (1 newton = $1\,\mathrm{kgms^{-2}}$.)

QUESTION 3.2 What value does that imply for c? What are its units?

It is important to remember that, for the reasons given earlier, c is not constant. Nevertheless, a value of c of about 2×10^{-3} gives values of τ that are accurate to within a factor of 2, and often considerably better than that.

Another empirical observation is that the surface current is typically about 3% of the wind speed, so that a $10\,\mathrm{ms^{-1}}$ wind might be expected to give rise to a surface current of about $0.3\,\mathrm{ms^{-1}}$. Again, this is only a rough 'rule of thumb', for reasons that should become clear shortly.

3.1.1 FRICTIONAL COUPLING WITHIN THE OCEAN

The effect of the wind stress at the surface is transmitted downwards as a result of internal friction within the upper ocean. This internal friction results from turbulence and is not simply the viscosity of a fluid moving in a laminar fashion (see Figure 3.4).

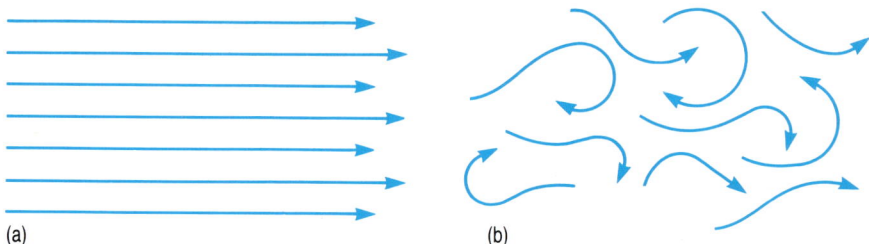

Figure 3.4 The difference between (a) laminar flow and (b) turbulent flow. The arrows represent the paths taken by individual parcels of water.

Friction in a moving fluid results from the transfer of momentum (mass × velocity) between different parts of the fluid. In a fluid moving in a laminar manner, momentum transfer occurs as a result of the transfer of *molecules* (and their associated masses and velocities) between adjacent layers, as shown schematically in Figure 3.5(a), and should therefore strictly be called **molecular viscosity**. At the sea-surface, as in the rest of the ocean, motion is never laminar, but instead turbulent so that parcels of water, rather than individual molecules, are exchanged between one part of the moving fluid and another (Figure 3.5(b)). The internal friction that results is much greater than that caused by exchange of individual molecules, and is known as **eddy viscosity**.

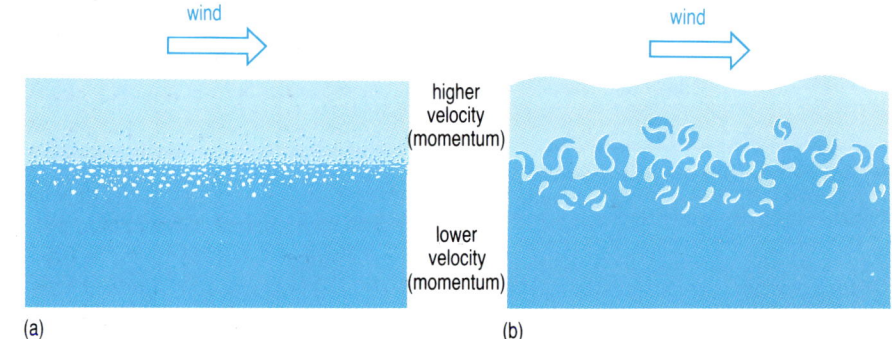

Figure 3.5 Schematic diagram to illustrate the difference between (a) molecular viscosity and (b) eddy viscosity. In (a), the momentum transferred between layers is that associated with individual molecules, whereas in (b) it is that associated with parcels of fluid. (For simplicity, we have only shown two layers of differing velocity; in reality, of course, there are an infinite number of layers.)

Turbulent eddies in the upper layer of the ocean act as a 'gearing' mechanism that transmits motion at the surface to deeper levels. The extent to which there is turbulent mixing, and hence the magnitude of the eddy viscosity, depend on how well stratified the water column is. If the water column is well mixed and hence fairly homogeneous, density will vary little with depth and the water column will be easily overturned by turbulent mixing; if the water column is well stratified so that density increases relatively sharply with depth, the situation is stable and turbulent mixing is suppressed.

QUESTION 3.3 Between the warm well-mixed surface layer and the cold waters of the main body of the ocean is the **thermocline**, the zone within which temperature decreases markedly with depth. Explain whether you would expect eddy viscosity to be greater in the thermocline or in the mixed surface layer.

An extreme manifestation of the answer to Question 3.3 is the phenomenon of the **slippery sea**. If the surface layers of the ocean are exceptionally warm or fresh, so that density increases abruptly not far below the sea-surface, there is very little frictional coupling between the thin surface layer and the underlying water. The energy and momentum of the wind are then transmitted only to this thin surface layer, which effectively slides over the water below.

As you might expect from the foregoing discussion, values of eddy viscosity in the ocean vary widely, depending on the degree of turbulence. Eddy viscosity (strictly, the *coefficient* of eddy viscosity) is usually given the symbol A, and we may distinguish between A_z which is the eddy viscosity resulting from vertical mixing (as discussed above) and A_h which is the eddy viscosity resulting from horizontal mixing—for example, that caused by the turbulence between two adjacent currents, or between a current and a coastal boundary. Values of A_z are typically 10^{-2}–$10^2 \text{kg m}^{-1}\text{s}^{-1}$, while those for A_h are generally much greater, ranging from 10^4 to $10^8 \text{kg m}^{-1}\text{s}^{-1}$.

Why are values of A_h so much higher than values of A_z?

The high values of A_h and the relatively low values of A_z reflect the differing extents to which mixing can occur in the vertical and horizontal directions. The ocean is stably stratified nearly everywhere, and stable stratification acts to suppress vertical mixing; motion in the ocean is nearly always in a horizontal or near-horizontal direction. In addition, the oceans are many thousands of times wider than they are deep and so the spatial extent of horizontal eddying motions is much less constrained than is vertical mixing.

The fact that frictional coupling in the oceans occurs through turbulence rather than through molecular viscosity has great significance for those seeking to understand wind-driven currents. Such currents increase to their maximum strength many times faster than would be possible through molecular processes alone. For example, in the absence of turbulence, the effect of a 10ms^{-1} wind would hardly be discernible 2m below the surface, even after the wind had been blowing steadily for two days. Similarly, when a wind driving a current ceases to blow, the current is slowed down by turbulent mixing many times faster than would otherwise be possible. Turbulence redistributes and dissipates the kinetic energy of the current; ultimately, however, it is converted into heat through molecular viscosity.

The types of current motion that are easiest to study, and that will be considered here, are those that result when the surface ocean has had time to adjust to the wind, and ocean and atmosphere have locally reached a state of equilibrium. When a wind starts to blow over a motionless sea-surface, the surface current which is generated takes some time to attain the maximum speed that can result from that particular wind speed; in other words, it first accelerates. The situations that are generally studied are those in which acceleration has ceased and the forces acting on the water are in balance.

The first satisfactory theory for wind-driven currents was developed by V. W. Ekman in the 1890s. It is to his ideas, and their surprising implications, that we now turn.

3.1.2 EKMAN MOTION

In the 1890s, the Norwegian scientist and explorer Fridtjof Nansen led an expedition across the Arctic ice. His specially designed vessel the *Fram* was allowed to freeze into the ice and drift with it for over a year. During this period, Nansen observed that ice movements in response to wind were *not* parallel to the wind, but at an angle of 20–40° to the right of it. Ekman developed his theory of wind-driven currents in order to explain this observation.

Ekman considered a steady wind blowing over an ocean that was infinitely deep, infinitely wide, and with no variations in density. He also assumed that the surface of the ocean remained horizontal, so that the pressure at any given depth was constant. This hypothetical ocean may be considered to consist of an infinite number of horizontal layers, of which the topmost is subjected to friction by the wind (wind stress) at its upper surface and to friction (eddy viscosity) with the next layer down at its lower surface; this second layer is acted upon at its upper surface by friction with the top layer, and by friction with layer three at its lower surface; and so on. In addition, because they are moving, all layers are acted upon by the Coriolis force.

By considering the balance of forces—friction and the Coriolis force—on the infinite number of layers making up the water column, Ekman deduced that the speed of the wind-driven current decreases exponentially with depth. He further found that the direction of the current deviates 45° *cum sole* from the wind direction at the surface, and that the angle of deviation increases with increasing depth. The current vectors therefore form a spiral pattern (Figure 3.6(a)) and this theoretical current pattern is now known as the **Ekman spiral**.

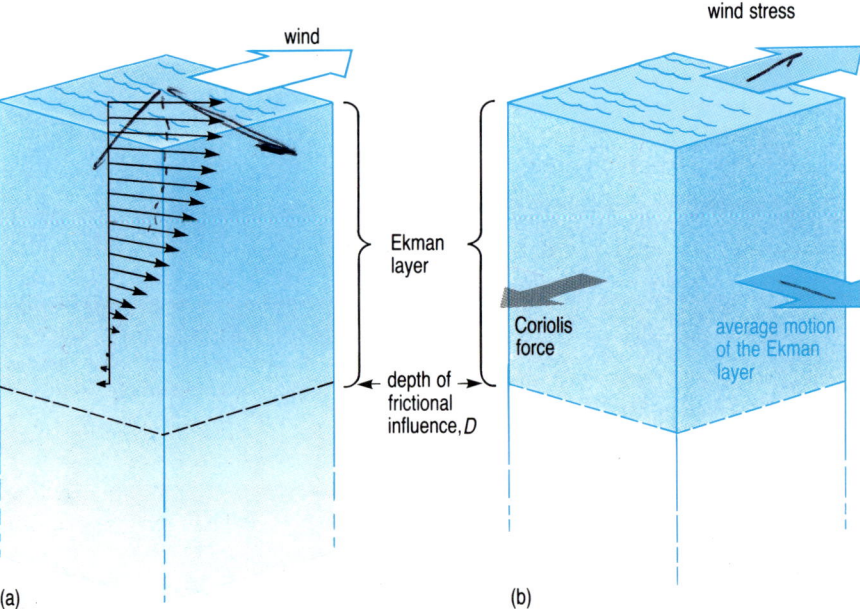

Figure 3.6 (a) The Ekman spiral current pattern believed to result from the action of wind on surface waters. The lengths and directions of the black arrows represent the speed and direction of the wind-driven current. (b) For the Ekman layer as a whole, the force due to the wind is balanced by the Coriolis force, which in the Northern Hemisphere is 90° to the right of the average motion of the layer (broad blue arrow).

We do not need to go into the calculations whereby Ekman deduced the spiral current pattern, but we can consider the forces involved in a little more detail. As discussed in Chapters 1 and 2, the Coriolis force acts 90° *cum sole* of the direction of current flow, and increases with latitude.

More specifically, the Coriolis force is proportional to the sine of the latitude, and for a particle of mass m moving with speed u it is given by:

$$\text{Coriolis force} = m \times 2\Omega \sin \phi \times u \tag{3.2a}$$

where Ω (capital omega) is the angular velocity of the Earth about its axis and ϕ (phi) is the latitude. The term $2\Omega \sin \phi$ is known as the **Coriolis parameter** and is often abbreviated to f, so that the expression given above becomes simply:

$$\text{Coriolis force} = mfu \tag{3.2b}$$

Thus, the Coriolis force—the force experienced by a moving particle of water by virtue of it being on the rotating Earth—is at right angles to the direction of motion and *increases* with the speed of the particle. This is important for the balances of forces that determine motion in the ocean— for example, the balance between frictional forces and the Coriolis force for each of the infinite number of layers of water making up the Ekman spiral.

Ekman deduced that in a homogeneous infinite ocean the speed of the surface current, u_0, is given by

$$u_0 = \frac{\tau}{\sqrt{A_z \rho f}} \tag{3.3}$$

where τ is the surface wind stress, A_z is the coefficient of eddy viscosity for vertical mixing, ρ (rho) is the density of seawater and f is the Coriolis parameter.

QUESTION 3.4 (a) What is the magnitude of f, the Coriolis parameter, at 40° S, given that $\Omega = 7.29 \times 10^{-5} \text{s}^{-1}$?

(b) (i) A westerly wind with a speed of 5ms^{-1} blows over the surface of the ocean at 40° S. Assuming that the wind stress that results is 0.1 Nm^{-2}, and that $A_z = 10^2 \text{kgm}^{-1}\text{s}^{-1}$ and $\rho = 10^3 \text{kgm}^{-3}$, what is the speed of the wind-generated current at the surface, according to equation 3.3?

(ii) In which direction does it flow?

(c) How well does the value you have calculated for the surface current speed correspond with the 'rule of thumb' mentioned earlier, given that the wind speed is 5ms^{-1}?

Observations of wind-driven currents in the real oceans, away from coastal boundaries, have shown that surface current speeds are similar to those predicted by Ekman and that the surface current direction deviates *cum sole* of the wind direction. However, the deviation is generally less than the 45° predicted, and the full spiral pattern has not been recorded. There are several reasons for this. For example, in many regions the ocean is too shallow, so that the full spiral cannot develop, and friction with the sea-bed becomes significant.

Bearing in mind the calculation you made in Question 3.4(b), can you suggest another more fundamental reason?

For the purpose of his calculations, Ekman assumed that the coefficient of eddy viscosity A_z remains constant with depth. As discussed in Section 3.1.1, values of A_z vary over several orders of magnitude and are generally much higher in the surface layers of the ocean than at depth.

The significant prediction of Ekman's theory is not the spiral current pattern but the fact that *the mean motion of the wind-driven layer is at right angles to the wind direction*, to the right in the Northern Hemisphere and to the left in the Southern Hemisphere. This may be understood by considering the averaged effect of the current spiral (Figure 3.6(a)), or by considering the forces acting on the wind-driven layer as a whole. The only forces acting on the wind-driven layer are wind stress and the Coriolis force (ignoring friction at the bottom of the wind-driven layer), and once an equilibrium situation has been reached these two forces must be equal and must act in opposite directions. Such a balance is obtained by the average motion, or **depth mean current**, being at right angles to the wind direction; Figure 3.6(b) illustrates this for the Northern Hemisphere.

The magnitude of the depth mean current, $\bar{u}$ (where the bar means 'averaged'), is given by

$$\bar{u} = \frac{\tau}{D\rho f} \tag{3.4}$$

where D is the depth at which the direction of the wind-driven current is directly opposite to its direction at the surface (Figure 3.6(a)). At this depth the current has decreased to 1/23 of its surface value, and the effect of the wind may be regarded as negligible. D, which depends on the eddy viscosity and the latitude, is therefore often equated with the **depth of frictional influence** of the wind. If it is assumed that the eddy viscosity $A_z = 10 \, \text{kg} \, \text{m}^{-1} \text{s}^{-1}$, D works out to about 40m at the poles and about 50m for middle latitudes; close to the Equator it rapidly approaches infinity. In the real oceans the thickness of the wind-driven layer is typically 100–200m. As you may expect from Section 3.1.1, it is largely determined by the thickness of the mixed surface layer, as frictional coupling between this and deeper water may be relatively weak.

The total *volume* of water transported at right angles to the wind direction per second may be calculated by multiplying $\bar{u}$ by the thickness of the wind-driven layer. This volume transport (in $\text{m}^3 \text{s}^{-1}$) is generally known as the **Ekman transport**, and the wind-driven layer (theoretically of depth D) as the **Ekman layer**. Ekman transports in response to prevailing wind fields contribute significantly to the general ocean circulation, as will become clear in Sections 3.4 and 4.2.2.

3.2 INERTIA CURRENTS

We should now briefly consider what happens when the wind that has been driving a current suddenly ceases to blow. Because of its momentum, the water will not come to rest immediately, and as long as it is in motion, both friction and the Coriolis force will continue to act on it. In the deep ocean away from any boundaries, frictional forces may be very small so that the energy imparted to the water by the wind takes some time to be dissipated; meanwhile the Coriolis force continues to turn the water *cum sole*. The resulting curved motion, under the influence of the Coriolis force, is known as an **inertia current** (Figure 3.7(a)). If the Coriolis force is the *only* force acting in a horizontal direction, and the motion involves only a small change in latitude, the path of the inertia current will be circular (Figure 3.7(b)).

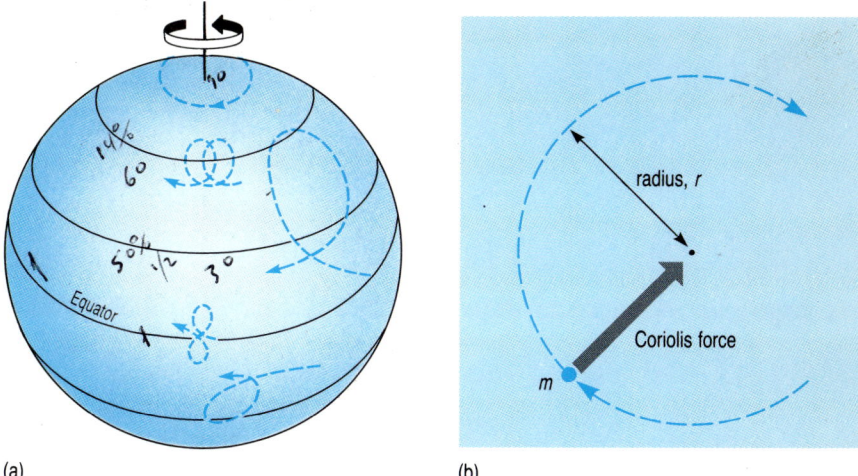

Figure 3.7 (a) Various possible paths for inertia currents. (b) Plan view showing inertial motion in the Northern Hemisphere. For details, see text.

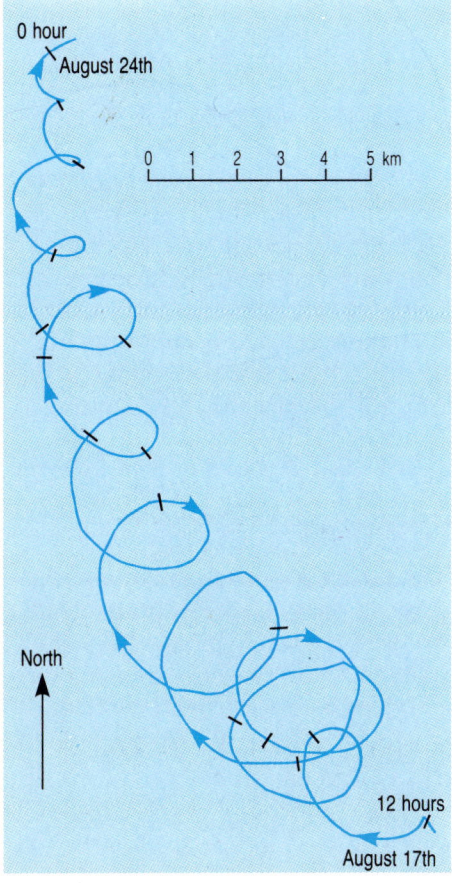

Figure 3.8 Plan view showing inertial motion observed in the Baltic Sea. The diagram shows the path of a parcel of water; if this path was representative of the general flow, the surface water in the region was both rotating and moving north–north-west. The observations were made between 17 and 24 August, 1933, and the tick marks on the path indicate intervals of 12 hours.

In an inertia current, the Coriolis force is acting as a centripetal force, towards the centre of the circle (Figure 3.7(b)). Now, if a body of mass m moves around a circle of radius r at a speed u, the centripetal force is given by

$$\text{centripetal force} = \frac{mu^2}{r} \tag{3.5}$$

In this case, the centripetal force is the Coriolis force (given by mfu, where f is the Coriolis parameter), so we can write:

$$\text{centripetal force} = \text{Coriolis force}$$

i.e. $\quad \dfrac{mu^2}{r} = mfu$

This equation simplifies to:

$$\frac{u}{r} = f \tag{3.6}$$

If the motion is small-scale and does not involve any appreciable changes in latitude, f is constant and so the water will follow a circular path of radius r, at a constant speed u. The time, T, taken for a water parcel to complete one circuit, i.e. the *period* of the inertia current, is given by the circumference of the circle divided by the speed:

$$T = \frac{2\pi r}{u}$$

and so, from equation 3.6,

$$T = \frac{2\pi}{f} \tag{3.7}$$

This equation shows that in the 'ideal' situation the only variable on which T depends is latitude. For example, at latitude 45°, T is approximately 17 hours, while at the Equator it becomes infinite.

Inertia currents have been identified by current measurement in many parts of the ocean. A classic example, observed in the Baltic Sea, is illustrated in Figure 3.8. Here a north–north-westerly wind-driven current has been superimposed on the inertial motion which had a period of about 14 hours (the theoretical value of T for this latitude is 14 hours 8 minutes), and died out after some nine rotations.

In Question 3.4(a) we used the fact that Ω, the angular speed of rotation of the Earth about its axis, is $7.29 \times 10^{-5} s^{-1}$. This may also be written as 7.29×10^{-5} radians*s^{-1}, and is calculated by dividing the angle the Earth turns through each day (2π) by the time it takes to do so (one day). Strictly, we should use one sidereal day, i.e. the time it takes the Earth to complete one revolution relative to the fixed stars. This is 23 hr 56 min. or 86 160 s.

Thus $\Omega = \dfrac{2\pi}{86\,160} s^{-1}$

$\qquad = \dfrac{2 \times 3.142}{86\,160} s^{-1}$

$\qquad = 7.29 \times 10^{-5} s^{-1}$

For many purposes, however, we can use $\Omega = 2\pi/24 \, hr^{-1}$.

QUESTION 3.5 Bearing this in mind, show that the period of an inertia current at the North Pole would be about 12 hours, while at a latitude of 30° it would be about 24 hours.

3.3 GEOSTROPHIC CURRENTS

So far, we have been assuming that the ocean is infinitely wide, as Ekman did in formulating his theory. When the effect of coastal boundaries is brought into the picture it becomes rather more complicated, because boundaries impede current flow and lead to slopes in the sea-surface—put simply, water tends to 'pile up' against boundaries. This is important because if the sea-surface is sloping, the hydrostatic pressure acting on horizontal surfaces at depth in the ocean will vary accordingly; in other words there will be horizontal pressure gradients. In the same way that winds blow from high to low pressure, so water tends to flow so as to even out lateral differences in pressure; the force that gives rise to this motion is known as the **horizontal pressure gradient force.**

We have seen how wind gives rise to the movement of surface waters and how consequent equilibrium motion depends on a balance between wind stress and the Coriolis force. If the Coriolis force acting on moving water is balanced by a horizontal pressure gradient force, the current is said to be in geostrophic equilibrium and is described as a **geostrophic current.**

Before we discuss geostrophic currents further, we should consider oceanic pressure gradients in a little more detail.

3.3.1 PRESSURE GRADIENTS IN THE OCEAN

As stated above, horizontal pressure gradients result from lateral variations in pressure. Pressure in the ocean is affected to some extent by the motion of the water. However, because ocean currents are relatively

---

* A radian is the angle subtended at the centre of a circle by an arc of the circle equal in length to the radius, r. As the circumference of a circle is given by $2\pi r$, there are $2\pi r/r = 2\pi$ radians in 360°. Because a radian is one length divided by another it is dimensionless and therefore not a unit in the usual sense.

slow, particularly in a vertical direction, for most purposes the pressure at depth may be taken to be the hydrostatic pressure—i.e. the pressure resulting from the overlying water, which is assumed to be static.

Hydrostatic pressure

The hydrostatic pressure at any depth, z, in the ocean is simply the weight of the water acting on unit area (say $1\,m^2$). This is given by:

hydrostatic pressure at depth $z = p =$ (mass of overlying seawater) $\times g$

where g is the acceleration due to gravity. If the density of seawater is ρ, this can be written as:

$p =$ (volume of overlying seawater) $\times \rho \times g$

which for hydrostatic pressure acting on unit area is simply

$$p = \rho g z \qquad (3.8a)$$

This is known as the **hydrostatic equation**.

It is usually written:

$$p = -\rho g z \qquad (3.8b)$$

because in oceanography the vertical z-axis has its origin at sea-level and is positive upwards and negative downwards.

In the real ocean, the density ρ varies with depth. A column of seawater may be seen as consisting of an infinite number of layers, each of an infinitesimally small thickness dz and contributing an infinitesimally small pressure dp to the total hydrostatic pressure at depth z (Figure 3.9(a)). In this case,

$$dp = -\rho g dz \qquad (3.8c)$$

and the total pressure at depth z is the sum of all the dp's. However, if we make the assumption that density is independent of depth, or we take an average value for ρ, the hydrostatic pressure at any depth may be calculated using equation 3.8a or 3.8b (Figure 3.9(b)).

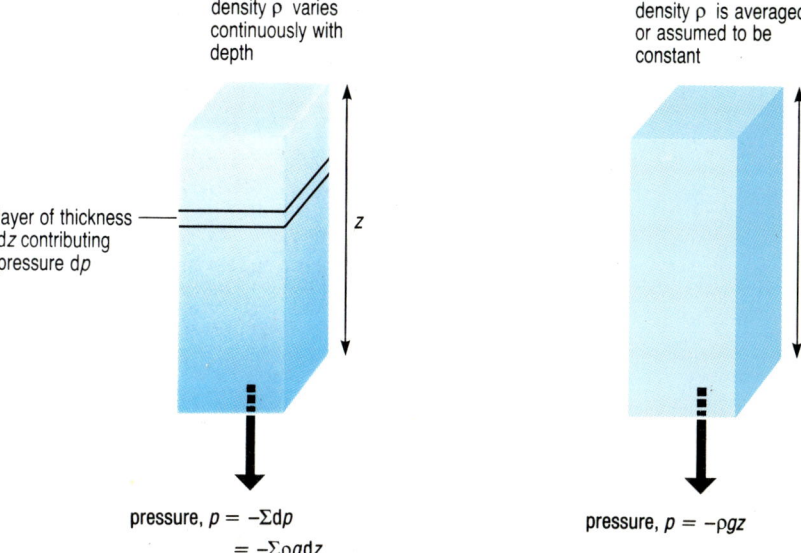

Figure 3.9 Diagrams to illustrate how pressure at depth in the ocean may be calculated. (a) The pressure p at depth z is the sum (Σ 'sigma') of all the pressures dp. (b) If an average value is taken for the density ρ, or it is assumed to be constant, the hydrostatic pressure at depth z is simply $-\rho gz$.

Horizontal pressure gradients

Horizontal pressure gradients are easiest to imagine by first considering a highly unrealistic situation. Seawater of constant density ρ occupies an ocean basin (or some other container) and a sea-surface slope is maintained across it without giving rise to any current motion (see Figure 3.10). The hydrostatic pressure acting at point A in Figure 3.10 is given by the hydrostatic equation (3.8a):

$$p_A = \rho g z$$

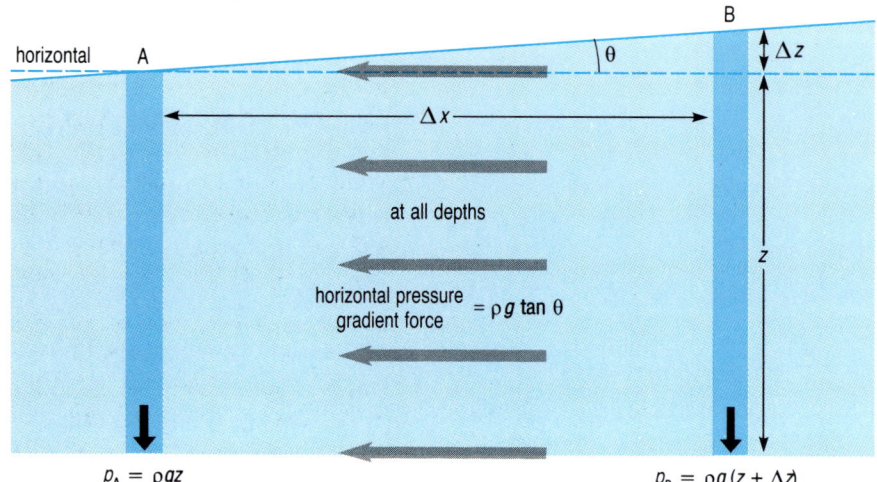

Figure 3.10 Schematic diagram to illustrate how a sloping sea-surface results in a horizontal pressure gradient. If the seawater density is constant, the horizontal pressure gradient force (grey arrows) is $\rho g \tan \theta$ (where $\tan \theta = \Delta z / \Delta x$) at all depths. (For further information, see text.)

where z is the height of the overlying column of water (and we are omitting the minus sign). At point B, where the sea-level is higher by an amount Δz (delta z),

$$p_B = \rho g (z + \Delta z)$$

The pressure at B is therefore greater than that at A by a small amount, which we will call Δp, so that:

$$\Delta p = p_B - p_A = \rho g (z + \Delta z) - \rho g z = \rho g \Delta z$$

If A and B are a distance Δx apart, the horizontal pressure gradient between them is given by:

$$\frac{\Delta p}{\Delta x} = \rho g \frac{\Delta z}{\Delta x}$$

and because $\frac{\Delta z}{\Delta x} = \tan \theta$,

$$\frac{\Delta p}{\Delta x} = \rho g \tan \theta \quad (3.9)$$

If we assume that Δp and Δx are extremely small, we can call them dp and dx, where dp is the incremental increase in pressure over a horizontal distance dx. Equation 3.9 then becomes

$$\frac{dp}{dx} = \rho g \tan \theta \quad (3.10)$$

This is the rate of change of pressure, or the horizontal pressure gradient force, in the x-direction. More precisely, it is the horizontal pressure gradient force acting on unit *volume* of seawater. In order to obtain the force acting on unit mass of seawater, we simply divide by the density, ρ:

$$\text{horizontal pressure gradient force per unit mass} = \frac{1}{\rho}\frac{dp}{dx} = g \tan\theta \qquad (3.10a)$$

If the ocean is homogeneously dense, the horizontal pressure gradient force from B to A is the same however deep A and B are. If there are no other horizontal forces acting, the entire ocean will be uniformly accelerated from the high pressure side to the low pressure side.

3.3.2 BAROTROPIC AND BAROCLINIC CONDITIONS

In the idealized situation discussed in the previous Section, the ocean across which a sea-surface slope had been set up consisted of seawater of constant density. Consequently, the resulting horizontal pressure gradient was simply a function of the degree of slope of the surface (tan θ)—the greater the sea-surface slope, the greater the horizontal pressure gradient. In such a situation, the surfaces of equal pressure within the ocean—the **isobaric surfaces**—are parallel to the sea-surface, itself the topmost isobaric surface.

In real situations where ocean waters are well-mixed and therefore fairly homogeneous, density nevertheless increases with depth because of compression caused by the weight of overlying water. As a result, the isobaric surfaces are parallel not only to the sea-surface but also to the surfaces of constant density or **isopycnic surfaces**. Such conditions are described as **barotropic** (see Figure 3.11(a)).

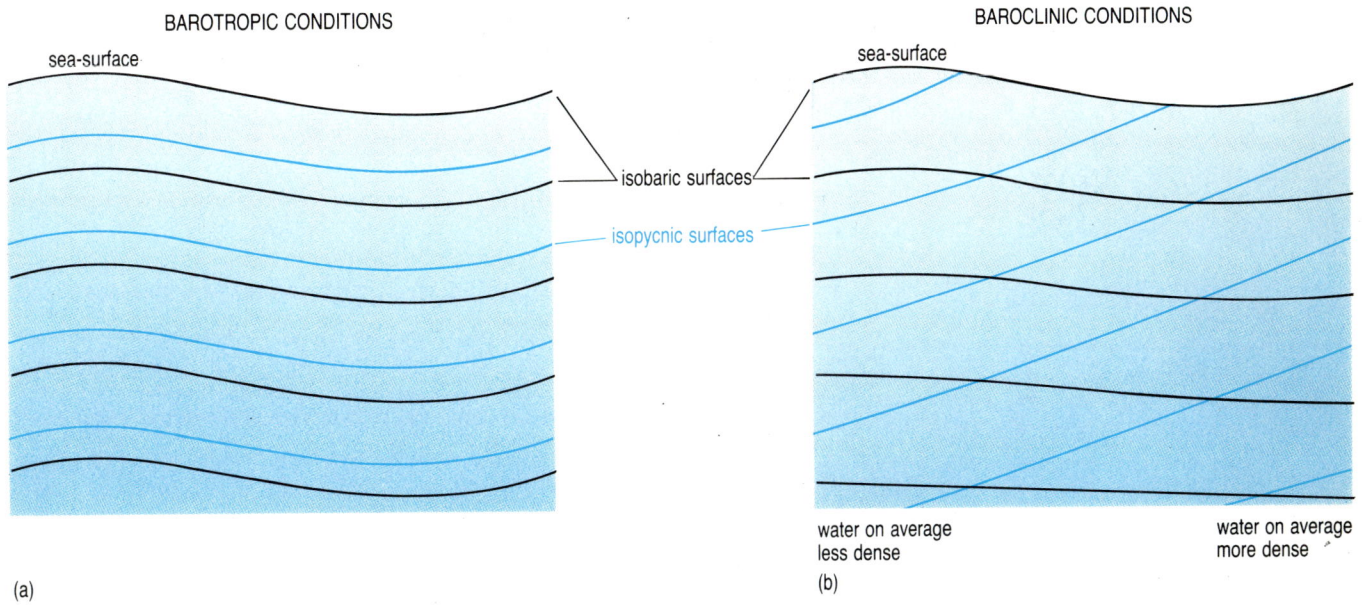

Figure 3.11 The relationship between isobaric and isopycnic surfaces in (a) barotropic conditions and (b) baroclinic conditions. In barotropic conditions, the density distribution (indicated by the intensity of blue shading) does not influence the shape of isobaric surfaces. By contrast, in baroclinic conditions, lateral variations in density *do* affect the shape of isobaric surfaces.

As discussed in Section 3.3.1, the hydrostatic pressure at any given depth in the ocean is determined by the weight of overlying seawater. In barotropic conditions, the variation of pressure over a horizontal surface at depth is determined *only* by the slope of the sea-surface, which is why isobaric surfaces are *parallel* to the sea-surface. However, any variations in the density of seawater will also affect the weight of overlying seawater, and hence the pressure, acting on a horizontal surface at depth.

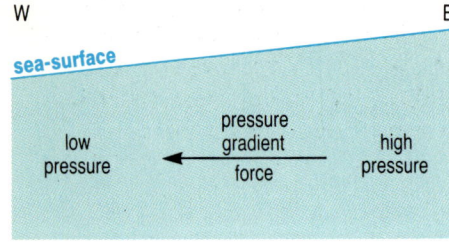

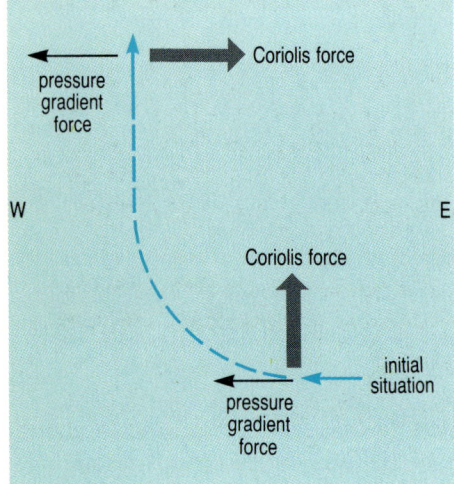

Figure 3.12 (a) In the Northern Hemisphere, a sea-surface slope up towards the east results in a horizontal pressure gradient force towards the west.
(b) Initially, this causes motion 'down the pressure gradient', but because the Coriolis force acts at right angles to the direction of motion, the equilibrium situation is one in which the direction of flow (blue line) is at right angles to the pressure gradient.

Therefore, *in situations where there are lateral variations in density*, *isobaric surfaces* are *not* parallel to the sea-surface; isobaric surfaces *intersect isopycnic surfaces* and the two slope in opposite directions (see Figure 3.11(b)). Because isobaric and isopycnic surfaces are *inclined* with respect to one another, such conditions are known as **baroclinic**.

Geostrophic currents—currents in which the horizontal pressure gradient force is balanced by the Coriolis force—may occur whether conditions in the ocean are barotropic (homogeneous), or baroclinic (with lateral variations in density). At the end of the previous Section we noted that in the hypothetical ocean, where the pressure gradient force was the only horizontal force acting, motion would occur in the direction of the pressure gradient. In the real ocean, as soon as water begins to move in the direction of a horizontal pressure gradient, it becomes subject to the Coriolis force. Imagine, for example, a region in a Northern Hemisphere ocean, in which the sea-surface slopes up towards the east, so that there is a horizontal pressure gradient force acting from east to west (Figure 3.12(a)). Water moving westwards under its influence immediately begins to be deflected towards the north by the Coriolis force, and eventually an equilibrium situation may be attained in which the water flows northwards and the Coriolis force acts towards the east, balancing the horizontal pressure gradient force towards the west (Figure 3.12(b)). Thus, in a geostrophic current, instead of moving *down* the horizontal pressure gradient, water moves *at right angles* to it.

Moving fluids tend towards situations of equilibrium, and so flow in the ocean is often geostrophic or nearly so. This is also true in the atmosphere; geostrophic winds may be recognized on weather maps by the fact that the wind-direction arrows are parallel to relatively straight isobars. Even when the motion of the air is strongly curved, and centripetal forces are important (as is the case in cyclones and anticyclones), wind-direction arrows cross the isobars obliquely rather than at right angles (*cf.* Figure 2.5).

So far, we have only been considering forces in a horizontal plane. Figure 3.13 is a partially completed diagram showing the forces acting on a parcel of water of mass m, in a region of the ocean where isobaric surfaces make an angle θ with the horizontal. If equilibrium has been attained so that the current is steady and not accelerating, the forces acting in the horizontal direction must balance one another, as must those acting in a vertical direction. The vertical arrow labelled mg represents the weight of the parcel of water, and the two horizontal arrows represent the horizontal pressure gradient force and the Coriolis force.

QUESTION 3.6 (a) Given the direction of slope of the isobars on Figure 3.13, which of the horizontal arrows represents the horizontal pressure gradient force and which the Coriolis force?

(b) If the current is flowing '*into* the page', is the situation illustrated in Figure 3.13 in the Northern or Southern Hemisphere?

In Section 3.3.1 we considered a somewhat unrealistic ocean in which conditions were barotropic. We deduced that if the sea-surface (and all other isobars down to the bottom) made an angle of θ with the horizontal, the horizontal pressure gradient force acting on *unit* mass of seawater is given by $g \tan \theta$ (equation 3.10a). The horizontal pressure

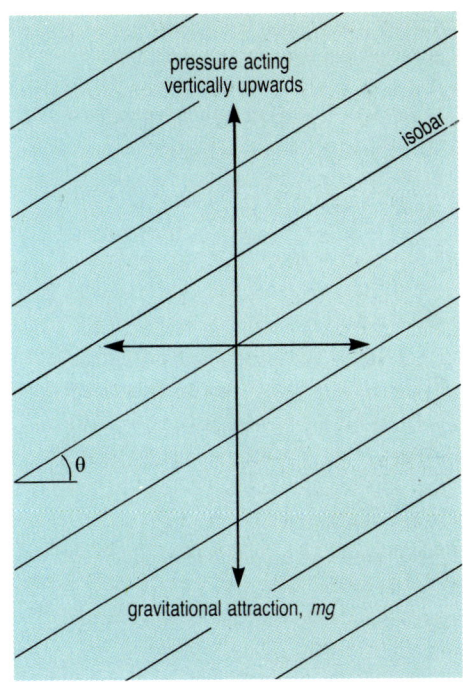

Figure 3.13 Cross-section showing the forces acting on a parcel of water, of mass *m* and weight *mg*, in a region of the ocean where the isobars make an angle θ with the horizontal. (Note that if the parcel of water is not to move vertically downwards under the influence of gravity, there must be an equal force acting upwards; this is supplied by pressure.)

gradient force acting on a water parcel of mass *m* is therefore given by *mg* tan θ. We also know that the Coriolis force acting on such a water parcel moving with velocity *u* is *mfu*, where *f* is the Coriolis parameter (equation 3.2). In conditions of geostrophic equilibrium, the horizontal pressure gradient force and the Coriolis force balance one another, and we can therefore write:

$$mg \tan \theta = mfu$$

or $\quad \tan \theta = \dfrac{fu}{g} \quad$ (3.11)

Equation 3.11 is known as the **gradient equation**, and in geostrophic flow *is true for every isobaric surface* (Figure 3.14).

It is worth noting that the Coriolis force and the horizontal pressure gradient force are extremely small: they are generally less than 10^{-4} N kg^{-1} and therefore several orders of magnitude smaller than the forces acting in a vertical direction. Nevertheless, in much of the ocean, the Coriolis force and the horizontal pressure gradient force are the largest forces acting in a horizontal direction.

As shown in Figure 3.11, in barotropic conditions the slope of the isobaric surfaces follows that of the sea-surface, whereas in baroclinic conditions the slope of the isobaric surfaces changes with depth.

Suppose that, at some depth, the isobaric surface becomes horizontal. What implication does this have for the velocity of the geostrophic current, given the relationship between isobaric slope and geostrophic velocity, *u*, expressed by the gradient equation (equation 3.11)?

As geostrophic velocity, *u*, is proportional to tan θ, the greater the slope of the isobars, the greater the geostrophic velocity. Therefore, in barotropic conditions the geostrophic velocity is constant with depth,

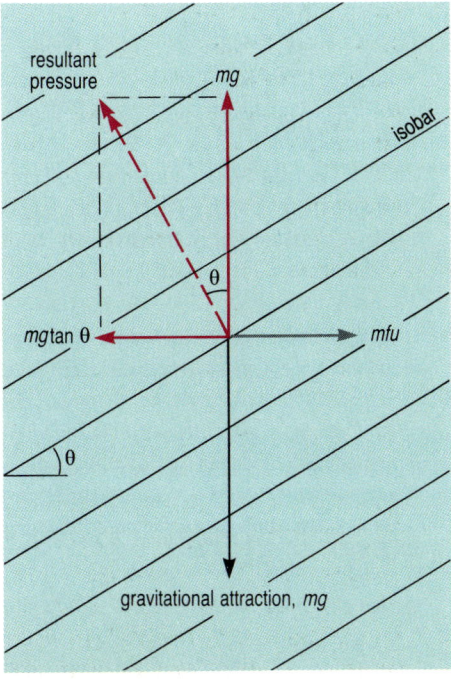

Figure 3.14 Diagram to illustrate the dynamic equilibrium embodied in the gradient equation. In geostrophic flow, the horizontal pressure gradient force (*mg* tan θ) is balanced by the equal and opposite Coriolis force (*mfu*). If you wish to verify for yourself trigonometrically that the horizontal pressure gradient force is given by *mg* tan θ, you will find it helpful to regard the horizontal pressure gradient force as the horizontal component of the resultant pressure which acts at right angles to the isobars; the vertical component is of magnitude *mg*, balancing the weight of the parcel of water (*cf.* Figure 3.13).

while in baroclinic conditions it varies with depth, becoming zero where isobaric surfaces become horizontal (i.e. where $\tan \theta = 0$).

Figure 3.15 summarizes the differences between barotropic and baroclinic conditions, and illustrates both how the distribution of density affects the slopes of the isobaric surfaces and how these, in turn, affect the variation of geostrophic current velocity with depth.

Barotropic conditions may be found in the well-mixed surface layers of the ocean, and in shallow shelf seas, particularly when they have been well-mixed by tidal currents. They also characterize the deep ocean, below the permanent thermocline, where density and pressure are generally only a function of depth so that isopycnic surfaces and isobaric surfaces are parallel. Conditions are most strongly baroclinic (i.e. the angle between isobaric and isopycnic surfaces is greatest) in regions of fast surface current flow.

Currents are described as geostrophic when the Coriolis force acting on the moving water is balanced by the horizontal pressure gradient force. This is true whether the water movement is being maintained by wind stress and the waters of the upper ocean have 'rearranged themselves' so that the density distribution is such that geostrophic equilibrium is attained, *or* whether the density distribution is itself the *cause* of water movement. Indeed, it is often not possible to determine whether a horizontal pressure gradient within the ocean is the cause or the result of current flow, and in many cases it may not be appropriate to try to make this distinction.

For there to be exact geostrophic equilibrium, the flow should be steady and the pressure gradient and the Coriolis force should be the only forces acting on the water, other than the attraction due to gravity. In the real oceans, other influences may be important; for example, there may be friction with nearby coastal boundaries or adjacent currents, or with the sea-floor. In addition, there may be local accelerations and fluctuations, both vertical and horizontal, resulting perhaps from internal waves. Nevertheless, many ocean currents, including all the major surface current systems—e.g. the Gulf Stream, the Antarctic Circumpolar Current and the equatorial currents—are, to a first approximation, geostrophic currents.

It is important to remember that the slopes shown in diagrams like Figures 3.10 to 3.15 are greatly exaggerated. Sea-surface slopes associated with geostrophic currents are broad, shallow, topographic irregularities. They may be caused by prevailing winds 'piling up' water against a coastal boundary, by variations in pressure in the overlying atmosphere, or by lateral variations in water density resulting from differing temperature and salinity characteristics (in which case, conditions are baroclinic), or by some combination of these. The slopes have gradients of about 1 in 10^5 to 1 in 10^8, i.e. a few metres in 10^2–10^5 km, so that they are extremely difficult to detect, let alone measure. However, under baroclinic conditions the isopycnic surfaces may have slopes several hundred times this, and these *can* be determined. How this is done is outlined in Section 3.3.3.

Figure 3.15 Diagrams to summarize the difference between (a) barotropic and (b) baroclinic conditions. The intensity of blue shading corresponds to the density of the water.

(a) In barotropic flow, isopycnic surfaces (surfaces of constant density) and isobaric surfaces are parallel and their slopes remain constant with depth. Because the slope of the isobaric surfaces remains constant with depth, the horizontal pressure gradient from B to A, and hence the geostrophic current, is constant with depth.

(b) In baroclinic flow, the isopycnic surfaces intersect (or are *inclined* to) isobaric surfaces. At shallow depths, isobaric surfaces are parallel to the sea-surface, but with increasing depth their slope becomes smaller, because the average density of a column of water at A is more than that of a column of water at B (in barotropic conditions the average density of two such columns is the same). As the isobaric surfaces become increasingly near horizontal, so the horizontal pressure gradient decreases and so does the geostrophic current, until at some depth the isobaric surfaces are horizontal and the geostrophic current is zero.

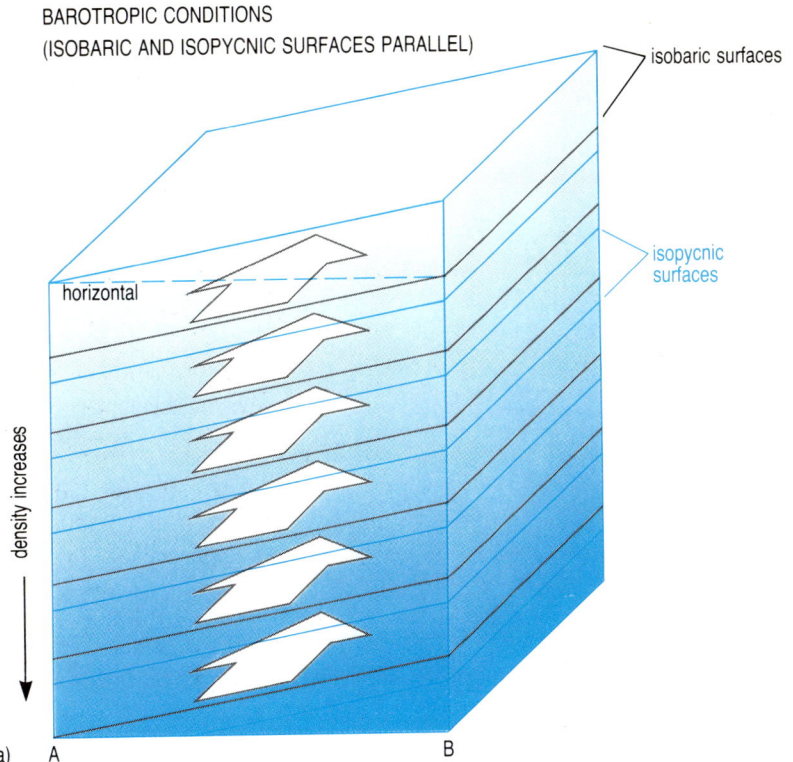

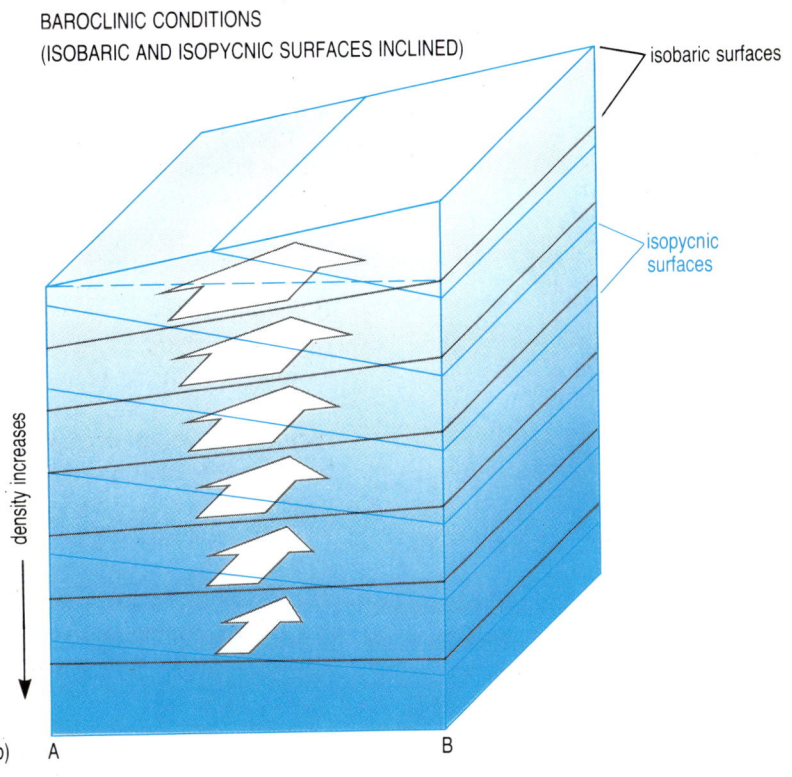

3.3.3 DETERMINATION OF GEOSTROPHIC CURRENT VELOCITIES

Note: From now on, we will usually follow common practice and use the terms isobar and isopycnal—lines joining points of equal pressure and of equal density, respectively—instead of the more cumbersome 'isobaric surface' and 'isopycnic surface'.

We have seen that in geostrophic flow the slopes of the sea-surface, isobars and isopycnals are all related to the current velocity. This means that measurement of these slopes can be used to determine the current velocity.

In barotropic conditions, how might the geostrophic current velocity be estimated?

In theory, if the slope of the sea-surface could be measured, the current velocity could be determined using the gradient equation (equation 3.11). In practice, it is only convenient to do this for flows through straits where the average sea-level on either side may be calculated using tide-gauge data; the measurement of the slope of the sea-surface in the open ocean would be extremely difficult, if not impossible, by traditional oceanographic methods.

In practice, it is the density distribution, as reflected by the slopes of the isopycnals, that is used to determine geostrophic current flow in the open oceans.

Why should it be relatively straightforward to determine the density distribution in a volume of ocean?

Because density is a function of temperature and salinity, both of which are routinely and fairly easily measured with the necessary precision. It is also a function of pressure or, to a first approximation, depth, which is also fairly easy to measure precisely.

Measurement of the density distribution also has the advantage—as we observed earlier—that in baroclinic conditions the slopes of the isopycnals are several hundred times those of the sea-surface and other isobars. However, it is important not to lose sight of the fact that it is the slopes of the *isobars* that we are ultimately interested in, because they control the horizontal pressure gradient; determination of the slopes of the isopycnals is, in a sense, only a means to an end.

Figure 3.16(a) and (b) are two-dimensional versions of Figure 3.15(a) and (b). In the situation shown in (a), conditions are barotropic and the current velocity, u, is constant with depth.

How may we determine u from the density distribution?

We know that for any isobaric surface making an angle θ with the horizontal, the velocity u along that surface may be calculated using the gradient equation:

$$\tan \theta = \frac{fu}{g} \qquad \text{(equation 3.11)}$$

where f is the Coriolis parameter and g is the acceleration due to gravity.

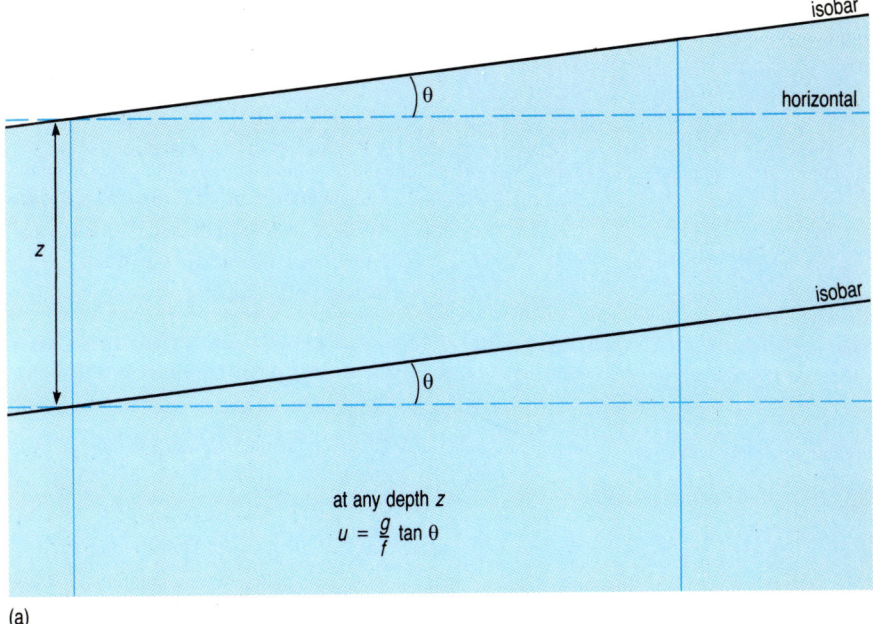

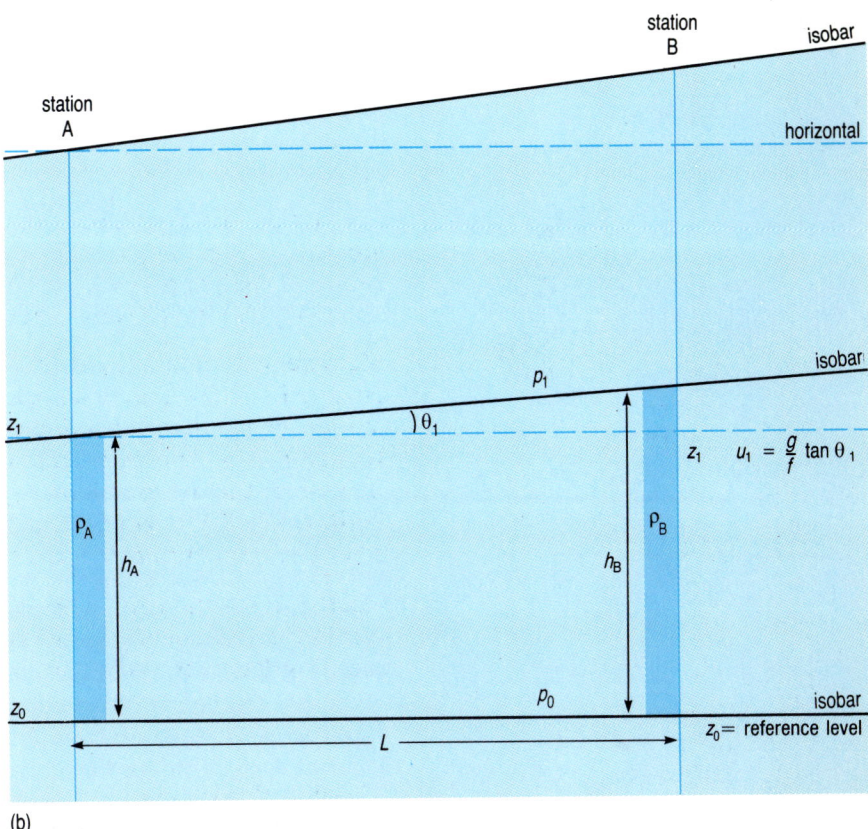

Figure 3.16 (a) In barotropic conditions, the slope of the isobars is tan θ at all depths; the geostrophic current velocity u is therefore (g/f) tan θ at all depths.

(b) In baroclinic conditions, the slope of the isobars varies with depth. At depth z_1, the isobar corresponding to pressure p_1 has a slope of tan θ_1. At depth z_0 (the reference level), the isobar corresponding to pressure p_0 is assumed to be horizontal. (For further details, see text.)

As conditions here are barotropic, the isopycnals are parallel to the isobars and we can therefore determine θ by measuring the slope of the isopycnals. The velocity, u, will then be given by:

$$u = \frac{g}{f} \tan \theta \qquad (3.11a)$$

at all depths in the water column. The depth-invariant currents that flow in barotropic conditions (isobars parallel to the sea-surface slope) are sometimes described as 'slope currents'; they are often too small to be measured directly.

By contrast, the geostrophic currents that flow in baroclinic conditions *vary* with depth (Figure 3.15(b)). Unfortunately, from the density distribution alone we can only deduce *relative* current velocities; that is, we can only deduce *differences* in current velocity between one depth and another. However, if we know the isobaric slope or the current velocity at some depth, we may use the density distribution to calculate how much greater (or less) is the isobaric slope and hence geostrophic current velocity at other depths. For convenience, it is often assumed that at some fairly deep level (known as a 'reference level') the isobars are horizontal (i.e. the horizontal pressure gradient force is zero) and the geostrophic velocity is therefore zero. Relative current velocities calculated with respect to this reference level may be assumed to be absolute velocities. This is the approach we will take here.

Look again at Figure 3.16(b), which we will assume represents a cross-section of ocean at *right angles* to the geostrophic current. A and B are two oceanographic stations a distance L apart. At each station, measurements of temperature and salinity have been made at various depths, and used to deduce how *density* varies with depth. However, if we are to find out what the geostrophic velocity is at depth z_1 (say), we really need to know how *pressure* varies with depth at each station, so that we can calculate the slope of the isobars at depth z_1.

Why do we want to know the slope of the isobars?

So that we can apply the gradient equation (3.11a), and hence obtain a value for u.

Assume that in this theoretical region of ocean the reference level has been chosen to be at depth z_0. From our measurements of temperature and salinity, we know that the average density ρ of a column of water between depths z_1 and z_0 is greater at station A than at station B; i.e. $\rho_A > \rho_B$. The distance between the isobars p_1 and p_0 must therefore be greater at B than at A, because hydrostatic pressure is given by $\rho g h$, where h is the height of the column of water (*cf.* equation 3.8). Isobar p_1 must therefore slope up from A to B, making an angle θ_1 with the horizontal, as shown in Figure 3.16(b).

How can $\tan \theta_1$ be expressed in terms of distances shown on Figure 3.16(b)?

It is given by

$$\tan \theta_1 = \frac{h_B - h_A}{L}$$

Substituting for $\tan \theta_1$ in the gradient equation (3.11a), we get

$$u = \frac{g}{f}\left(\frac{h_B - h_A}{L}\right) \tag{3.12}$$

The difference in hydrostatic pressure between isobars p_1 and p_0 is the same at both A and B, and so

$$\rho_A g h_A = \rho_B g h_B$$

i.e. $h_A = h_B \dfrac{\rho_B}{\rho_A}$

Therefore, substituting for h_A in equation 3.12 we get

$$u = \frac{g}{f}\left(\frac{h_B - h_B \dfrac{\rho_B}{\rho_A}}{L}\right)$$

i.e. $u = \dfrac{g h_B}{fL}\left(1 - \dfrac{\rho_B}{\rho_A}\right)$ \hfill (3.13)

We now have an equation that enables us to deduce the geostrophic current velocity u from the density distribution; we will refer to it as the 'geostrophic equation'. Use the geostrophic equation, in conjunction with Figure 3.16(b), to answer Question 3.7.

QUESTION 3.7 (a) Stations A and B are 100km apart at about 30° of latitude. Calculate the geostrophic velocity u at depth $z_1 = 1000$m, taking your reference level (z_0) as 2 000m; temperature and salinity measurements at the two stations indicate that $\rho_A = 1.0265$ kgm^{-3} and $\rho_B = 1.0262$ kgm^{-3}.

Note: In order to apply equation 3.13 you will have to make the assumption that h_B and $z_0 - z_1$ are equal. The value for g is 9.8 ms^{-2}.

(b) Use equation 3.12 to calculate the difference $h_B - h_A$.

(c) If stations A and B are in the Southern Hemisphere, with B due east of A, in which direction is the geostrophic current flowing?

By setting h_B equal to $z_0 - z_1$ in Question 3.7(a) you were effectively using average densities ρ_A and ρ_B that had been calculated assuming that the columns of water between isobars p_1 and p_0 were the same height at A and B. As you will have found in part (b), the difference in height between the two columns ($h_B - h_A$) is a very small fraction (0.03%) of $z_0 - z_1$, and the resulting error in ρ_B, and hence u, is also extremely small.

In Question 3.7(a) you calculated the velocity of the geostrophic current at one depth, z_1. A complete profile of the variation of the geostrophic current velocity with depth may be obtained by choosing a reference level z_0 and then applying equation 3.13 at successively higher levels, each time calculating the geostrophic current velocity in relation to the level below (see for example Figure 3.17(a)). This is how geostrophic current velocities have been calculated since the beginning of the century; the full version of equation 3.13 used for this purpose is often referred to as Helland-Hansen's equation, after the Scandinavian oceanographer B. Helland-Hansen who, with J. W. Sandström, did much pioneering work in this field.

Until relatively recently, oceanographers had to calculate seawater densities from measured values of temperature, salinity and depth, using standard tables. Temperature, salinity and depth may now be recorded electronically and the necessary calculations done by computer, but it is still important for the oceanographer to be aware of the limitations of the geostrophic method and the various assumptions and approximations that are implicit in the calculations.

The first point to note is that, so far, we have been assuming that the geostrophic current we are attempting to quantify is flowing *at right angles* to the section A–B. In reality, there is no way of ensuring that this is the case. If the current direction makes an angle with the section, the geostrophic equation (3.13) will only provide a value for the component of the flow at right angles to it. The calculated geostrophic velocities will therefore be underestimates. To take an extreme example, if the average flow in a region is north-easterly and the section taken is from the south-west to north-east, the geostrophic velocity calculated will be only a very small proportion of the actual geostrophic velocity to the north-east. To get over this problem, a second section may be made at right angles to the first, and the total current calculated by combining the two components. In practice, the geostrophic flow in an area may be determined in this way using data from a grid of stations.

The second important limitation of this method is that, as discussed earlier, it may only be used to determine *relative* velocity. Thus in calculating the geostrophic current in Question 3.7(a), it was assumed that at a depth of 2 000m—the reference level—the current was zero. If in fact current measurements revealed that there was a current of, say, 0.05ms^{-1} at this depth, this amount could be added on to the velocity at all depths.

One approach is to regard any current that persists below the chosen reference level as the 'barotropic' part of the flow (Figure 3.17(b)). As mentioned in Section 3.3.2, in the deep ocean below the permanent thermocline, density and pressure usually vary as a function of depth only, and so even if they are not horizontal, isobars and isopycnals are likely to be parallel to one another.

Would the barotropic part of the flow show up in calculations like those you made in Question 3.7(a)?

No. The geostrophic velocity u obtained using equation 3.13 is that resulting from lateral variations in density (i.e. attributable to the difference between ρ_A and ρ_B). The effect of any horizontal pressure gradient that remains constant with depth is not included in equation 3.13. In reality, 'barotropic flow' resulting from a sea-surface slope caused by wind, and 'baroclinic flow' associated with lateral variations in density—or, put another way, the 'slope current' and the 'relative current'—may not be as easily separated from one another as Figure 3.17 suggests.

Another point that must be borne in mind is that the geostrophic equation only provides information about the *average* flow between two stations (which may be many tens of kilometres apart) and gives no information about the details of the flow. However, if the investigator is interested only in the large-scale mean conditions, this does not matter.

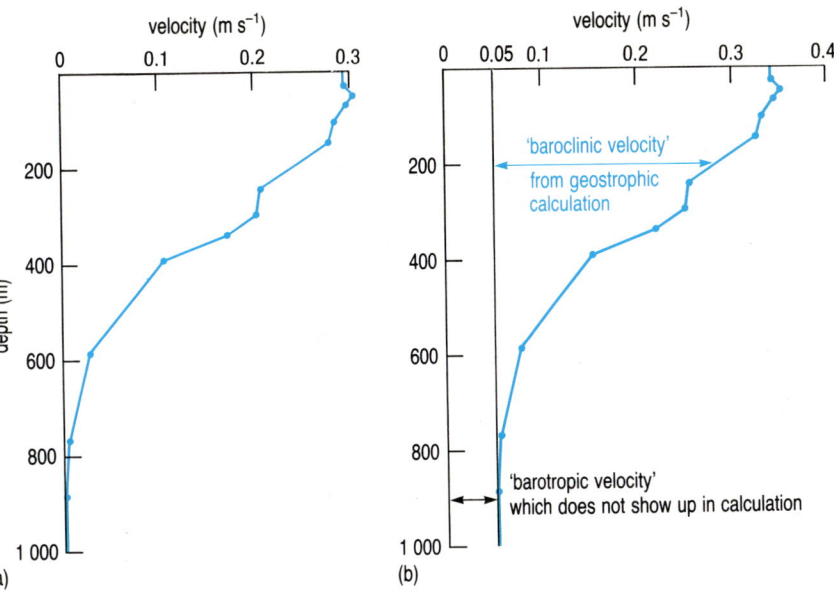

Figure 3.17 (a) Example of a profile of geostrophic current velocity, calculated on the assumption that the horizontal pressure gradient, and hence the geostrophic current velocity, are zero at 1 000 m depth (the reference level).

(b) If direct current measurements reveal that the current velocity below about 1 000 m is not zero as assumed, but some finite value (say 0.05 m s^{-1}), the geostrophic current velocity profile would look like this. The geostrophic velocity at any depth may therefore be regarded as a combination of baroclinic and barotropic components.

Indeed, in some ways it is even an advantage because it means that the effects of small-scale fluctuations are averaged out, along with variations in the flow that take place during the time the measurements are being made (which may be from a few days to a few weeks).

We have seen how information about the distribution of density with depth may be used to determine a detailed profile of geostrophic current velocity with depth. Although both density and current velocity generally vary continuously with depth (e.g. Figure 3.18(a) and Figure 3.17), for some purposes it is convenient to think of the ocean as a number of homogeneous layers, each with a constant density and velocity. This simplification is most often applied in considerations of the motion of the mixed surface layer, which is assumed to be a homogeneous layer separated from the deeper, colder waters by an abrupt density discontinuity (Figure 3.18(b)). In this situation, the slopes of the sea-surface and the interface will be as shown in Figure 3.19(a) and, for convenience, the geostrophic velocity of the upper layer may be calculated on the assumption that the lower layer is motionless. Figure 3.19(b) is an example of a more complex model, which may also sometimes approximate to reality; here there are three homogeneous layers, with the intermediate layer flowing in the opposite direction to the other two.

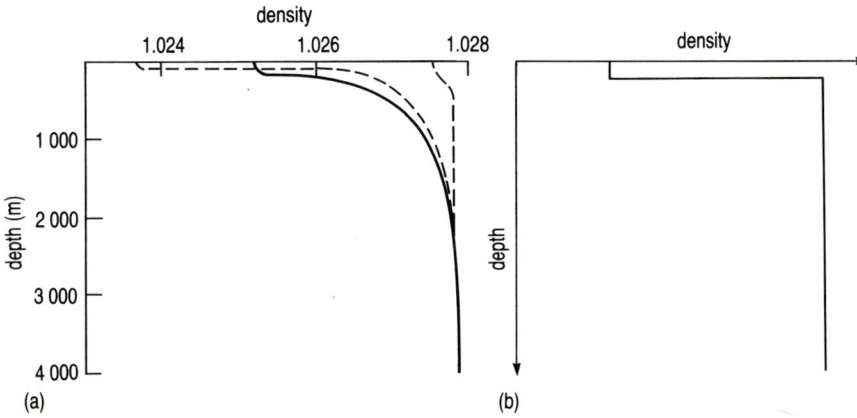

Figure 3.18 (a) Typical density–depth profiles for different latitudes: *solid line* = tropical latitudes; *1st dashed line* = equatorial latitudes; *2nd dashed line* = high latitudes. (b) The type of simplified density distribution sometimes assumed in order to estimate geostrophic currents in the mixed surface layer.

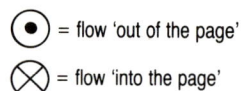

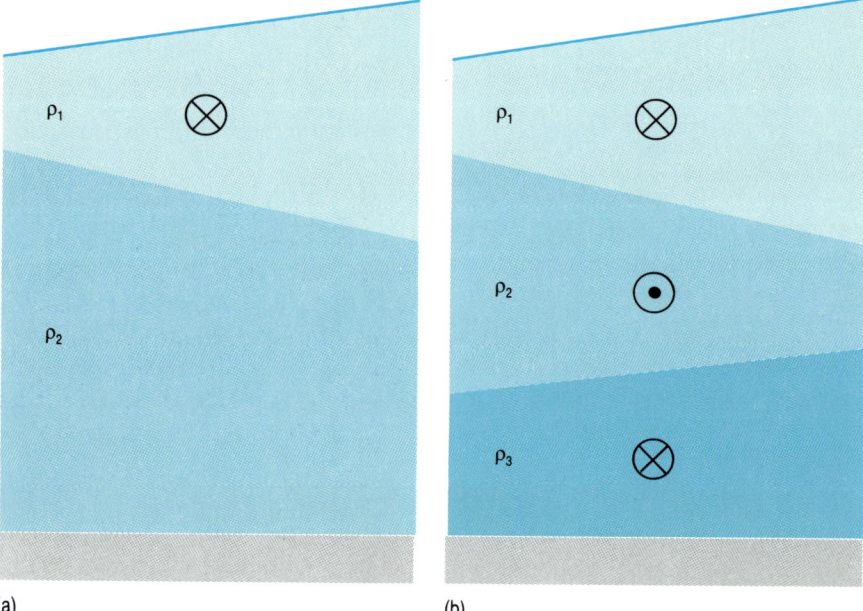

Figure 3.19 Diagrams to show how the interfaces between layers slope in the case of (a) two and (b) three homogeneous layers, where $\rho_1 < \rho_2 < \rho_3$. The velocity symbols are drawn on the assumption that this is for the Northern Hemisphere.

In Chapter 5, you will see how such simplifications may help us to interpret the density and temperature structure of the upper ocean in terms of geostrophic current velocity.

3.3.4 PRESSURE, DENSITY AND DYNAMIC TOPOGRAPHY

We stated earlier that seawater densities used in the calculation of geostrophic velocities are computed from data on temperature, salinity and depth. Of course, it is not depth *per se* that affects the density of a sample of seawater but the pressure it is under, which is related to the depth by the hydrostatic equation (3.8). For the purpose of making geostrophic calculations it is usual to interconvert hydrostatic pressure and depth by using the following approximate relationship:

$$\text{pressure} \approx 10^4 \times \text{depth} \qquad (3.14)$$

where pressure is in newtons per square metre (N m^{-2}) and depth is in metres.

Can you see where the factor of 10^4 comes from?

Relationship 3.14 is simply the hydrostatic equation $p = \rho g z$, in which it has been assumed that g is exactly 10 m s^{-2} (instead of 9.8 m s^{-2}) and that the density of seawater is a constant $1 \times 10^3 \text{ kg m}^{-3}$ (rather than ranging between 1.025 and $1.029 \times 10^3 \text{ kg m}^{-3}$). This approximation is very useful for 'back-of-envelope' calculations involving pressure at depth in the ocean, as we shall now demonstrate.

One possible source of error in geostrophic calculations that we have not considered is the effect of atmospheric pressure. An oceanographic section may be of the order of 100 km across and it is quite likely that there will be a difference in atmospheric pressure between the two oceanographic stations.

Would such a difference in atmospheric pressure be taken into account by equation 3.13?

No, it would not. In deducing equation 3.13 we used the *hydro*static equation (3.8), which gives pressures due to the weight of overlying *water* only. Nevertheless, lateral variations in atmospheric pressure would contribute to any slope in the sea-surface and hence to the slopes of isobars at depth. As the ultimate purpose of equation 3.13 is to obtain geostrophic current velocities from isobaric slopes, it would be useful to know the extent to which variations in atmospheric pressure affect these slopes. Now, the pressure resulting from a standard atmosphere is 1 bar or 10^5 N m^{-2}.

Given relationship 3.14 above, what depth of seawater would give rise to the same pressure?

The answer must be $10^5/10^4 = 10$ m. If a standard atmosphere is equivalent to 10 m of seawater, a difference in atmospheric pressure of a few millibars—which is what we might expect over distances of the order of 100 km—would be equivalent to a few centimetres of seawater. Thus, variations in atmospheric pressure typically give rise to isobaric slopes of about a few centimetres in 100 km or about 1 in 10^7, and hence contribute between about 1 and 10% of the total horizontal pressure gradient (*cf.* Section 3.3.2).

Would geostrophic current flow resulting from variations in atmospheric pressure vary with depth?

No, it would not, because it would be a 'slope current' and as such be in the 'barotropic' component of the flow (*cf.* Figure 3.17(b)). Thus, although the effects of variations in atmospheric pressure cannot be completely ignored, in theory they could be easily corrected for. However, weather systems may travel several hundred kilometres per day and so differences in atmospheric pressure between two stations are likely to vary over the period in which measurements are made. In practice, therefore, the effect of atmospheric pressure on geostrophic current flow is difficult to take account of and in many situations is considered sufficiently small to be ignored.

We deduced above that 10 m of seawater is equivalent to a pressure of 1 bar, i.e. that 1 m of seawater is equivalent to 1 decibar (dbar). For many purposes this is a very useful approximation. However, like all fluids, seawater is compressible, and for greater depths quite significant errors result from converting pressures in decibars directly to depths in metres. For example, the pressure recorded at the bottom of the Marianas Trench is 11 240 decibars, but soundings of the area indicate that the maximum depth is about 10 880 m.

Dynamic topography

You have seen how, to determine geostrophic current velocities, we need to quantify departures of isobaric surfaces from the horizontal. But what does 'horizontal' really mean? The simplest answer is that a horizontal surface is any surface at right angles to a vertical plumb-line, i.e. a plumb-line hanging so that it is parallel to the direction in which the force of gravity acts. The reason for this apparently perverse approach is that the upper layers of the solid Earth are neither level nor uniformly dense, so that a surface over which gravitational potential energy is constant is not smooth but has a topography, with bumps and dips on a horizontal scale of tens to thousands of kilometres and with a relief of up to 200m. If there were no currents, the sea-surface would be coincident with an equipotential surface. The particular equipotential surface that corresponds to the sea-surface of a hypothetical motionless ocean is known as the marine **geoid**.

In the context of geostrophic current flow, the important aspect of a 'horizontal' or equipotential surface is that the potential energy of a parcel of water moving over such a surface remains constant. If a parcel of water moves from one equipotential surface to another, it gains or loses potential energy, and the amount of potential energy gained or lost depends on the vertical distance moved and the value of g at that location.

Now imagine a situation in which a steady wind has been blowing sufficiently long for the slopes of isobaric and isopycnic surfaces to have adjusted so that there is a situation of geostrophic equilibrium. If the wind speed now increases up to another steady value, more energy is supplied to the upper ocean; the speed of the current increases and the slopes of isobaric and isopycnic surfaces become even steeper—i.e. they depart even further from the horizontal, the position of least energy. In other words, the ocean gains potential energy as well as kinetic energy. You might like to think of this as being analogous to a motorcyclist riding around a 'wall of death'. As the motorcyclist travels faster and faster, so his circuit moves higher and higher up the 'wall', thus increasing the potential energy of the rider and machine.

Because of the relationship between isobaric slope and potential energy, departures of isobaric surfaces from the horizontal are often discussed in terms of 'dynamic height', i.e. vertical distances are quantified in terms of changes in potential energy (or 'work') rather than simply distance. The units of work that have been adopted for this purpose are known as 'dynamic metres' because they are numerically very close to actual metres. How closely dynamic metres and geometrical metres correspond depends on the local value for g: if g is 9.80ms^{-2}, 1 dynamic metre is equivalent to 1.02 geometric metres.

You have effectively encountered dynamic height already, in Question 3.7(b). When you calculated the value of $h_B - h_A$ as 0.3m, you could equally well have deduced that the difference in the dynamic height of isobar p_1 between stations A and B was about 0.3 dynamic metres. Such variations in dynamic height are described as **dynamic topography**. Because of the way geostrophic calculations are made (*cf.* Question 3.7), dynamic heights are always given relative to some depth or pressure at which it is assumed, for the purposes of the calculations, that the isobaric surface is horizontal. In Question 3.7 and Figure 3.16(b) the depth was z_0 and the pressure p_0.

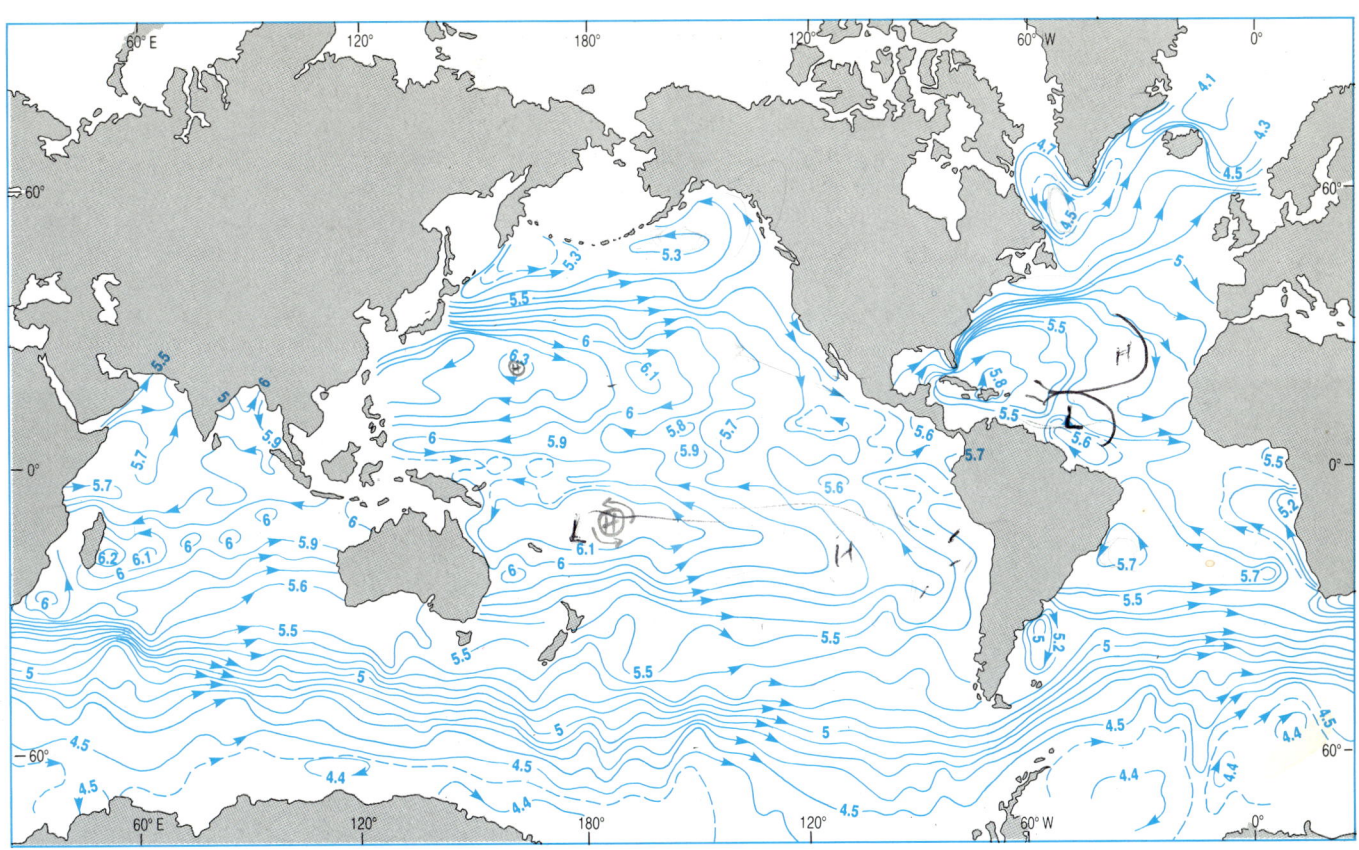

Figure 3.20 The dynamic topography of the sea-surface relative to 1 500 dbar. The contours are labelled in dynamic metres, and values range from 4.4 to 6.4, corresponding to a relief of about 2m.

It is possible to determine the dynamic topography of any isobaric surface relative to another. The simplest situation to imagine is that in which the isobaric surface under consideration is the uppermost one—the sea-surface. Figure 3.20 shows the dynamic topography of the sea-surface relative to the 1500 dbar isobaric surface, for the world ocean. Clearly, maps like Figure 3.20 may be used to determine geostrophic current flow in the upper ocean.

How does the direction of geostrophic current flow relate to the contours of dynamic height?

The current flows *at right angles* to the slope of the isobaric surface, and therefore flows *along* the contours of dynamic height. The direction of flow will be such that the isobaric surface slopes up towards the right in the Northern Hemisphere and up towards the left in the Southern Hemisphere.

Bearing in mind equation 3.11 (the gradient equation), can you suggest how the contours of dynamic height reflect the geostrophic current velocity?

The greater the geostrophic current velocity, the greater the isobaric slope, and—by analogy with ordinary topographic contours—the closer together the contours of dynamic height.

With these points in mind, try Question 3.8.

QUESTION 3.8 (a) Identify on Figure 3.20 the regions corresponding to flow in (i) the Gulf Stream, and (ii) the Antarctic Circumpolar Current, referring to Figure 3.1 if necessary.

(b) Where is the fastest part of the Antarctic Circumpolar Current, according to Figure 3.20?

(c) At 120°E and about 65°S there is a closed contour corresponding to 4.4 dynamic metres. In the region of this contour, does the sea-surface form a 'hill' or a depression? Given the direction of current flow, is this what you would expect?

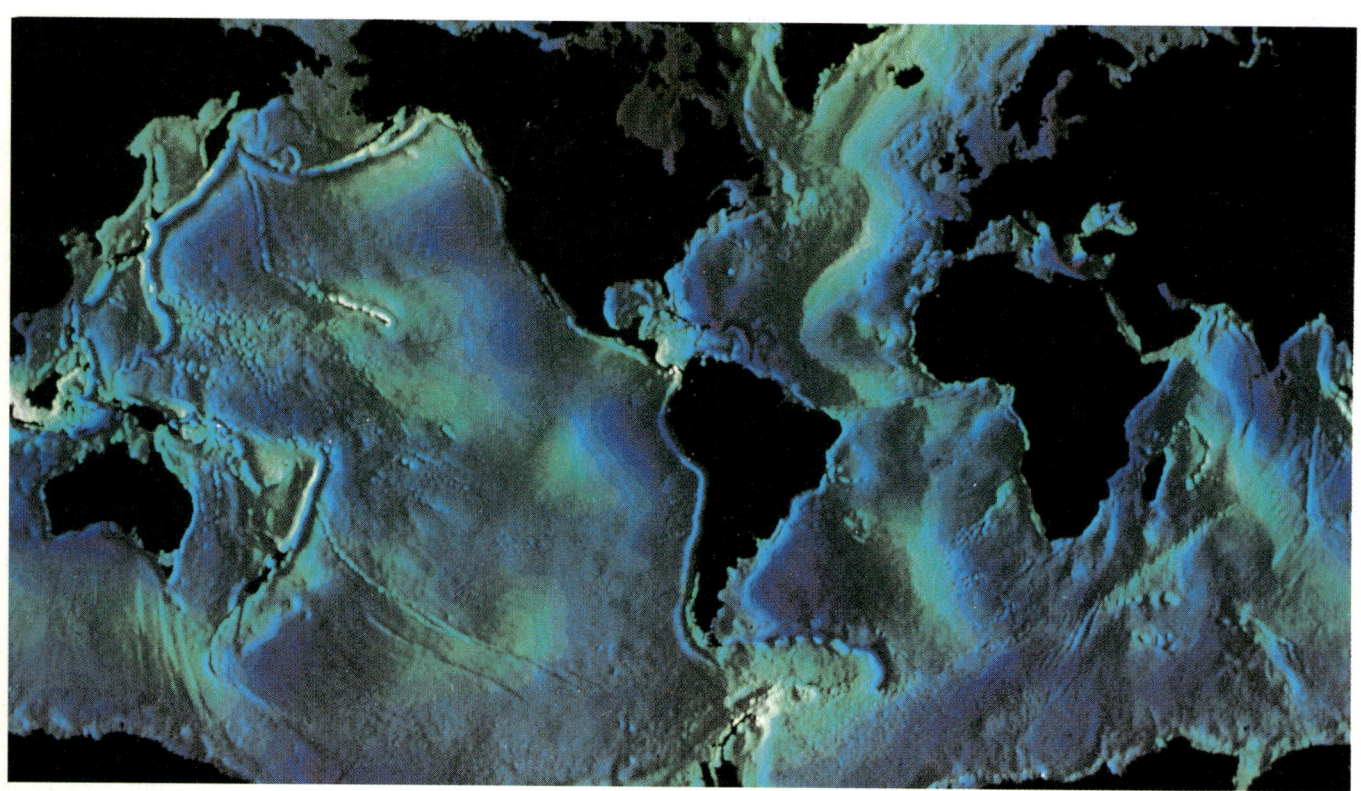

Figure 3.21 Topographic map of the mean sea-surface, as determined from a satellite-borne radar altimeter. The sea-surface topography reflects the topography of the sea-floor rather than geostrophic current flow, as the effect of the latter is about two orders of magnitude smaller, even in regions of strong current flow.

Figure 3.21 is a topographic map of the *mean* sea-surface as obtained by satellite altimetry. The mean sea-surface is a very close approximation to the marine geoid because even in regions where there are strong and relatively steady currents (e.g. the Gulf Stream or the Antarctic Circumpolar Current) the contribution to the topography from current flow is only about one-hundredth of that resulting from variations in the underlying solid crust (*cf*. Figure 3.20). Hence the topographic surface in Figure 3.21 may be regarded as the 'horizontal' surface, departures from which are shown by the topographic surface in Figure 3.20.

3.4 DIVERGENCES AND CONVERGENCES

Wind stress at the sea-surface not only causes horizontal movement of water, it also leads to vertical motion. When the wind stress leads to a **divergence** of surface water, deeper water rises up to take its place (Figure 3.22(a)); conversely, when there is a **convergence** of the surface water, sinking occurs (Figure 3.22(b)). **Upwelling** of subsurface water and sinking of surface water occur throughout the oceans, both at coastal boundaries and away from them. We will consider coastal upwelling in Chapter 4; here we will concentrate on the upwelling that occurs in the open ocean.

Figure 3.23(a) shows the effect on surface waters of a cyclonic wind in the Northern Hemisphere. The Ekman transport—the average movement of the wind-driven layer—is to the right of the wind, causing divergence of surface water and upwelling. Figure 3.23(b) shows how, under these conditions, the sea-surface is lowered and the thermocline raised. This upward movement of water in response to wind stress is sometimes called **Ekman pumping**.

Figure 3.23(c) and (d) show the situation that results from anticyclonic wind stress in the Northern Hemisphere: under these conditions there is convergence and sinking.

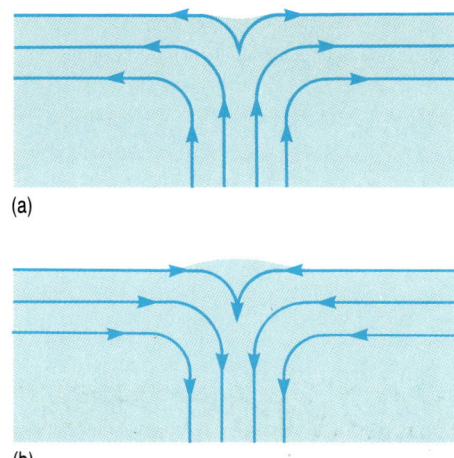

Figure 3.22 (a) Divergence of surface waters leads to upwelling while (b) convergence of surface waters leads to sinking.

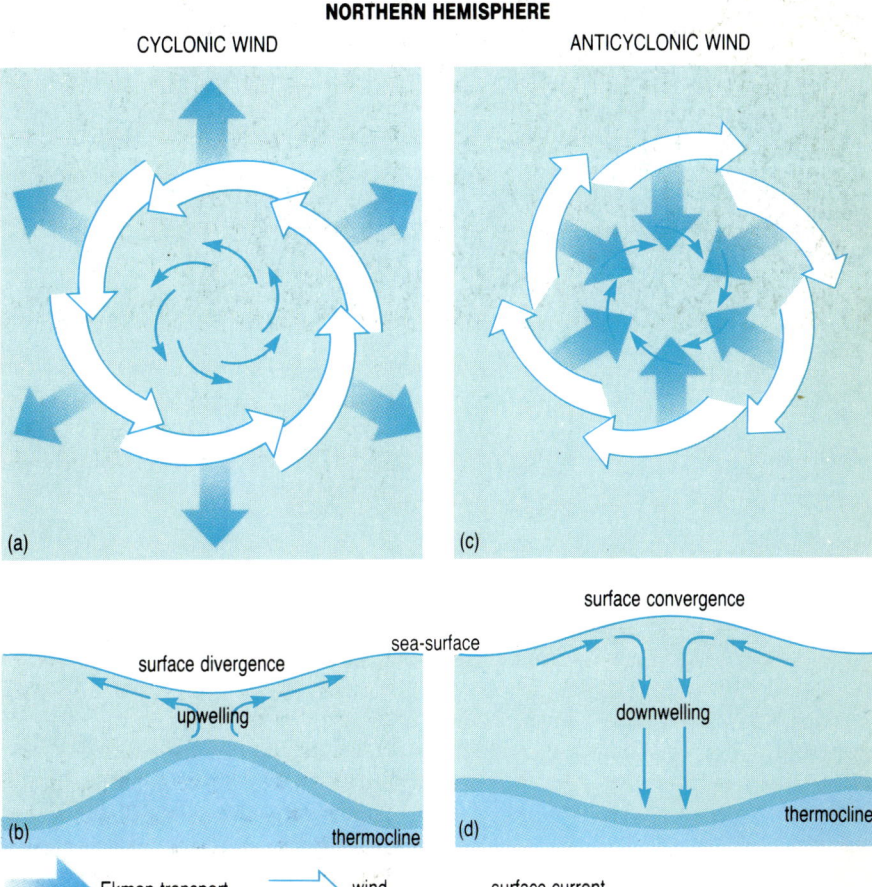

Figure 3.23 The effect of a cyclonic wind in the Northern Hemisphere (a) on surface waters, (b) on the shape of the sea-surface and thermocline. Diagrams (c) and (d) show the effects of an anticyclonic wind in the Northern Hemisphere.

QUESTION 3.9 Sketch diagrams analogous to those in Figure 3.23 to show the conditions that would lead to convergence and sinking of surface water in the Southern Hemisphere.

The convergence of water as a result of anticyclonic winds thus causes the sea-surface to slope upwards towards the middle of the gyre. As a result, the circulating water will be acted upon by a horizontal pressure gradient force.

In which direction will the horizontal pressure gradient force act?

It will act outwards from the centre, from the region of higher pressure to the region of lower pressure (see Figure 3.24). Under steady conditions the horizontal pressure gradient force will be balanced by the Coriolis force, and a geostrophic ('slope') current will flow in the same direction as the wind. The gyral current systems of the Atlantic and Pacific Oceans between about 10 and 40 degrees of latitude—the **subtropical gyres**—are large-scale gyres of the type we have been discussing.

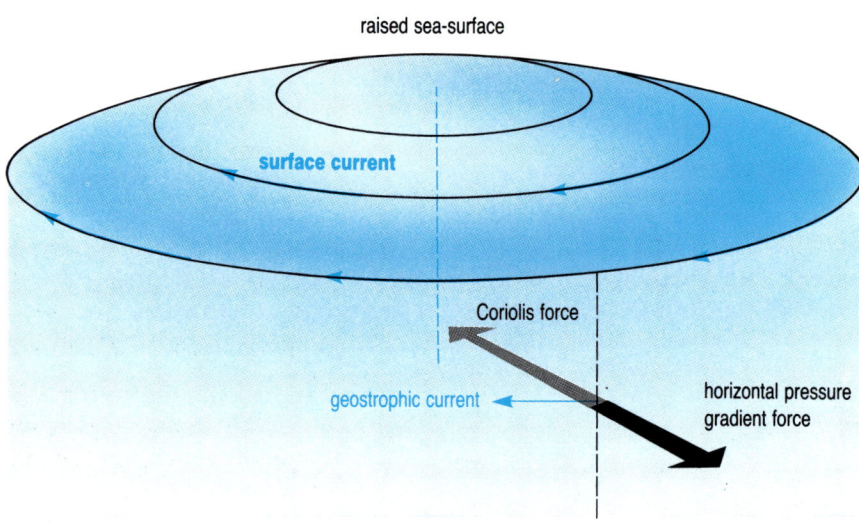

Figure 3.24 The generation of geostrophic current flow in a gyre driven by anticyclonic winds in the Northern Hemisphere. This current is driven by the wind only indirectly and persists below the wind-driven (Ekman) layer.

The surface waters of the ocean move in complex patterns, and divergences and convergences occur on small scales as well as on the scale of the subtropical gyres. Figure 3.25 illustrates schematically several types of flow that would lead to vertical movement of water. Such divergences and convergences may be seen in rivers, lakes and shelf seas, as well as in the open ocean.

The position of a large-scale oceanic divergence may sometimes be inferred from the colour of the surface water: upwelled water is often greener than the surrounding water because it is more nutrient-rich and therefore able to support larger populations of phytoplankton. Upwelled

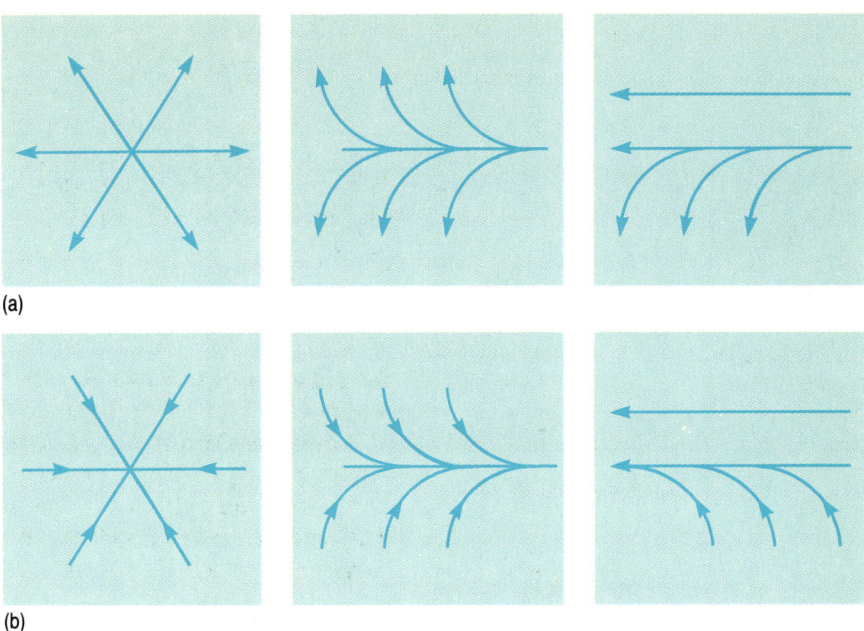

Figure 3.25 Schematic representations of (a) divergent surface flow patterns which would lead to upwelling of subsurface water and (b) convergent flow patterns that would lead to sinking of surface water.

water is also colder than surface water and so divergences are sometimes marked out by fog banks. Linear convergences are often known as **fronts**, especially when water properties (e.g. temperature and productivity) are markedly different on either side of the convergence.

Small-scale convergences are often marked by a collecting together of surface debris, seaweed or foam (Figure 3.26(a) overleaf). In certain circumstances, linear convergences form *parallel* to the wind (Figure 3.26(b)). These 'windrows' were first studied by Langmuir, who in 1938 noticed large amounts of seaweed arranged in lines parallel to the wind, in the North Atlantic Ocean. He proposed that the wind had somehow given rise to a series of helical vortices, with axes parallel to the wind. This circulation system—called **Langmuir circulation**—is illustrated schematically in Figure 3.26(c). It is thought to result from instability in the well-mixed, and therefore fairly homogeneously dense, surface water. 'Longitudinal roll vortices' are now known to be common in both the upper ocean and the lower atmosphere, where they give rise to the linear arrangements of clouds, known as 'cloud streets', that are often seen from airliners (Figure 3.26(d)).

The spatial scales of the circulatory systems of windrows and of basin-wide features like the subtropical gyres are very different. The latter involve horizontal distances of thousands of kilometres; the former extend only for a kilometre or so. It is characteristic of dynamic systems that phenomena with small length scales also have short time-scales, while phenomena with long length scales also have long time-scales. This aspect of ocean circulation will be developed further in the next Section.

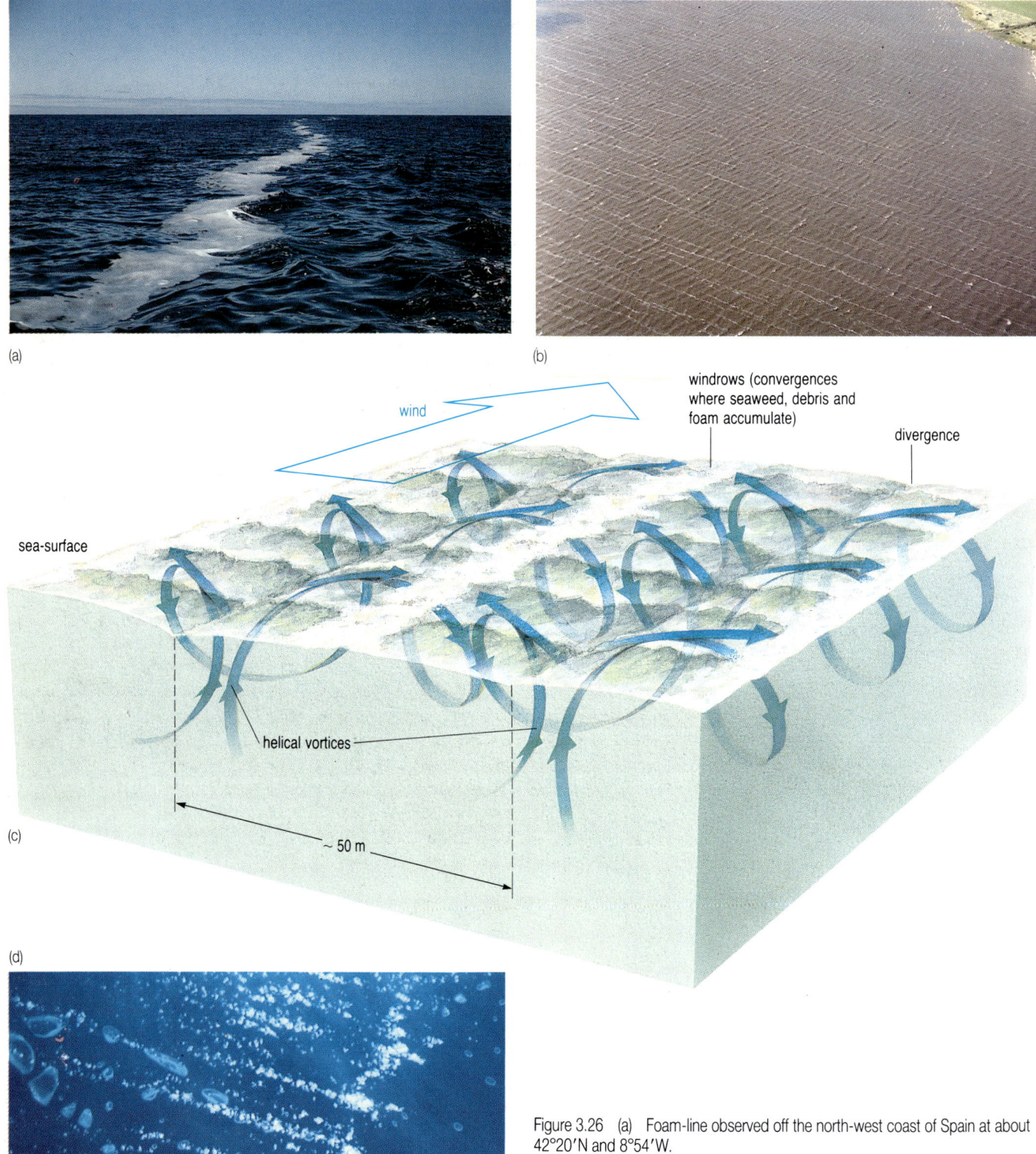

Figure 3.26 (a) Foam-line observed off the north-west coast of Spain at about 42°20'N and 8°54'W.

(b) Windrows observed on a Welsh lake. The streaks, which were 5–10m apart, were observed to respond quickly to changes in the wind direction (inferred from the orientation of the wave crests).

(c) Schematic diagram to show Langmuir circulation in the upper ocean. The distance between the surface streaks may be as much as a few hundred metres, but is typically a few tens of metres.

(d) Cloud streets over the reefs of the Maldive Islands in the Indian Ocean.

3.5 THE ENERGY OF THE OCEAN: SCALES OF MOTION

Energy is imparted to the ocean from the Sun, directly through solar radiation (which leads to heating, evaporation and, ultimately, precipitation) and indirectly through winds.

As discussed in Section 3.3.4, the ocean possesses both kinetic and potential energy: kinetic energy by virtue of its motion and potential energy as a result of isopycnic and isobaric surfaces being displaced from their position of least energy parallel to the geoid (i.e. horizontal). The potential energy of the world ocean is about *one hundred times* greater than its kinetic energy. It has been calculated that if all the isopycnic surfaces that are presently sloping were allowed to become horizontal, 10^6 joules of potential energy would be released for every square metre of sea-surface. This huge amount of stored potential energy ensures that if the global winds stopped blowing, the ocean circulation would take a decade or so to run down.

3.5.1 KINETIC ENERGY SPECTRA

We have seen that oceanic circulation consists of many types of motion, acting over a range of scales within space and time.

How, then, is it possible to calculate the total kinetic energy of the ocean?

At any one point in the ocean, at a given time, the current may be regarded as the combination of a number of components—perhaps a southward tidal current of (say) $0.2 \, \text{ms}^{-1}$ has been superimposed on a local wind-driven current of (say) $0.1 \, \text{ms}^{-1}$ to the north, in which case the actual current will be $0.1 \, \text{ms}^{-1}$ to the south. Perhaps both of these occur in the region of an established current system like the Gulf Stream, which shows fluctuations but is nevertheless a permanent feature of the oceanic circulation. Such a long-term, large-scale flow is referred to as the **mean motion**. If the varying components of the flow are resolved mathematically into regular oscillations, the relationship of the fluctuating components to the mean flow might look like Figure 3.27.

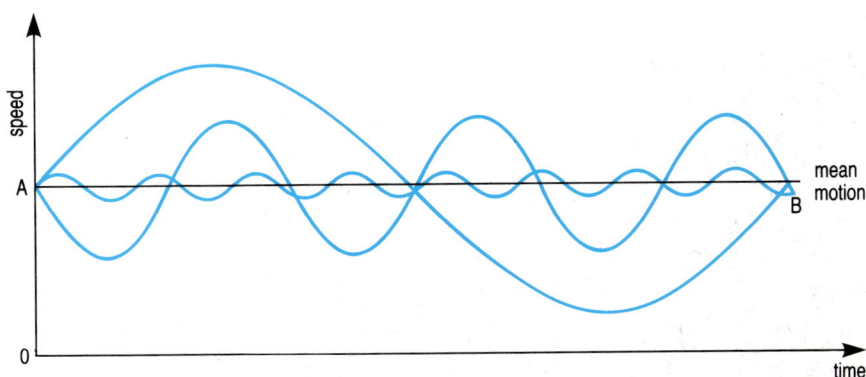

Figure 3.27 A diagrammatic representation of how speed might vary with time for current components which fluctuate about the mean motion. Over a long period (e.g. from A to B), the fluctuating motions average out, leaving only the mean motion represented by the straight line.

The kinetic energy possessed by a parcel of water at any instant is the sum of the energies of all the components contributing to its motion, *as determined by the squares of their speeds*. This is because the kinetic energy of a mass m moving with speed v is given by $\frac{1}{2}mv^2$. For any particular region of the ocean, the component oscillations may be further analysed to produce a **kinetic energy density spectrum**.

Figure 3.28 is an energy density spectrum based on a 48-day record of current measurements made at 120 m depth at a station in the north-west Atlantic. It represents the kinetic energy of the fluctuating components of the motion with periods of less than 20 days (480 hours). The fluctuating components account for most of the total kinetic energy of the flow, with only a very small proportion being contained in the mean motion.

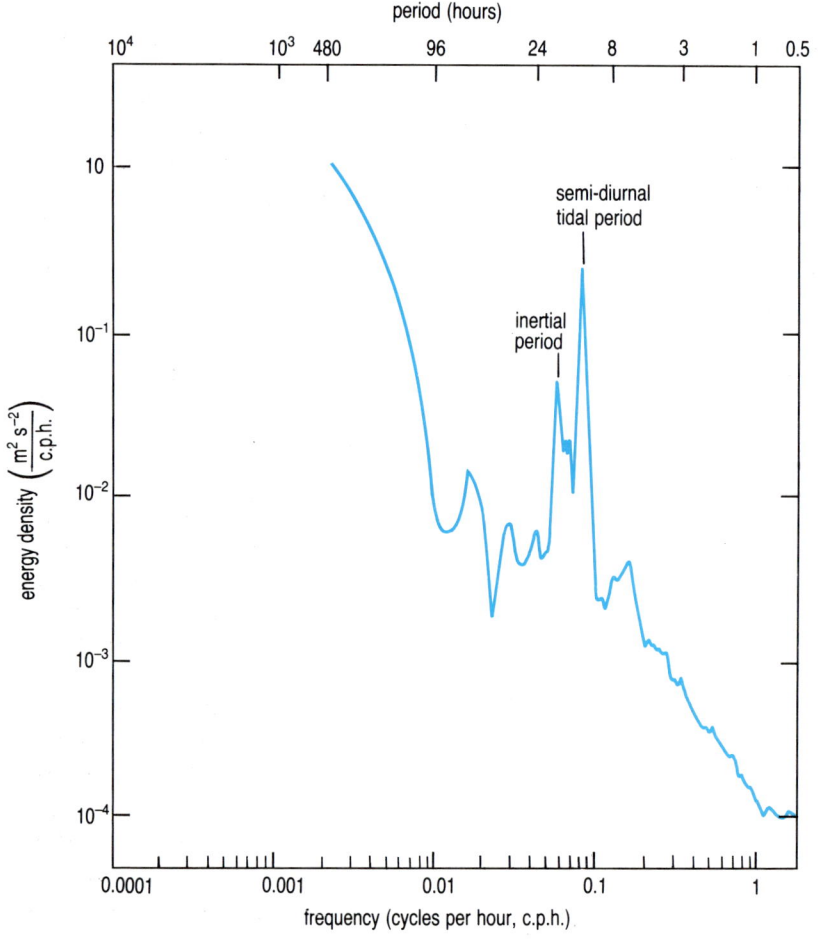

Figure 3.28 The kinetic energy density spectrum for current flow at a depth of 120 m in the north-west Atlantic off Woods Hole, Mass., USA, as determined from current meter measurements made from June 24 to 11 August 1965. The spectrum has only been determined for periods less than 480 hours. (Note that the scales on the axes are logarithmic.)

The components of a flow that have a specific period (specific frequency) typically show up on an energy density spectrum as individual peaks. In Figure 3.28 a peak corresponding to the semi-diurnal tidal motion is clearly seen at a period of about 12 hours, while the peak to the left of that has been attributed to inertia currents. In general, however, there is a continuous spectrum containing an infinite number of frequencies because variations in current flow occur over the full range of possible time-scales. This means that it is usually not possible to determine the energy corresponding to a single frequency, as one frequency can only possess an infinitesimal proportion of the total energy. It *is* possible, however, to determine the amount of energy corresponding to a frequency *interval* and the units of energy density are generally given in terms of energy/frequency which is shorthand for 'energy per frequency interval'. The total *area* under an energy density curve corresponds to the total kinetic energy contributed by all the frequencies within the range considered.

3.5.2 MESOSCALE EDDIES

The mean motion—i.e. long-term, large-scale current systems like the Gulf Stream or the Antarctic Circumpolar Current—is the oceanic equivalent of climate. It was not until relatively recently that oceanographers began to get to grips with the study of the ocean's variability over shorter time-scales—i.e. with the ocean's 'weather'.

During 1971 to 1973, an international expedition was mounted to study intensively an area of ocean several hundred kilometres across, in the western North Atlantic to the east of the Gulf Stream. The aim of the expedition was to reveal relatively small-scale variability within the ocean by combining the results produced from fixed current meters, free-drifting floats, and dynamic topography calculated from temperature and salinity data. This experiment, which was a turning point in oceanographic experimentation, was named the **Mid-Ocean Dynamics Experiment (MODE)**.

The most exciting result to come out of analysis of the MODE records was that motions with periods greater than the tidal and inertial periods are dominated by what are now called **mesoscale eddies** (the prefix 'meso-' means 'intermediate'). They have length scales of 50–200 km and periods of one to a few months. In many ways, mesoscale eddies are the oceanic analogues of weather systems in the atmosphere, although the density distribution in the atmosphere is such that cyclones (depressions) and anticyclones have length scales of about 1000 km and periods of about a week. Mesoscale eddies travel at a few kilometres per day (compared with about 1000 km per day for atmospheric weather systems), and have rotatory currents with speeds of the order of $0.1\,\mathrm{ms}^{-1}$.

Figure 3.29 shows the path taken by a freely drifting float tracked during the MODE experiment. For most of the monitoring period the float circulated within a cyclonic eddy which moved due west. It has since been observed that mesoscale eddies tend to move westwards and/or equatorwards rather than polewards and/or eastwards. Their lifetimes seem to depend to a large extent on the path they take: in Chapter 4 you will see how mesoscale eddies formed in the region of the Gulf Stream may only survive for a matter of months before being 'entrained' back into the Stream; by contrast, eddies moving southwards in the eastern Atlantic have been observed to survive for about two years.

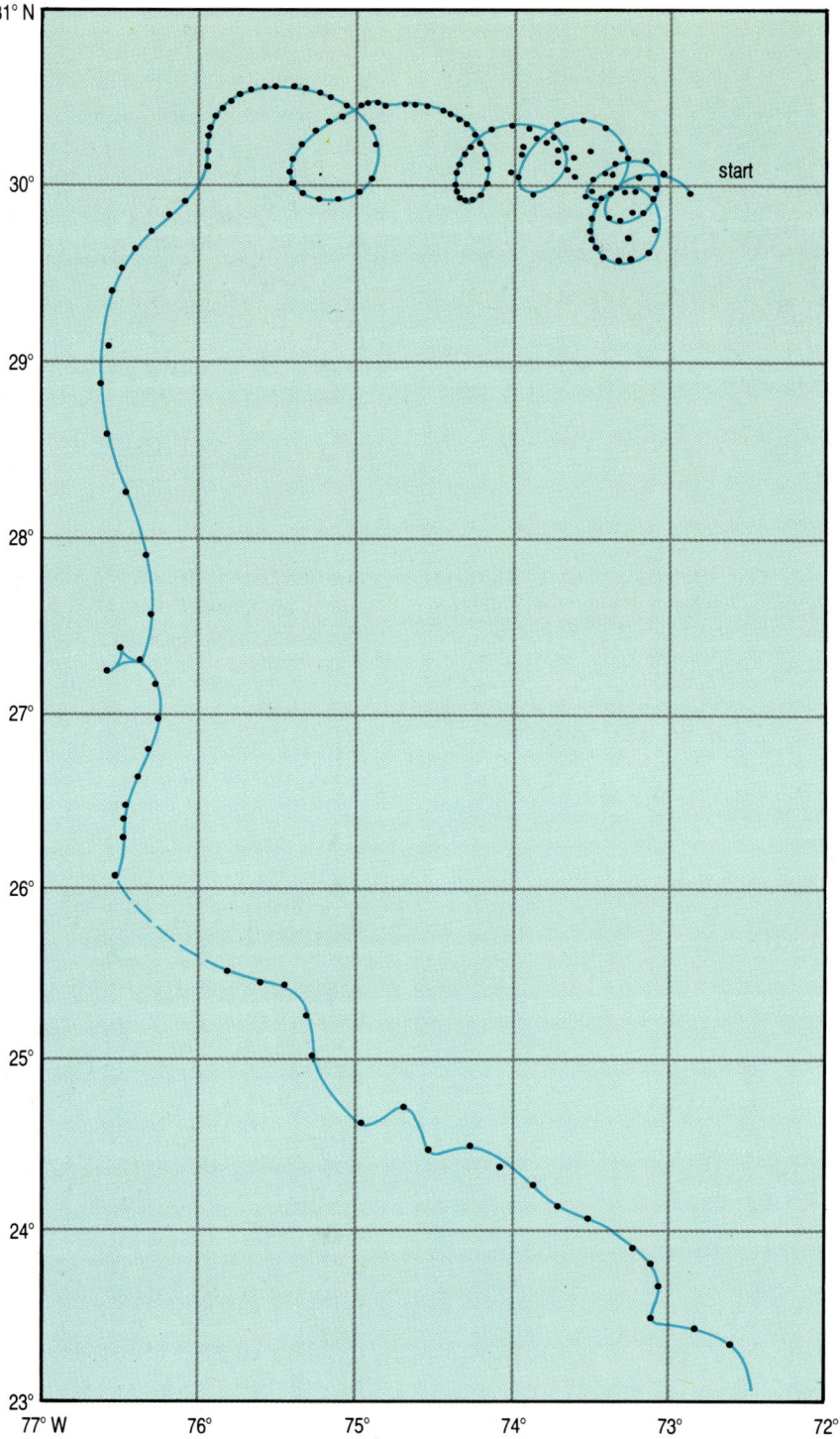

Figure 3.29 The path of a Sofar (SOund Fixing And Ranging) float drifting in the sound channel at 2 000 m, one of many such floats tracked during the MODE experiment. Dots show daily positions, so the track shows the path taken by the float over 5½ months. The float first circulates at speeds of about 0.5 m s^{-1} within a westward-moving eddy. It is then caught in a fast southward-flowing current. (Sofar floats will be discussed further in Chapter 4.)

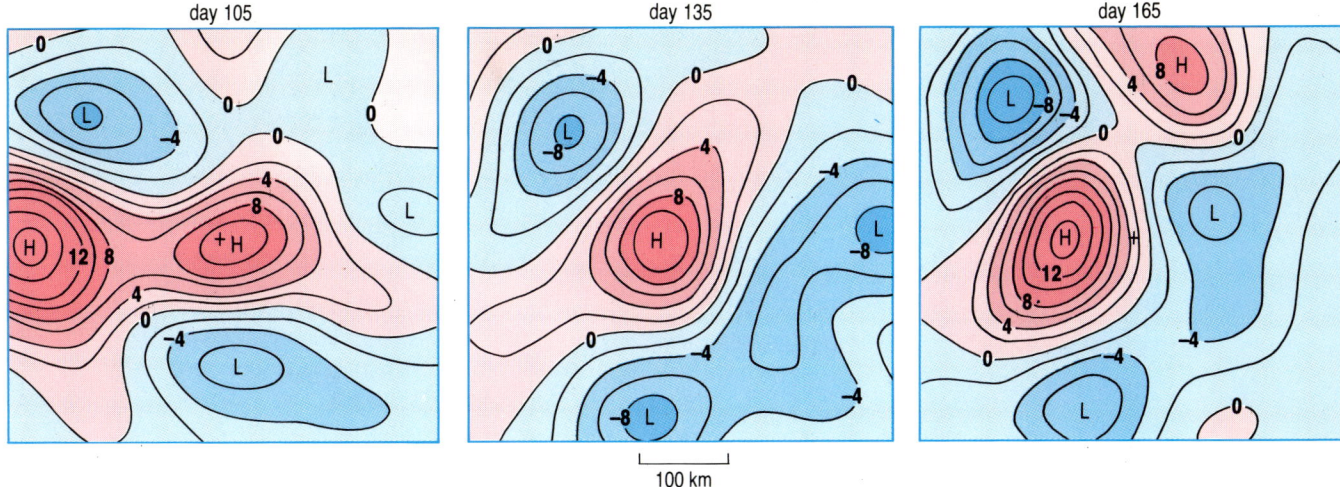

Figure 3.30 The mesoscale structure in the region of 69° 40′ W, 28° N (marked by the cross in the centre of each map), as revealed by the dynamic topography at 150m depth, calculated from temperature and salinity data collected during MODE. The numbers on the contours are a measure of the total flow between the contour in question and the zero contour. The 'highs' (H) correspond to warmer water and the 'lows' (L) to cooler water. The three pictures show the situation at intervals of 30 days.

A general westward movement may also be discerned in the position of the best-defined eddy in Figure 3.30. Here, density determinations made during MODE have been used to estimate geostrophic flow at a depth of 150m.

QUESTION 3.10   Given that the contours on Figure 3.30 represent dynamic topography, use your knowledge of geostrophic currents to determine the directions of flow around the 'highs' and 'lows', respectively.

Maps similar to Figure 3.30 were drawn for different depths in the water column and it was found that although the fastest speeds occurred near the surface (being about $0.15\,\mathrm{m\,s^{-1}}$ at 150m depth) the eddies persisted down to at least 1500m. Furthermore, it was found that the 'axes of rotation' of the eddies were not all vertical.

The kinetic energy of mesoscale eddies is believed to be derived from the potential energy of the mean flow, i.e. from the 'relaxation' of sloping isopycnals (*cf.* Section 3.5). The eddies seem to be most commonly generated in regions of the ocean where there are marked lateral density discontinuities, i.e. boundaries between different water masses, or fronts. Like fronts in the atmosphere, oceanic boundaries slope and are intrinsically unstable. If there is a significant change of velocity with depth, i.e. strong vertical velocity shear, waves form on the fronts and may evolve into eddies. These waves—known as 'baroclinic instabilities' because they are associated with marked lateral variations in density—are only one of several possible mechanisms whereby mesoscale eddies may be generated. Regions where mesoscale eddies have been observed to form include the Gulf Stream, and the Antarctic Polar Frontal Zone in the Antarctic Circumpolar Current (which is discussed in Section 5.5.2).

QUESTION 3.11   (a)   The mean circulation of the ocean has an average speed of the order of $0.01\,\mathrm{m\,s^{-1}}$. Mesoscale eddies typically have currents of about $0.1\,\mathrm{m\,s^{-1}}$. Explain why the kinetic energy per square kilometre of ocean in the region of an eddy is about 100 times that associated with the mean flow.

(b) Look at Figure 3.31. What atmospheric phenomenon has approximately the same amount of energy as a mesoscale eddy?

(c) Which ocean current has the greatest kinetic energy, according to Figure 3.31?

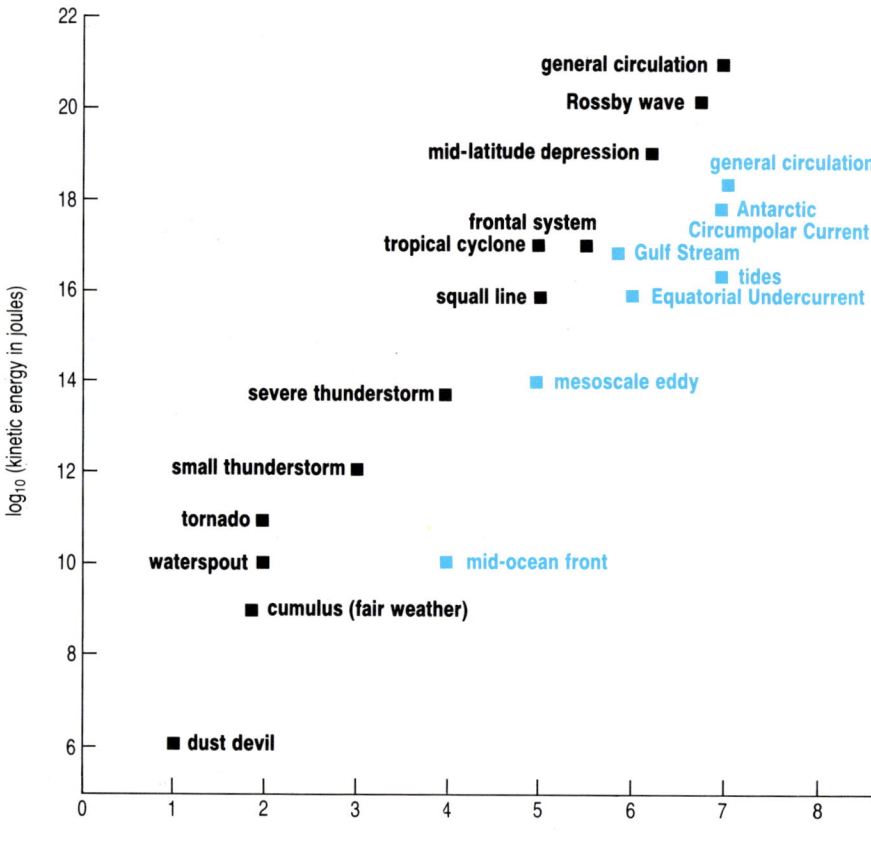

Figure 3.31 The kinetic energy possessed by various oceanic (blue) and atmospheric (black) phenomena. Rossby waves are discussed in Section 5.3.2. (Note that for convenience both vertical and horizontal scales are logarithmic.)

Since the 1970s it has become clear that mesoscale eddies are an intrinsic part of the ocean circulation. They are particularly numerous in certain regions of the ocean (such as those mentioned above) but no part of the ocean has been shown to be without them. Kinetic energy spectra that extend into periods greater than 30 days nearly always show a bulge in the mesoscale part of the curve (see Figure 3.32). Indeed, it is believed that *most* of the kinetic energy of the ocean—perhaps as much as 99%—is contained in the ocean's 'weather'. The extent to which the energy of mesoscale eddies feeds back into the general large-scale circulation is not clear. Insight into this and other related problems can only be obtained through computer models with the ability to predict basin-wide flow patterns on time-scales and space scales that are small enough to 'resolve' mesoscale motions. Such computer models (described as 'eddy-resolving') only became possible in the mid-1980s, when the powerful computers required to run them became available.

As might be expected, mesoscale eddies play an important role in the transport of heat and salt across frontal boundaries, from one water mass to another. It has been estimated that they transport heat polewards across the Antarctic Circumpolar Current at a rate of about 0.4×10^{15} J s^{-1} (watts) (*cf*. Figure 1.5). Through their 'stirring' motions, they also act to homogenize water characteristics within water masses (this will be discussed further in Chapter 6).

Mesoscale eddies are an exciting and relatively new discovery. Understanding them, and in particular understanding how they interact with the mean flow, will immeasurably improve our understanding of the oceanic circulation as a whole. Nevertheless, it is important to remember that mesoscale eddies are only one part of a continuous spectrum of possible motions in the oceans.

3.6 SUMMARY OF CHAPTER 3

1   The global surface current system to some extent reflects the surface wind field, but ocean currents are constrained by continental boundaries and current systems are often characterized by gyral circulations.

2   Maps of wind and current flows of necessity represent average conditions only; at any one time the actual flow at a given point might be markedly different from that shown.

3   The frictional force caused by the action of wind on the sea-surface is known as the wind stress. Its magnitude is proportional to the square of the wind speed; it is also dependent on the roughness of the sea-surface and conditions in the overlying atmosphere.

4   Wind stress acting on the sea-surface causes motion in the form of waves and currents. The surface current is typically 3% of the wind speed. Motion is transmitted downwards through frictional coupling caused by turbulence. As a result, the coefficient of friction that is important for studies in the ocean is the coefficient of eddy viscosity, which varies from $10^{-2} - 10^2 \text{kgm}^{-1}\text{s}^{-1}$ for motion in a vertical direction to $10^4 - 10^8 \text{kgm}^{-1}\text{s}^{-1}$ for motion in a horizontal direction.

5   Moving water tends towards a situation of equilibrium. Flows adjust to the forces acting on them so that those forces balance one another. Major forces that need to be considered with respect to moving water are wind stress at the sea-surface, internal friction, the Coriolis force and horizontal pressure gradient forces; in some situations, friction with the sea-bed and/or with coastal boundaries also needs to be taken into account.

6   Ekman showed theoretically that under idealized conditions the surface current resulting from wind stress will be 45° *cum sole* of the wind, and that the direction of the wind-induced current will rotate *cum sole* with depth, forming the Ekman spiral current pattern. An important consequence of this is that the mean flow of the wind-driven (or Ekman) layer is 90° to the right of the wind in the Northern Hemisphere and 90° to the left of the wind in the Southern Hemisphere.

7   When forces that have set water in motion cease to act, the water will continue to move until the energy supplied has been dissipated, mainly by internal friction. During this time the water is still acted upon by the Coriolis force, and the rotational flows that result are known as inertia currents. The period of rotation of an inertia current varies with the Coriolis parameter $f = 2\Omega \sin \phi$, and hence with latitude, ϕ.

8   The currents that result when the horizontal pressure gradient force is balanced by the Coriolis force are known as geostrophic currents. The horizontal pressure gradient force may result only from the slope of the sea-surface, and in these conditions isobaric and isopycnic surfaces are parallel and conditions are described as barotropic. When the water is not homogeneous, but instead there are lateral variations in temperature and salinity, part of the variation in pressure at a given depth level is due to the density distribution in the overlying water. In these situations, isopycnic surfaces slope in the opposite direction to isobaric surfaces; thus, isobars and isopycnals are inclined to one another and conditions are described as baroclinic.

9   In geostrophic flow, the angle of slope (θ) of each isobaric surface may be related to u, the speed of the geostrophic current in the vicinity of that isobaric surface, by the gradient equation: $\tan \theta = fu/g$. In barotropic flow, the slope of isobaric surfaces remains constant with depth and so also does the velocity of the geostrophic current; in baroclinic conditions, the slope of isobaric surfaces and the velocity of the geostrophic current vary with depth. The types of geostrophic current that occur in the two situations are sometimes known as 'slope currents' and 'relative currents', respectively. In the oceans, flow is often a combination of the two types of flow, with a relative current superimposed on a slope current.

10   In baroclinic conditions, the slopes of the isopycnals are very much greater than the slopes of the isobars. As a result, the gradient equation may be used to construct a relationship which gives the average velocity of the geostrophic current flowing between two oceanographic stations in terms of the density distributions at the two stations. This relationship is known as the geostrophic equation or (in its full form) as Helland-Hansen's equation. It may be used to determine *relative* current velocities (i.e. velocities relative to a selected depth or isobaric surface) at right angles to the section. It provides information about *average* conditions only, and is subject to certain simplifying assumptions. Nevertheless, much of what is known about oceanic circulation has been discovered through geostrophic calculations.

11   Departures of isobaric surfaces from the horizontal (i.e. from an equipotential surface) may be measured in terms of units of work known as dynamic metres. Variations in the dynamic height of an isobaric surface (including the sea-surface) are known as dynamic topography. On a map of dynamic topography, geostrophic flow is parallel to the contours of dynamic height in such a direction that the 'highs' are on the right in the Northern Hemisphere and on the left in the Southern Hemisphere. Dynamic topography represents departures of an isobaric surface from the geoid, which itself has a relief of the order of 100 times that of dynamic topography.

12   Surface wind stress gives rise to vertical motion of water, as well as horizontal flow. In particular, cyclonic wind systems give rise to a lowered sea-surface, raised thermocline and divergence and upwelling, while anticyclonic wind systems lead to a raised sea-surface, lowered thermocline and convergence and sinking. Relatively small-scale linear divergences and convergences occur as a result of Langmuir circulation in the upper ocean.

13 Flow in the ocean occurs over a large range of time-scales and space scales. The general circulation, as represented by the average position and velocity of well-established currents such as the Gulf Stream, is known as the mean motion.

14 The energy of the ocean, both kinetic and potential, derives ultimately from solar energy. The ocean's potential energy is about 100 times its kinetic energy and results from isobars and isopycnals being displaced from their position of least energy (parallel to the geoid) as a result of wind stress or changes in the density distribution of the ocean; if the ocean were at rest and homogeneous, all isobaric and isopycnic surfaces would be parallel to the geoid. The ocean's kinetic energy results from the motion in ocean currents. For any given area of ocean, it may be represented by a kinetic energy density spectrum. The kinetic energy associated with a current is proportional to the square of the current speed.

15 Mesoscale eddies, which have length scales of 50–200 km and periods of one to a few months, represent the ocean's 'weather'. They contain a significant proportion of the ocean's energy, but how they interact and exchange energy with the mean flow is not yet clear. Flow around mesoscale eddies is in approximate geostrophic equilibrium. Mesoscale eddies have been found in all parts of the oceans, but it seems that they are generated at intense frontal regions like the Gulf Stream and the Antarctic Circumpolar Current, where there is a marked lateral variation in density distribution and significant vertical current shear, which give rise to wave-like perturbations known as 'baroclinic instabilities'.

Now try the following questions to consolidate your understanding of this Chapter.

QUESTION 3.12 (a) The forces which act on water or any other fluid may be divided into three categories:

(i) external forces, that arise from outside the fluid;
(ii) internal (or body) forces, that act within the fluid;
(iii) secondary forces, that come into play only because the fluid is in motion relative to the Earth's surface.

Some of the main forces that act on ocean water are:

1 wind stress;
2 viscous forces;
3 the Coriolis force;
4 horizontal pressure gradient forces; and
5 the tide-producing forces.

How would you classify each of forces 1–5 in terms of categories (i)–(iii)?

(b) Motion in the oceans is in equilibrium when the flow has had time to adjust so that the forces acting on the water balance. In the following types of flow, which of the forces 1–5 (above) are balancing one another?

(i) geostrophic flow;
(ii) the mean flow of the whole Ekman layer at right angles to the wind.

QUESTION 3.13 Use information from the end of Section 3.1.2 to show that, theoretically, for a given wind stress, the total volume transport in the wind-driven layer is independent of the value of A_z, the coefficient of eddy viscosity.

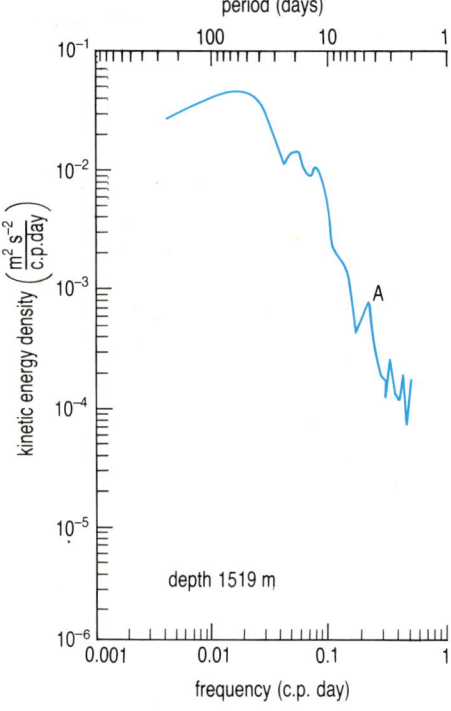

Figure 3.32 Kinetic energy density spectrum for flow in the Drake Passage, between South America and Antarctica.

QUESTION 3.14 In the Straits of Dover, water is well mixed by tidal currents and wind. The mean geostrophic current is towards the east.

(a) Are conditions in the Straits of Dover barotropic or baroclinic?

(b) If the mean geostrophic current through the Straits is 0.2 m s^{-1}, what is the slope of the sea-surface, i.e. what is tan θ? Is the mean sea-level higher on the French or the English side, and by how much? (You will need to use $\Omega = 7.29 \times 10^{-5}$ s^{-1}; the Straits of Dover are at 51° N and are about 35 km wide.)

QUESTION 3.15 Until MODE in the 1970s, a large proportion of studies of ocean currents had been made using the indirect geostrophic method (Section 3.3.3), combined with a few direct current measurements. Bearing this in mind, explain briefly why it is not surprising that mesoscale eddies had not previously been well documented.

QUESTION 3.16 In Chapter 2, we mentioned that the surface tracks of tropical cyclones are marked out by cooler water upwelled from a depth of 100 m or so. Using information in Sections 3.1 and 3.4, can you now explain why this happens?

QUESTION 3.17 Study the kinetic energy density spectrum in Figure 3.32.

(a) Why might we expect the contribution of mesoscale eddies to the spectrum to be above average?

(b) Explain, using calculations where necessary, why the peak at A on Figure 3.32 might be attributable to inertia currents.

QUESTION 3.18 Figure 3.33 shows the *variability* in the level of the sea-surface as computed from satellite measurements.

(a) Compare Figure 3.33 with Figure 3.1. What do the regions showing the greatest variability in sea-surface height have in common? Can you explain this?

(b) Why are the positions of the equatorial current systems not strongly visible on Figure 3.33? (*Hint*: Think about *all* the variables in the gradient equation.)

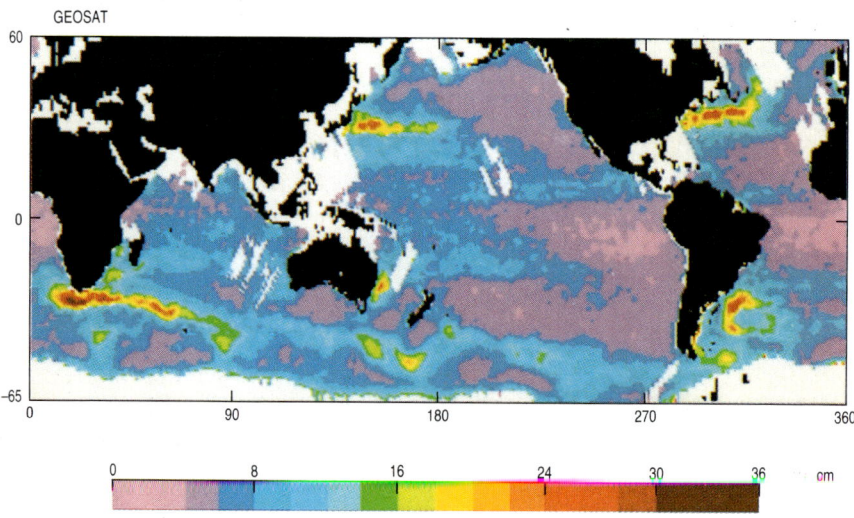

Figure 3.33 Variability in sea-surface height, as computed from satellite altimeter measurements made over the course of a year from November 1986 to November 1987. The colours represent different deviations from the mean sea-level.

CHAPTER 4

THE NORTH ATLANTIC GYRE: OBSERVATIONS AND THEORIES

The North Atlantic is the most studied—and most theorized-about—area of ocean in the world. Over the course of many centuries ships have crossed and recrossed it. The effects of currents on sailing times became well known, and no current was more renowned than the Gulf Stream. The Gulf Stream has also fascinated natural philosophers, both because of the mystery of its very existence and because of the benign influence it appeared to have on the European climate.

The aim of this Chapter is to use the example of the North Atlantic gyral current system, and of the Gulf Stream in particular, to introduce some actual observations and measurements, and some of the ideas and theories that have been put forward to explain them.

4.1 THE GULF STREAM

Before starting Section 4.1.1, look again at Figure 3.1 to remind yourself of the geographical position of the Gulf Stream and its relationship to the general circulation of the North Atlantic.

4.1.1 EARLY OBSERVATIONS AND THEORIES

European exploration of the eastern seaboard of the New World began in earnest during the sixteenth century. Coastal lands were investigated and much effort was put into the search for a North-West Passage through to the Pacific. The earliest surviving reference to the Gulf Stream was made by the Spaniard Ponce de Leon in 1513. His three ships sailed from Puerto Rico, crossed the Gulf Stream with great difficulty north of what is now Cape Canaveral (in Florida), and then turned south. By 1519 the Gulf Stream was well known to the ships' masters who sailed between Spain and America. On the outgoing voyage they sailed with the Trade Winds in the North Equatorial Current; on their return, they passed through the Straits of Florida and followed the Gulf Stream up about as far as the latitude of Cape Hatteras (~ 35° N), and then sailed for Spain with the prevailing Westerlies.

As early as 1515, there were well-considered theories about the origin of the Gulf Stream. Peter Martyr of Angheira used the necessity for conservation of mass to argue that the Gulf Stream must result from the deflection of the North Equatorial Current by the mainland. Explorers had found no passage through which the North Equatorial Current could flow to the Pacific and thence back around to the Atlantic; the only other possibility was that water piled up continuously against the Brazilian coast, and this had not been observed to happen. The North Equatorial Current itself was thought to result in some way from the general westward movement across the heavens of the celestial bodies, which drew the air and waters of the equatorial regions along with them.

During the 1600s, the eastern coast of North America was colonized by Europeans and the Gulf Stream was traversed countless times and at various locations. In the following century the experience gained in the great whaling expeditions added further to the knowledge of currents, winds and bottom topography. This accumulating knowledge was not, however, readily accessible in technical journals, but handed down by word of mouth. Charts indicating currents did exist—the first one to show the 'Gulf Stream' was published in 1665—but they were of varying quality and showed features that owed more to the imagination than observation.

The first authoritative chart of the Gulf Stream was made by William Gerard De Brahm, an immensely productive scientist and surveyor who, in 1764, was appointed His Majesty's Surveyor-General of the new colony of Florida. The chart of the Gulf Stream (Figure 4.1(a)) and De Brahm's reasoned speculation about its origin were published in *The Atlantic Pilot* in 1772. At about the same time, a chart of the Gulf Stream was engraved and printed by the General Post Office, on the instructions of the Postmaster-General of the Colonies, Benjamin Franklin. The chart was made for the benefit of the masters of the packet ships which carried mail between London and New England. A Nantucket sea captain, Timothy Folger, had drawn Franklin's attention to the Gulf Stream as one cause of delay of the packet ships, and had plotted the course of the Stream for him (Figure 4.1(b)).

The theories of oceanic circulation that had evolved during the seventeenth century were seldom as well constructed as the charts they sought to explain. During the eighteenth century, understanding of fluid dynamics advanced greatly. Moreover, the intellectual climate of the times encouraged scientific advances based on observations, rather than fanciful theories based in the imagination. Franklin, who had observed the effect of wind on shallow bodies of water, believed that the Trade Winds caused water to pile up against the South American coast; the head of pressure so caused resulted in a strong current flowing 'downhill' through the Caribbean islands, into the Gulf of Mexico, and out through the Straits of Florida.

Franklin was also one of those who hit upon the idea of using the thermometer as an instrument of navigation. Seamen had long been aware of the sharp changes in sea-surface temperature that occur in the region of the Gulf Stream. At the beginning of the seventeenth century Lescarbot had written:

> 'I have found something remarkable upon which a natural philosopher should meditate. On the 18th of June, 1606, in latitude 45° at a distance of six times twenty leagues east of the Newfoundland Banks, we found ourselves in the midst of very warm water despite the fact that the air was cold. But on the 21st of June all of a sudden we were in so cold a fog that it seemed like January and the sea was extremely cold too.'

Franklin made a series of surface temperature measurements across the Atlantic and also attempted to measure subsurface temperatures; he collected water for measurement from a depth of about 100 feet using a bottle, and later a cask, with valves at each end—a piece of equipment not unlike the modern Nansen bottle.

The next intensive study of the Gulf Stream was made by James Rennell, whose authoritative work was published posthumously in 1832. Rennell

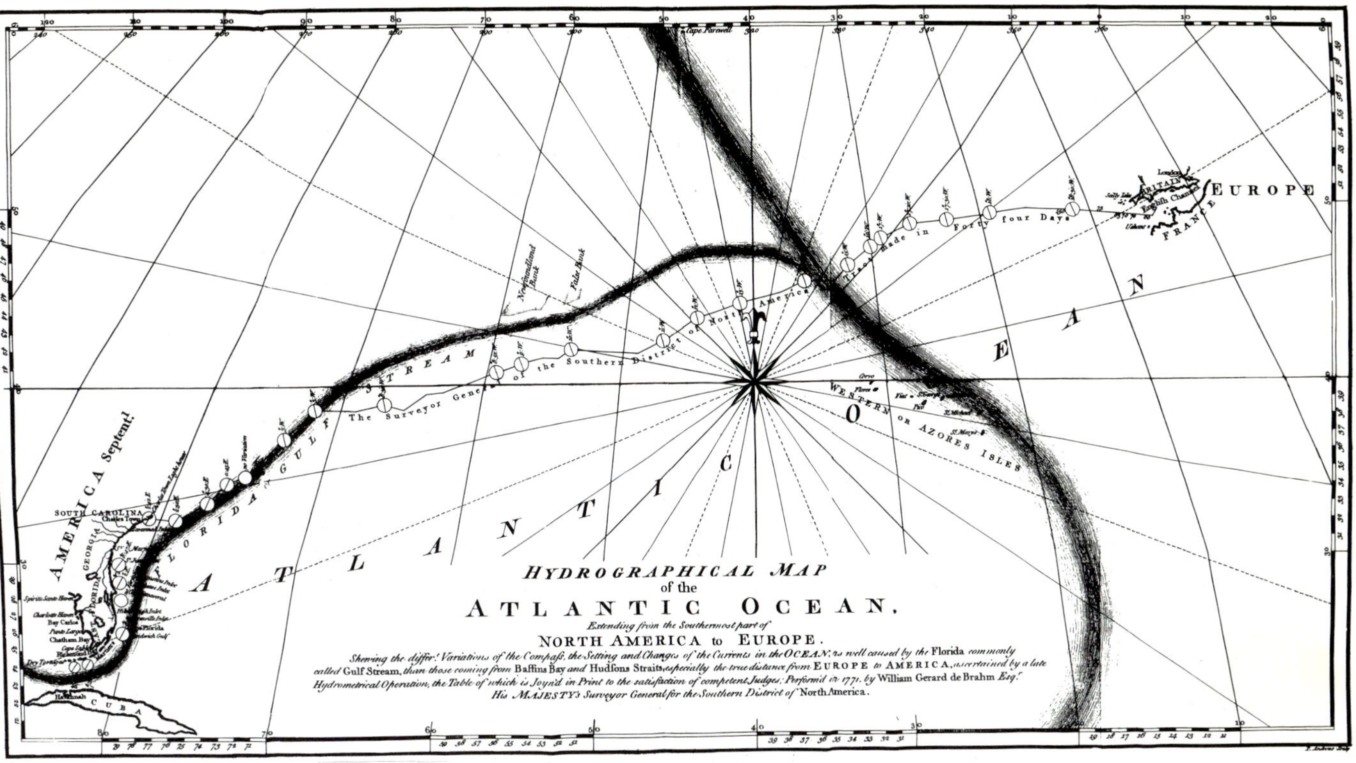

Figure 4.1 (a) De Brahm's chart of Florida and the Gulf Stream. On this projection, lines of latitude (not shown) are parallel to the top and bottom of the map, and the curved lines are lines of longitude. In the small circles along the path of the Gulf Stream are indicated the bearings of the current read against magnetic north.

(b) The chart of the Gulf Stream made by Timothy Folger and Benjamin Franklin.

compiled large amounts of data from the British Admiralty Office and carefully documented the variability found in the Gulf Stream. He distinguished between 'drift currents', produced by direct stress of the wind, and 'stream currents' produced by a horizontal pressure gradient in the direction of the flow (the term 'drift current' is still used occasionally). Rennell agreed with Franklin that the Gulf Stream was a 'stream current'.

The mean current charts shown in Figure 4.1 were produced mainly from measurements of ships' drift. By the time of Franklin and De Brahm, accurate chronometers had become available and ships' positions could be fixed with respect to longitude as well as latitude. Position fixes were made every 24 hours; between fixes, a continuous record of position was kept by means of *dead-reckoning*, i.e. the ship's track was *de*duced from its speed and the course steered. The accumulated discrepancy after 24 hours between the dead-reckoning position and the accurately fixed position gave an indication of the average drift of the surface water through which the vessel had passed (Figure 4.2). However, sailing vessels were themselves strongly influenced by the wind, and there could be considerable difficulties in maintaining course and speed in heavy seas. These inherent inaccuracies, combined with any inaccuracies in position-fixing, meant that estimates of current speeds so obtained could be very unreliable.

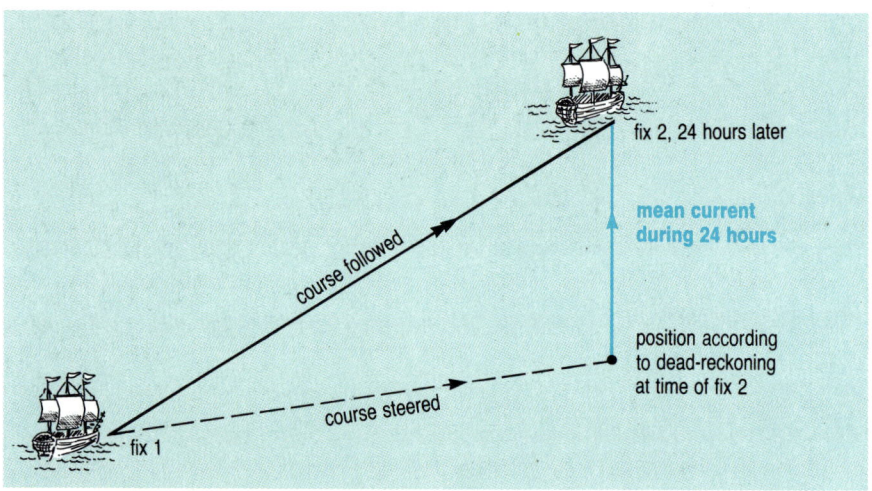

Figure 4.2 Deduction of the mean current during a 24-hour period from the ship's drift.

Although single estimates of current speed obtained from ships' drift were unreliable, accumulations of large numbers of measurements, as compiled by Rennell, could be used to construct charts of current flow, averaged over time and area. Collection of such data from commercial shipping has continued to the present day and this information has been used to construct maps of mean current flow, such as that in Figure 3.1. Current speeds estimated by dead-reckoning are now considerably more accurate, thanks to position-fixing by radar and navigational satellites, and the use of automatic pilots.

Systematic collection of oceanographic data was given a great impetus by the American naval officer and hydrographer, Matthew Fontaine Maury. Maury arranged for the US Hydrographic Office to supply mariners with charts of winds and currents, accompanied by sailing directions; in return, the mariners agreed to observe and record weather and sea conditions and to provide Maury with copies of their ships' logs. In 1853, on Maury's initiative, a conference of delegates from maritime nations was convened in Brussels to devise a standard code of observational practice. It was agreed that observations of atmospheric and oceanic conditions (including sea-surface temperature, specific gravity and temperature at depth) should be made at intervals of two hours. This was soon changed to four hours, but apart from that the system has remained largely the same up to the present time.

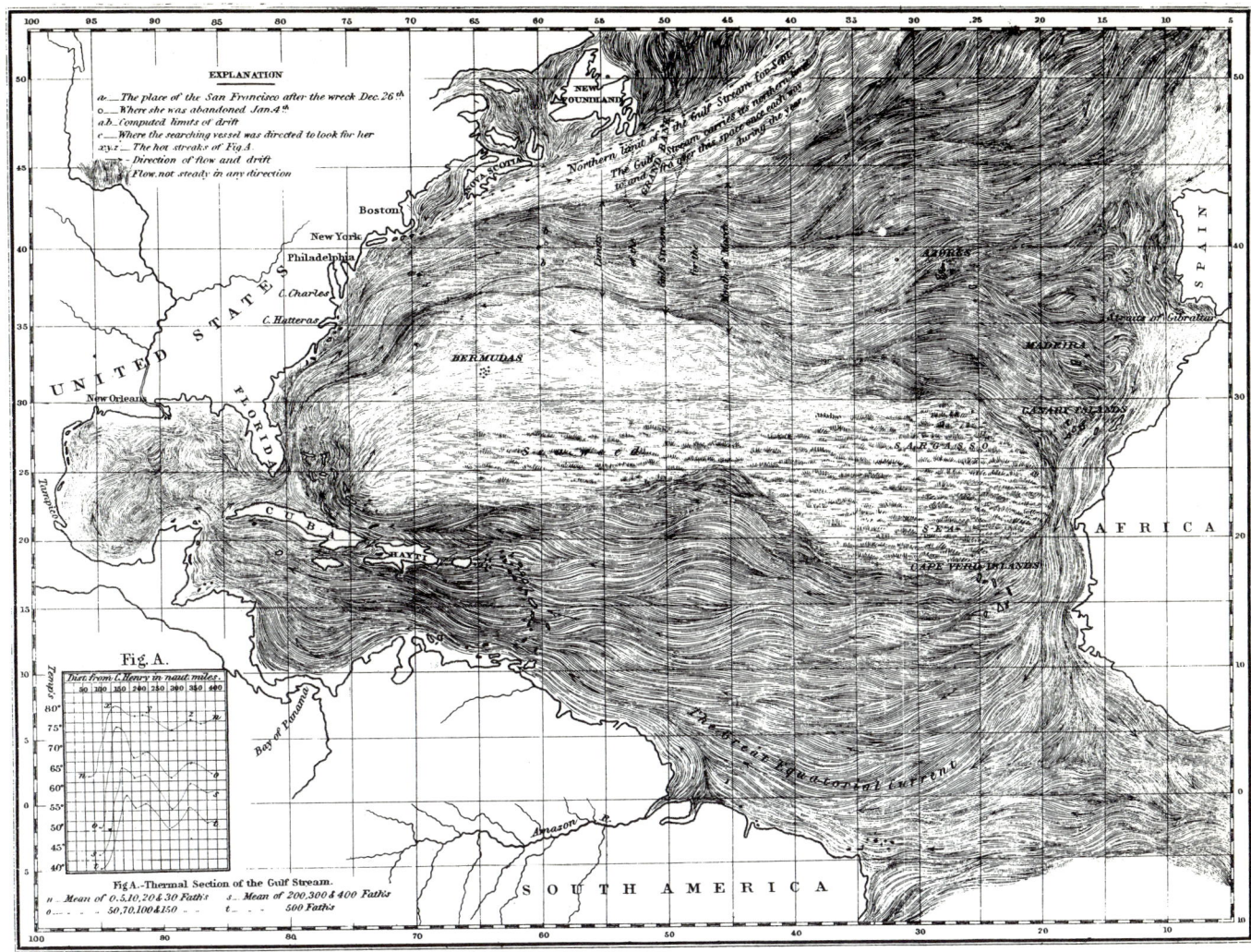

Figure 4.3 Maury's chart of the Gulf Stream and North Atlantic Drift. Note also the temperature traverses of the Gulf Stream at different latitudes, shown at bottom left.

Maury's contribution to practical navigation can be immediately appreciated by looking at his chart of the Gulf Stream and North Atlantic Drift (Figure 4.3). He was not, however, a great theoretician. He rejected the theory that the Gulf Stream owes its origin to the wind on the basis of the following argument. The Gulf Stream is wider at Cape Hatteras than in the Straits of Florida, so given that its total volume transport stays the same, and that the flow extends to the bottom, the sea-bed must be shallower at Cape Hatteras. The Gulf Stream therefore has to flow uphill and must be maintained by some mechanism other than wind stress.

QUESTION 4.1 (a) By reference to the idea about Gulf Stream generation shared by Franklin and Rennell, explain briefly why Maury's argument is seriously flawed. About what basic principle was he confused?

(b) In fact, Maury was not justified in assuming that the volume of water transported in the Gulf Stream is the same at Cape Hatteras as it is in the Straits of Florida. By reference to Figure 3.1, can you explain why?

Maury suggested that the Gulf Stream was driven by a kind of peristalsis, and believed that the general ocean circulation resulted from the difference in seawater density between the Equator and the Poles. This latter idea had been put forward in 1836 by Arago, as an alternative to the theory that the Gulf Stream owed its existence ultimately to the wind. Arago pointed out that the measured drop in the level of the Gulf Stream from the Gulf of Mexico side of Florida to the Atlantic side, as measured by a levelling survey across Florida, was only 7½ inches at most, and this was surely much too small to drive the Gulf Stream. In fact, more recent calculations (assuming greater frictional forces in the Gulf Stream than were thought of then) have shown that 7½ inches *is* enough of a head to drive the Gulf Stream. The Stream is sometimes thought of as a 'jet' being squirted out through the Straits of Florida.

Modern hydrographic surveying of the Gulf Stream can be said to have begun in 1844 with the work of the United States Coast and Geodetic Survey, under the direction of Franklin's great-grandson, A. D. Bache. Since that time there have been a number of surveys of both current velocity and temperature distribution. Perhaps the most impressive work was done by John Elliott Pillsbury who made observations along the Gulf Stream at a number of locations, including several in the Straits of Florida. He measured the temperature at a number of depths and also recorded the direction and speed of the current using a current meter of his own design. These observations made in the 1890s were not repeated until relatively recently and have been invaluable to oceanographers in the twentieth century (as you will see in Section 4.3.2).

Like most seagoing men of his time, Pillsbury was convinced that ocean currents were generated by the wind. There were, however, severe difficulties in explaining the mechanism whereby the wind could drive the ocean. Advances in understanding were held up for two main reasons. The first of these was that the significance of the turbulent nature of flow in the oceans was not fully appreciated.

QUESTION 4.2 One objection to the idea of wind-driven currents (put forward by a contemporary mathematical physicist) was that the wind would need to blow over the sea-surface for hundreds of thousands of years before a current like the Gulf Stream could be generated. From your reading of Chapter 3, can you suggest what misunderstanding this objection was based on?

The other great difficulty hindering progress was that the effect on ocean currents of the rotation of the Earth was not properly understood, with the result that the role of the Coriolis force in balancing horizontal pressure gradients in the oceans was overlooked. One of the first scientists to understand the effect of the Coriolis force on ocean currents was William Ferrel. Ferrel derived the relationship between atmospheric pressure gradient and wind speed, i.e. the equation for geostrophic flow in the atmosphere. Unfortunately, oceanographers were not aware of Ferrel's work and so its application to the oceans was not immediate. The formula for computing ocean current speeds from the slopes of isobaric surfaces—in other words, the gradient equation (3.11)—was derived by Henrik Mohn in 1885, a few years later. It was not until the first decade of the twentieth century that the Scandinavians Sandström and Helland-Hansen seriously investigated the possibility of using the density distribution in the oceans to deduce current velocities (*cf.* Section 3.3.3).

It is now clear that the Gulf Stream *is* wind-driven and is not so much a warm current as a ribbon of high-velocity water forming the boundary between the warm waters of the Sargasso Sea and the cooler waters over the continental margin. Armed with an understanding of geostrophic currents we can see that the sharp gradient in temperature (and hence density) across the Stream is an expression of a balance between the horizontal pressure gradient force and the Coriolis force, an equilibrium that has been brought about indirectly by the wind. (We will return to geostrophic equilibrium in the Gulf Stream in Section 4.3.2.)

4.2 THE SUBTROPICAL GYRES

As discussed in Section 3.4, current flow in the subtropical gyres is related to the overlying anticyclonic wind systems, which blow around the subtropical high pressure regions. The centres of the atmospheric and oceanic gyres are not, however, coincident: the centres of the atmospheric gyres tend to be displaced towards the eastern side of the oceans, while the centres of the oceanic gyres tend to be displaced towards the western side, especially in the Northern Hemisphere. As a result, the currents that flow along the western sides of oceans—the **western boundary currents**—are characteristically fast, intense, deep and narrow, while those that flow along the eastern sides—the **eastern boundary currents**—are characteristically slow, wide, shallow and diffuse.

The western boundary current in the North Atlantic is the Gulf Stream; its counterpart in the North Pacific is the Kuroshio. Both of these currents are typically only some 100km wide and in places have surface velocities in excess of $2\,\mathrm{m\,s^{-1}}$. By contrast, the Canary Current and the California Current are over 1000km wide and generally have surface velocities less than $0.25\,\mathrm{m\,s^{-1}}$. In the South Atlantic and the South Pacific, the difference between the western boundary currents (the Brazil Current and the East Australian Current, respectively) and the eastern boundary currents (the Benguela Current and the Humboldt, or Peru, Current) is not so marked. This may be because the South Atlantic and South Pacific are open to the Southern Ocean so that their gyres are strongly influenced by the Antarctic Circumpolar Current; furthermore, the South Pacific does not have a continuous barrier along its western boundary. However, some of the differences between the gyres of the Northern and Southern Hemispheres may be apparent rather than real. The northern gyres have been more fully investigated; and perhaps when more information is available, the southern gyres will be seen to resemble their northern counterparts more closely.

4.2.1 VORTICITY

'Big whirls have little whirls
which feed on their velocity.
Little whirls have lesser whirls
and so on to viscosity.'

L. F. Richardson (1881–1953), referring to atmospheric motions and paraphrasing Augustus de Morgan (who had himself paraphrased Jonathan Swift).

You will already be familiar with the idea—essential to an understanding of dynamic systems—that energy and mass must be conserved. Another

Figure 4.4 Examples of eddies in the oceans.
Left: Eddies in the Mediterranean off the coast of Libya. They were photographed from the Space Shuttle and show up because of variations in surface roughness. The picture shows an area about 75km across; the white line is a ship's wake or bilge dump.
Right: Small-scale eddies forming in the vicinity of a rocky coastline.

property that must be conserved is momentum—both linear momentum associated with motion in straight lines and angular momentum associated with rotatory movement. In oceanography, it is more convenient to view the conservation of angular momentum as the conservation of a tendency to rotate, or the conservation of **vorticity**, the tendency to form vortices.

Ocean waters have rotatory motions on all scales, from the basin-wide subtropical gyres down to the smallest swirls and eddies (Figure 4.4). Fluid motion does not have to be in closed loops to be rotatory; whenever there is **current shear** (a change in velocity at right angles to the direction of flow) there is a tendency to rotate and the water has vorticity. For convenience, *a tendency to rotate anticlockwise is referred to as positive and a tendency to rotate clockwise is referred to as negative*. These aspects of vorticity are illustrated schematically in Figure 4.5.

Note that we say 'a *tendency* to rotate' rather than simply 'rotatory motion'. This is because water may acquire positive vorticity by one mechanism at the same time as it acquires negative vorticity by another. For example, water may be acquiring positive vorticity as a result of

NEGATIVE (CLOCKWISE) VORTICITY   POSITIVE (ANTICLOCKWISE) VORTICITY

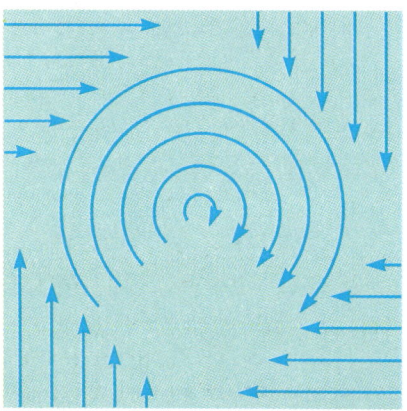
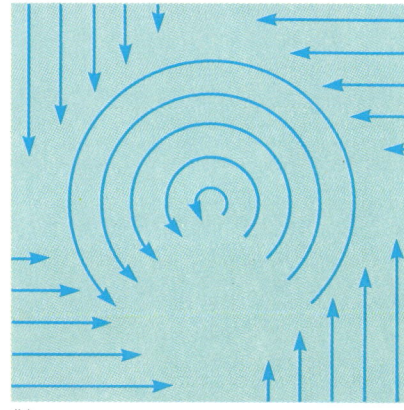

(a)   (b)

Figure 4.5 Diagrams to show examples of flow with (a) negative and (b) positive vorticity. The speed and direction of the flow are indicated by the lengths and directions of the arrows.

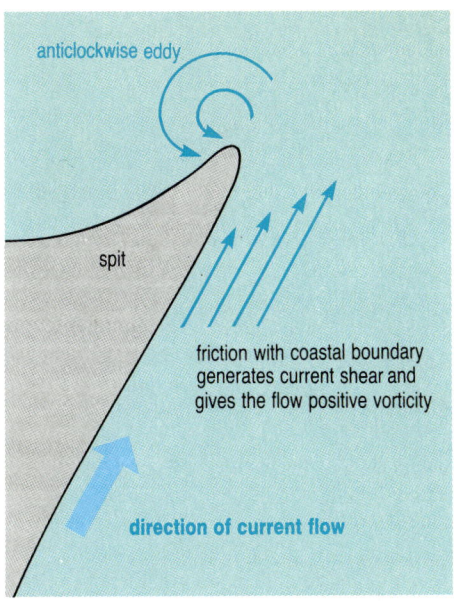

Figure 4.6 Diagram to illustrate how eddies are generated off spits; the lengths and directions of the arrows indicate the speed and direction of the current.

current shear caused by friction with a coastal boundary such as a spit (Figure 4.6) or adjacent water, while at the same time it is acquiring negative vorticity from the wind. The actual rotatory motion that results will depend on the relative sizes of the two effects. In theory, positive and negative vorticity tendencies could be exactly equal so that *no* rotatory motion would result.

Water that has a rotatory motion in relation to the surface of the Earth, caused by wind stress and/or frictional forces, is said to possess **relative vorticity**. However, the Earth itself is rotating. The vorticity possessed by a parcel of fluid by reason of its being on the rotating Earth is known as its **planetary vorticity**.

Planetary vorticity and the Coriolis force

In Chapter 1, you saw how the rotation of the Earth about its axis results in the deflection of currents and winds by the Coriolis force. These deflections were explained in terms of the poleward decrease in the eastwards velocity of the surface of the Earth. In addition to a linear eastwards velocity, the surface of the Earth has an *angular velocity*. Like the linear eastwards velocity, this angular velocity depends on latitude. At the poles, the angular velocity of the surface of the Earth is simply Ω (= $7.29 \times 10^{-5} s^{-1}$ (*cf.* Section 3.2)). At lower latitudes, it is a proportion of Ω: the bigger the angle between the Earth's axis of rotation and the local vertical axis, the smaller the angular velocity of the surface of the Earth about this vertical axis. At the Equator, where a vertical axis is at right angles to the axis of rotation of the Earth, the angular velocity of the surface is zero. In general, the angular velocity of the surface of the Earth about a vertical axis at latitude ϕ is given by $\Omega \sin \phi$ (Figure 4.7(a)).

You saw in Chapter 1 that a missile fired northwards from the Equator is deflected eastwards in relation to the surface of the Earth because, with increasing latitude, the surface of the Earth travels eastwards at a progressively decreasing rate. However, as just described, the surface of the Earth also has an *angular* velocity, so that in the Northern Hemisphere it rotates anticlockwise about a vertical axis, and in the Southern Hemisphere it rotates clockwise (Figure 4.7(b)). For this reason, there would be relative motion between a hypothetical missile and the surface of the Earth, regardless of the direction in which the missile is fired (as long as it is not fired along the Equator), and this relative motion would increase with increasing latitude. Similarly, winds and currents moving eastwards and westwards experience deflection, as well as those moving northwards and southwards. You have seen that the Coriolis force causing this deflection is weakest in the tropics and strongest at the poles. Using the fact that the angular velocity of the surface of the Earth is given by $\Omega \sin \phi$, it is possible to deduce the exact relationship of the Coriolis force to latitude, which you have already encountered in Section 3.1.2:

$$\text{Coriolis force} = m \times 2\Omega \sin \phi \times u$$
$$= mfu \quad \text{(equation 3.2)}$$

where m is the mass of a body moving with velocity u, and f (= $2\Omega \sin \phi$) is known as the Coriolis parameter.

Clearly, if the surface of the Earth at latitude ϕ has an angular velocity of $\Omega \sin \phi$, so too does any object or parcel of fluid on the Earth at that latitude. In other words, any parcel of fluid on the surface of the Earth

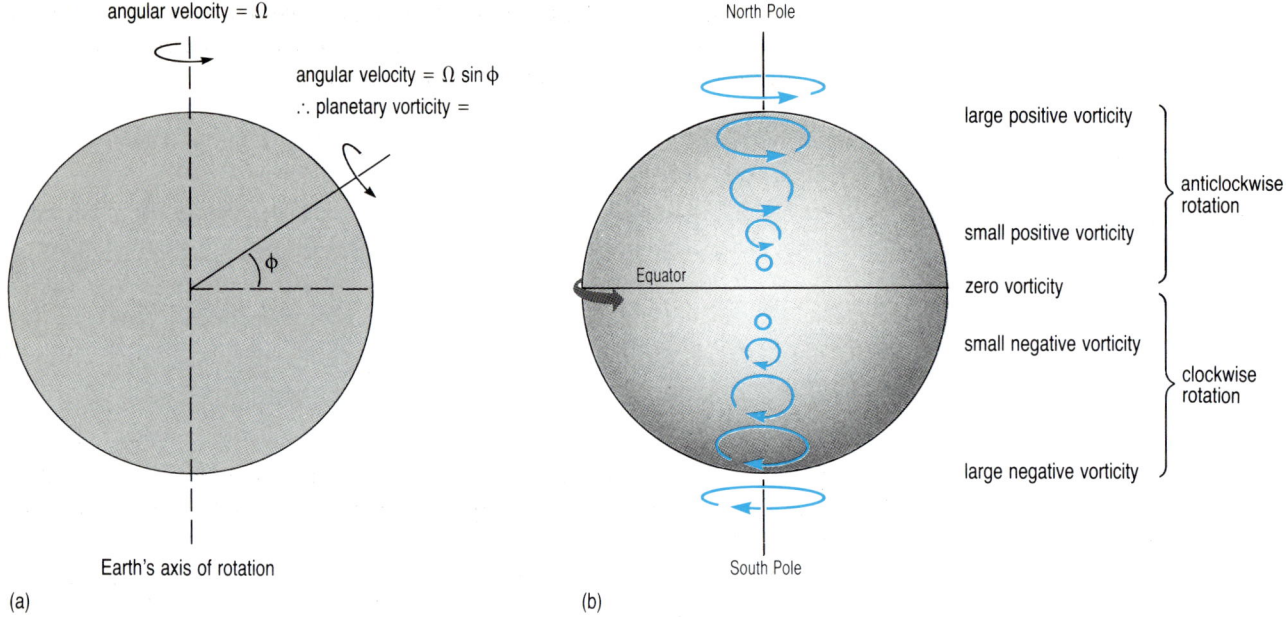

Figure 4.7 (a) Diagram to show the derivation of the expression for the planetary vorticity of a fluid parcel on the surface of the Earth at latitude φ. (For use with Question 4.3(a).)

(b) Schematic diagram to show the variation of planetary vorticity with latitude; the blue arrows represent the component of the Earth's rotation about a vertical axis (viewed from above) at different latitudes.

shares the component of the Earth's angular rotation appropriate to that latitude.

The vorticity of a parcel of fluid is defined mathematically to be equal to twice its angular velocity.

QUESTION 4.3 (a) What, then, is the planetary vorticity possessed by a parcel of fluid on the surface of the Earth at latitude φ? (To answer this Question, complete Figure 4.7(a).)

(b) What is the planetary vorticity (i) at the North Pole, (ii) at the South Pole, and (iii) at the Equator?

In answering Question 4.3(a), you will have seen that planetary vorticity is given by the same expression as the Coriolis parameter; for that reason, it also is given the symbol f. Relative vorticity is given the symbol ζ (zeta).

At the beginning of this Section it was stated that vorticity—effectively, angular momentum—must be conserved. But it is vorticity *relative to fixed space* (i.e. with respect to an observer outside the Earth) that must be conserved; in other words, it is the **absolute vorticity** of a parcel of fluid— the combination of the planetary and relative vorticity ($f + \zeta$)—that must remain constant, in the absence of external forces such as wind stress and friction. Consider, for example, a parcel of water that is being carried northwards from the Equator in a current. As it moves northwards it will be moving into regions of increasingly large positive (anticlockwise) planetary vorticity (see Figure 4.7(b)). However, as it is only weakly bound to the Earth by friction, it tends to be 'left behind' by the Earth rotating beneath it and, relative to the Earth, acquires an increasingly negative (clockwise) rotational tendency*. If there are no external influences imparting positive or negative relative vorticity to the parcel of water (e.g. it does not come under the influence of cyclonic or

---

*If you find this difficult to envisage, you can understand what we mean by being 'left behind' simply by filling a glass with water and twirling it. You will find that the water tends to remain stationary, as the glass rotates outside it. Of course, *in relation to the glass* the water *is* moving.

anticyclonic wind fields), the increase in positive planetary vorticity will be exactly equal to the increase in negative relative vorticity, and the absolute vorticity $(f + \zeta)$ will remain constant.

QUESTION 4.4 (a) A body of water is carried *southwards* from the Equator in a current.

(i)   As the water moves south, how is its planetary vorticity changing?

(ii) How does this affect its relative vorticity?

(b) What happens to the relative vorticity of a body of water if it is acted upon by:

(i)   winds blowing in a clockwise direction?

(ii) cyclonic winds in the Southern Hemisphere?

However, the situation is not quite as straightforward as this. In order to understand what happens in the oceans, we have to bear in mind that the vorticity of a body of water is the *sum* of the vorticity of all the constituent particles of water. Consider, for convenience, a column of water moving in a current. For the purposes of this discussion, we can imagine the column of water spinning about its own axis, although to have vorticity it could of course have any type of rotatory motion (Figure 4.5). For simplicity, we can also assume that the column is rotating anticlockwise in the Northern Hemisphere, i.e. it has positive relative vorticity.

What will happen if the rotating column of water becomes longer and thinner as a result of stretching, because it moves into a region of deeper sea-floor, for example? (Consider what happens when spinning skaters bring their arms in close to their sides.)

The simple answer is that the column of water will spin faster, i.e. its relative vorticity will increase (become more positive). Looked at from the point of view of angular momentum, the angular momentum of each particle of water is $mr^2\omega$, where r is the distance from the particle (of mass m) to the axis of rotation, and ω is the angular velocity (Figure 4.8(a)). When the column stretches, the average radius r goes down and so, for angular momentum to be conserved, ω the speed of rotation must increase (Figure 4.8(b)). Because of the effect of changes in the length

Figure 4.8 (a) The angular momentum of a particle of mass m moving with angular velocity ω in a circle of radius r is given by $mr^2\omega$.

(b) For the total angular momentum (vorticity) of a stretched column of water to remain constant, the angular velocity ω of the particles in the column must increase.

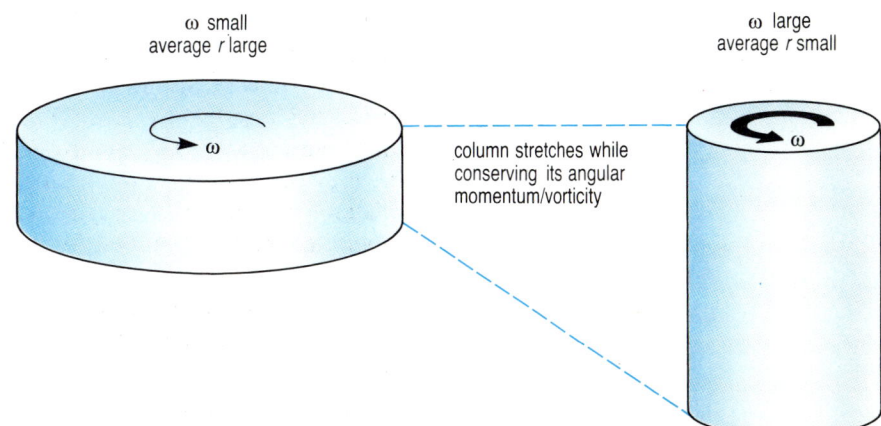

(D) of the water column, the property that is actually conserved is therefore not $f + \zeta$, the absolute vorticity, but $(f + \zeta)/D$, the **potential vorticity**.

The example given above is unrealistic because we have omitted consideration of variations in f, the planetary vorticity. In fact, in the real oceans, away from coastal boundaries and other regions of large current shear, f is very much greater than ζ. This means that $(f + \zeta)/D$ is effectively equal to f/D and it is f that must change in response to changes in D; and as f is simply a function of latitude, it can only be changed by water changing its latitude.

QUESTION 4.5 (a) Imagine a current flowing from east to west in the Northern Hemisphere. What will happen if the current flows over a shallow bank, so that the depth D is reduced? (You may assume that this occurs in the open oceans and that $f \gg \zeta$.)

(b) What will happen to a zonal* current flowing over a shallow bank in the Southern Hemisphere?

The fact that conservation of f/D causes currents to swing equatorwards and polewards over topographic highs and lows, respectively, is sometimes referred to as topographic steering.

In studies of potential vorticity in the real oceans, D, the 'depth of the water column', need not be the total depth to the sea-floor. It is the thickness of the body of water under consideration and so could, for example, be the depth from the sea-surface to the bottom of the permanent thermocline or, for deep-water movements, the depth from the bottom of the permanent thermocline to the sea-floor.

Since the mid-1980s, potential vorticity has become an important tool in the study of the movement of oceanic water masses; this application of vorticity will be discussed in Chapter 6.

4.2.2 WHY IS THERE A GULF STREAM?

We will now summarize briefly, with as little mathematics as possible, some of the developments in ideas that have led to an increased understanding of the dynamics of ocean circulation in general, and of the subtropical gyres in particular. In the course of this discussion, we will go a substantial way towards answering the question posed above: why *is* there a Gulf Stream?

Ekman's initial work on wind-driven currents, intended to explain current flow at an angle to the wind direction, was published in 1905. In 1947, the Scandinavian oceanographer H.U. Sverdrup used mathematics to demonstrate another surprising relationship between wind stress and ocean circulation. In constructing his theory, Ekman had assumed a hypothetical ocean which was not only infinitely wide and infinitely deep, but also had no horizontal pressure gradients because the sea-surface and all other isobaric surfaces were assumed to be horizontal. Sverdrup's aim was to determine current flow in response to wind stress *and* horizontal

---

*Zonal means *either* from east to west *or* from west to east; the equivalent word for north–south flow is 'meridional'.

pressure gradients. Unlike Ekman, Sverdrup was not interested in determining how flow varied with depth; instead, he derived an equation for the *total* or *net* flow resulting from wind stress.

Consider a simple situation in which the winds blowing over a hypothetical ocean are purely zonal and vary in strength with latitude sinusoidally, as shown in Figure 4.9(a).

QUESTION 4.6 (a) How closely does this hypothetical wind field resemble that over the real oceans between 15 and 45 degrees of latitude?

(b) Draw arrows on Figure 4.9(b) to indicate the Ekman transport that would result from the wind field shown in (a).

(c) What effect will this have on the shape of the sea-surface and on the horizontal pressure gradients at depth? Draw more arrows on Figure 4.9(b) to show the direction of the geostrophic currents that would result from the horizontal pressure gradients.

Figure 4.9 (a) A hypothetical wind field in which the wind is purely zonal (i.e. easterly or westerly) and varies in strength with latitude sinusoidally.

(b) Plan view showing the direction of the Ekman transports, horizontal pressure gradients and geostrophic currents that result from the wind field shown in (a). (To be completed for Question 4.6(b) and (c).)

In answering Question 4.6 you have identified the two aspects of current flow resulting from wind stress that Sverdrup combined mathematically:
(i) Ekman transport at right angles to the direction of the wind, and
(ii) geostrophic flow, in response to horizontal pressure gradients caused by the 'piling up' of water through Ekman transport.

The pattern of flow that results from combining these two components is shown in Figure 4.10(b). The 'gyre' is not complete because in deducing this flow pattern, Sverdrup considered the effect of an *eastern* boundary but could not also include the effect of a *western* boundary. As a result, there could be no northward flow in a 'Gulf Stream'.

The most interesting aspect of Sverdrup's results, however, was that the net amount of water transported by a given pattern of wind stress depends not on the *absolute* value of the wind stress but on its *torque*, i.e. its tendency to cause rotation—or, if you like, its ability to supply relative vorticity to the ocean. In particular, Sverdrup showed that the net amount of water transported meridionally (i.e. north–south or south–north) is

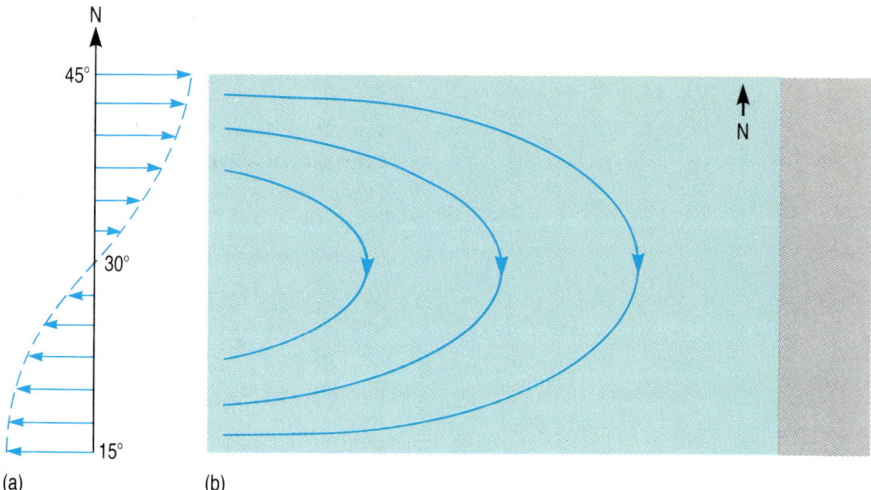

Figure 4.10 (a) As for Figure 4.9. (b) The 'depth-integrated' flow pattern that results from combining Ekman transports at right angles to the wind with geostrophic current flow in response to horizontal pressure gradient forces, according to Sverdrup.

directly proportional to the torque of the wind stress. You can see this for yourself, in a qualitative way, by studying Figure 4.10. The first thing to note is that the pattern of wind stress shown in part (a) of Figures 4.9 and 4.10 will result in a clockwise torque on the ocean (*cf.* Question 4.6(a)).

At what latitude is this torque greatest?

At latitude 30°—and it is here that the southwards motion, as shown by the flow lines on Figure 4.10(b), is greatest. The torque is least (effectively zero) at 45° and 15°, and here the flow is almost entirely zonal (eastwards and westwards, respectively).

In summary then, Sverdrup showed that at any location in the ocean, the total amount of water transported meridionally is proportional to the torque of the wind stress. If the meridional transport is given the symbol *M*, this can be written as:

$$M = \text{constant} \times \text{torque of } \tau \tag{4.1a}$$

where τ is the wind stress. The word used in mathematics for 'torque' is 'curl', so equation 4.1a can be written as:

$$M = \text{constant} \times \text{curl } \tau \tag{4.1b}$$

Sverdrup found that the constant in equation 4.1 is the reciprocal of the rate of change of the Coriolis parameter with latitude. The rate of change of *f* with latitude is commonly given the symbol β, so equation 4.1 is usually written as:

$$M = \frac{1}{\beta} \text{curl } \tau \tag{4.1c}$$

You are not expected to manipulate this equation, but you should appreciate that it is a convenient way of summarizing Sverdrup's theoretically determined relationship, namely that at any location in the ocean the *total* meridional flow is determined by the rate of change of *f* at the latitude concerned and the torque, or curl, of the wind stress. It is important to remember that Sverdrup's relationship concerns the *total* meridional transport of water, i.e. the *net* flow taking into account the speed and direction of flow at *all* depths, from the surface down to the depth at which even indirect effects of the wind cannot be felt.

Sverdrup's theory was an advance on Ekman's in that the ocean was not assumed to be unlimited but, as mentioned above, it could not take account of a western boundary, and certainly could not explain the existence of intense western boundary currents like the Gulf Stream (Figure 4.11).

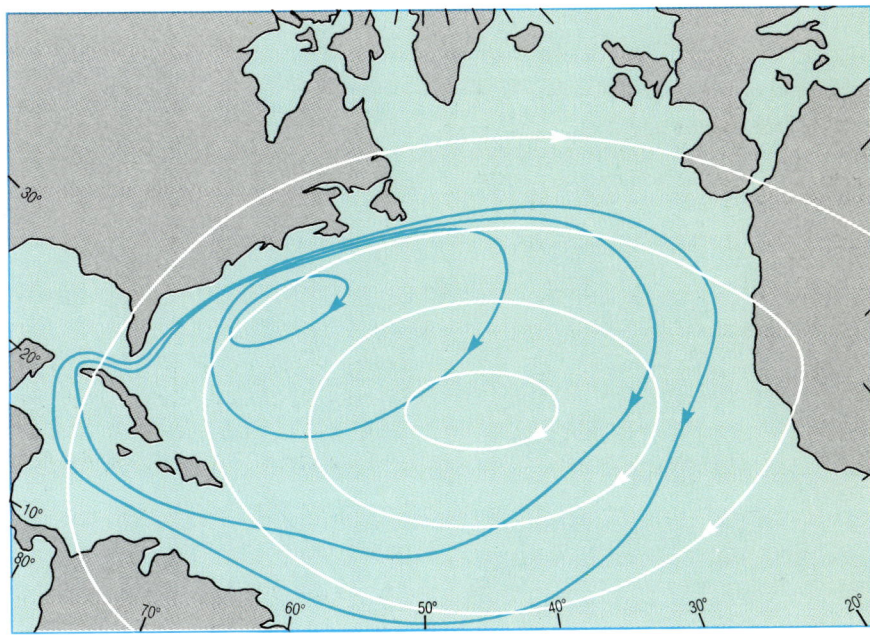

Figure 4.11 Schematic illustration of the asymmetrical North Atlantic gyre (blue) and the more or less symmetrical wind field which overlies it.

This problem was solved by the American oceanographer Henry Stommel, in a paper published in 1948. Stommel considered the effect of a symmetrical gyral wind field on a rectangular ocean in three different situations:

1   The ocean is not rotating, i.e. it is assumed to be on a non-rotating Earth.

2   The ocean is rotating but the Coriolis parameter f is constant.

3   The ocean is rotating and the Coriolis parameter varies with latitude (for the sake of simplicity, this variation is assumed to be linear).

Unlike Sverdrup, Stommel included friction in his calculations and he worked out the flow that would result when the wind stress and frictional forces balanced (i.e. when there was a steady state with no acceleration or deceleration) for each of the three situations described above.

Figure 4.12 summarizes the results of Stommel's calculations. The left-hand diagrams represent the patterns of flow that result from a symmetrical gyral wind field in each of situations (a)–(c), and the right-hand diagrams show the sea-surface shapes that accompany this flow.

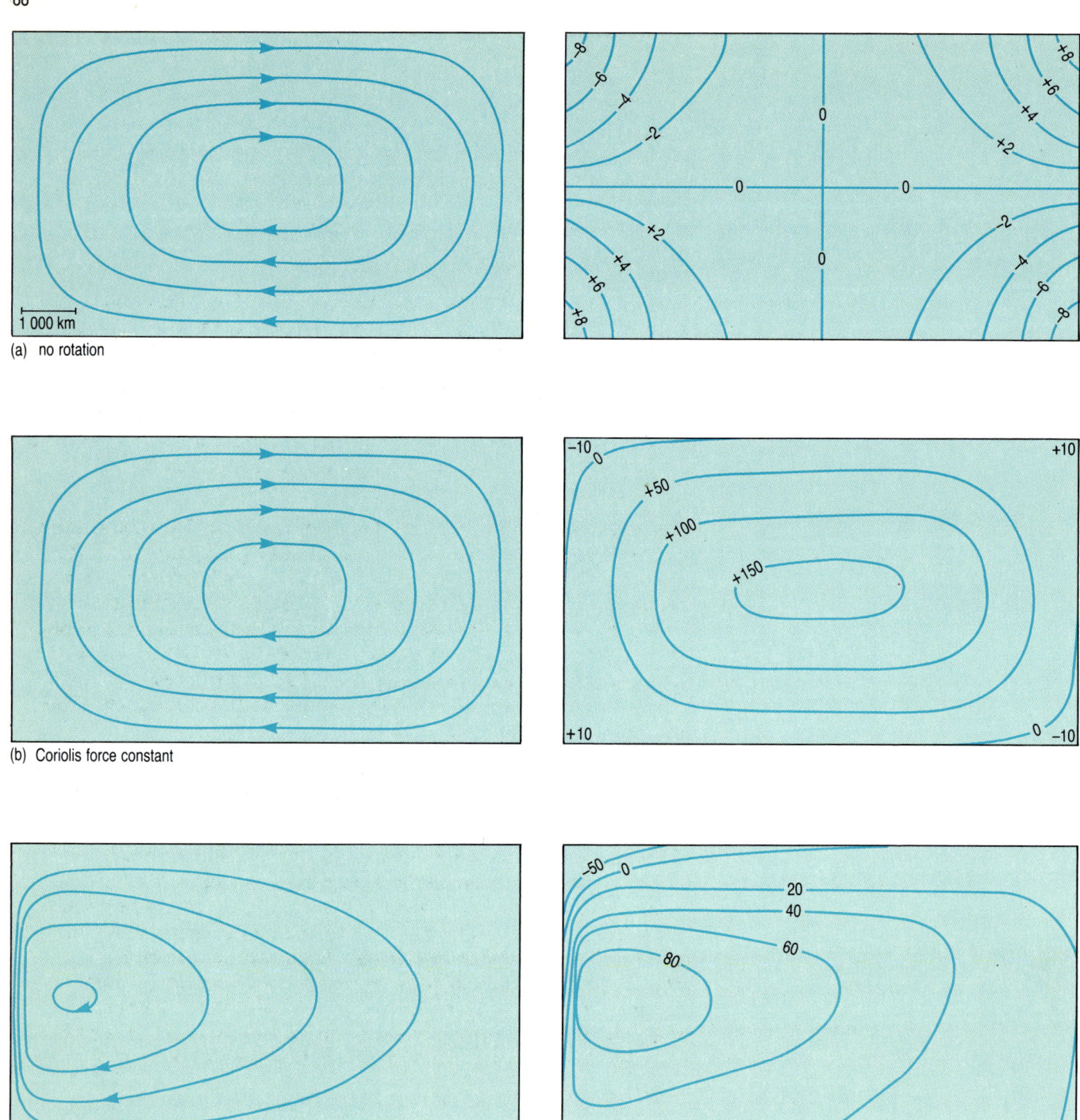

Figure 4.12 Summary of the results of Stommel's calculations. The diagrams on the left-hand side show the streamlines parallel to which water flows in the wind-driven layer (assumed to be 200 m deep). The volume of water transported around the gyre per second between one streamline and the next is 20% of the total flow. The diagrams on the right-hand side show contours of sea-surface height in cm. In (a), the ocean is assumed to be on a non-rotating Earth; in (b), the ocean is on a rotating Earth but the Coriolis force is assumed to be constant with latitude; in (c), the Coriolis parameter is assumed to vary linearly with latitude.

QUESTION 4.7   Study Figure 4.12 and its caption.

(a)   Concentrate first on the diagrams on the right-hand side. What is the striking difference in the shape of the sea-surface between situation (a) and situation (b)? Given that each diagram represents a situation of equilibrium, where forces balance, can you suggest why there is this marked difference?

(b)   Now concentrate on the diagrams on the left-hand side. What do they tell us about the cause of intensification of western boundary currents? Is it simply the result of the *existence* of the Coriolis force?

Thus, Stommel showed that the intensification of western boundary currents is, in some way, the result of the fact that the Coriolis force *varies with latitude*, increasing from the Equator to the poles.

Intensification of western boundary currents may also be explained in terms of vorticity balance. In some ways, it is more convenient to work with vorticity than with linear current flow, because horizontal pressure gradient forces, which lead to complications, do not need to be considered.

Imagine a subtropical gyre, acted upon by a symmetrical wind field like that in Figure 4.11. If the wind has been blowing for a long time, the ocean will have reached a state of equilibrium, tending to rotate neither faster nor slower with time; at every point in the ocean, the relative vorticity has a fixed value. This means that, over the ocean as a whole, those factors which act to change the relative vorticity of the moving water must cancel each other out. Assuming for convenience that the depth of the flow is constant, we will now consider in turn each of the factors that affect relative vorticity.

The most obvious factor affecting the relative vorticity of the water in the gyre is the *wind*. The wind field is symmetrical and acts to supply negative (clockwise) vorticity over the whole region (Figure 4.13(a)).

The next factor to consider is *change in latitude*. Water moving northwards on the western side of the gyre is moving into regions of larger positive planetary vorticity and hence acquires negative relative vorticity (Figure 4.7(b)); similarly, water moving southwards on the eastern side of the gyre is moving into regions of smaller positive planetary vorticity and so loses negative relative vorticity (or gains positive relative vorticity). However, because as much water moves northwards as moves southwards, the net change in relative vorticity of water in the gyre as a result of change in latitude is zero.

What, then, is acting to counteract the negative (clockwise) tendency supplied by the wind stress, so that the gyre does not speed up indefinitely?

The answer is *friction*. We may assume that the wind-driven circulation is not frictionally bound to the sea-floor, but there will be significant friction with the coastal boundaries as a result of horizontal eddies. Using appropriate values of A_h, the coefficient of eddy viscosity for horizontal motion (Section 3.1.1), we may make vorticity-balance calculations for a symmetrical ocean circulation. Such calculations show that for the frictional forces to be large enough to provide sufficient positive relative vorticity to balance the negative relative vorticity provided by wind stress, the gyral circulation would have to be many times faster than that observed in the real oceans.

Figure 4.13(a) is a pictorial representation of the vorticity balance of a symmetrical Northern Hemisphere gyre (*cf*. Figure 4.12(b)) that carries water around at the rate observed in the real oceans. In this gyre there is an approximate vorticity balance in the eastern part of the ocean. The negative relative vorticity supplied to the water by the wind is nearly cancelled out by the positive relative vorticity that results from friction with the eastern boundary combined with that gained by the water as a result of moving into lower latitudes.

In the western part of the ocean, however, the combined effect of the negative relative vorticity supplied by the wind and that resulting from the movement of water into higher latitudes far outweighs the positive relative vorticity provided by friction. There is a continual gain of negative relative vorticity in the western part of the ocean and the gyre will accelerate indefinitely.

Figure 4.13(b) shows the corresponding situation for a strongly asymmetrical current system in which water flows north in a narrow boundary current in the western part of the ocean and flows south over most of the rest of the ocean (*cf*. Figure 4.12(c)). In this asymmetrical gyre the vorticity balance may be maintained in both the eastern *and* the western part of the ocean. In the eastern part of the ocean, where friction with the boundary is small, the vorticity balance is hardly affected at all, but on the western side, flow must now be much faster and the effects of both friction and change in latitude increase significantly. The negative relative vorticity resulting from change in latitude is acquired at a greater rate because water is now changing latitude much faster, and the gain of positive relative vorticity through friction is increased because both velocity and velocity shear have increased.* Both the planetary and frictional relative vorticity tendencies (which oppose one another) increase by an order of magnitude at least, and a vorticity balance is attained in the western ocean.

You should not regard the result demonstrated in Question 4.7(b) as being mutually exclusive with the vorticity-balance explanation given above. Both are concerned with the effect of the latitudinal variation in the angular velocity of the surface of the Earth on ocean circulation. In the first case, this variation in angular velocity is represented by the variation of the Coriolis force with latitude, and in the second it is represented by the variation in planetary vorticity with latitude.

The theoretically derived pictures—or models—of ocean circulation derived by Sverdrup and Stommel (Figures 4.10 and 4.12 (c)) bear a strong resemblance to the circulatory systems of the subtropical gyres. These circulation models were extended and made even more realistic by the work of Munk (1950). Like Stommel, Munk used a rotating rectangular ocean and assumed that the Coriolis force varied linearly with latitude, but he extended the ocean to include latitudes up to 60° and the equatorial zone. He also balanced the relative vorticity supplied by wind, change of latitude and friction, but improved on the representation of

---

*The frictional force between moving water and a boundary is approximately proportional to the square of the current speed; this is analogous to the relationship between the frictional force of the wind on the sea-surface (the wind stress, τ) and the wind speed, W (equation 3.1).

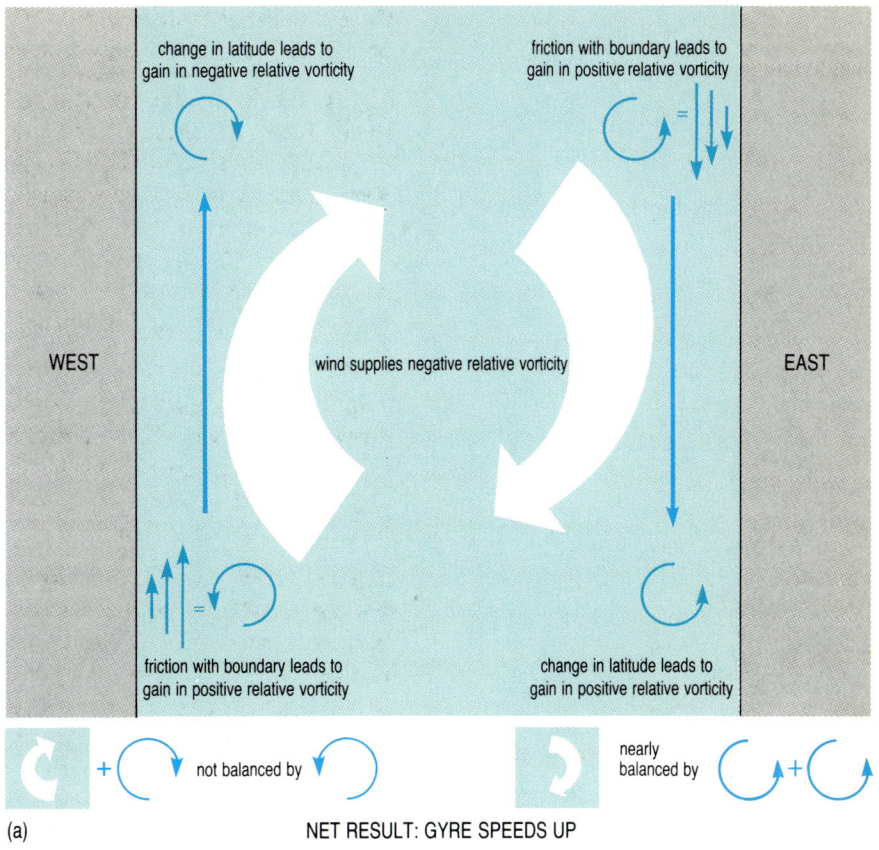
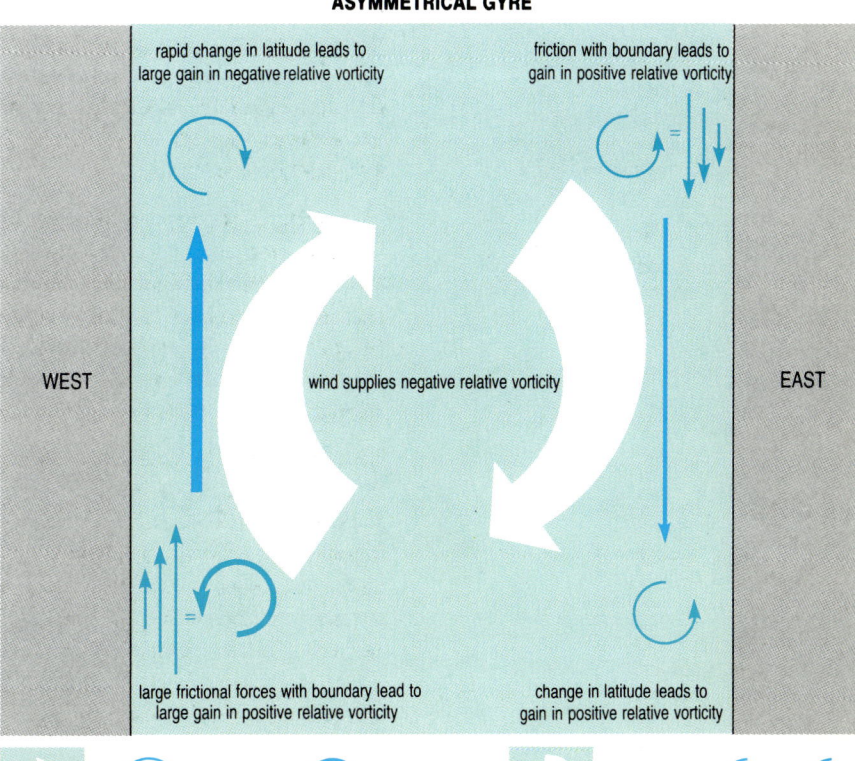

Figure 4.13 Pictorial representation of the various contributions to the vorticity of (a) a symmetrical subtropical gyre (cf. Figure 4.12(b)), and (b) a strongly asymmetrical subtropical gyre with an intensified western boundary current (cf. Figure 4.12(c)). The flow pattern in (b), in contrast to that in (a), enables a vorticity balance to be attained, so that the gyre does not rotate faster and faster indefinitely.

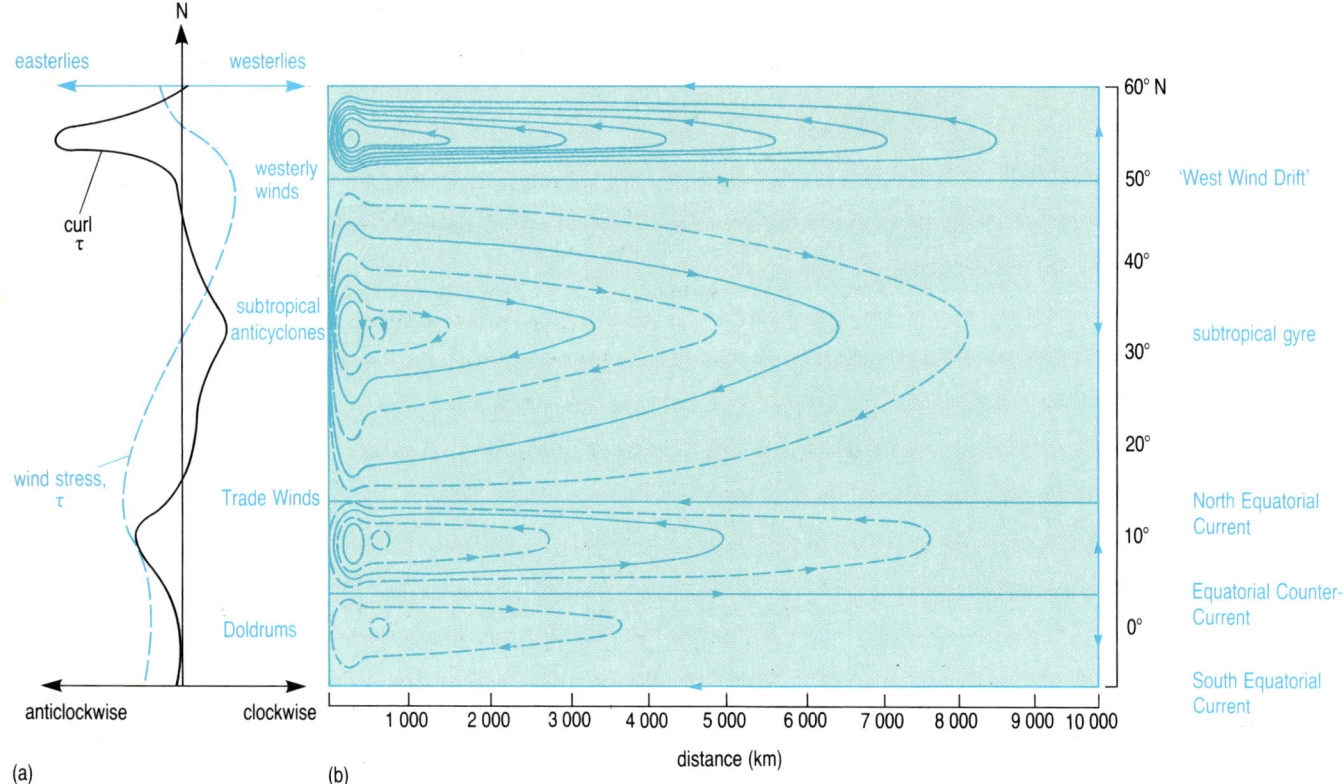

Figure 4.14 (a) The blue curve is the averaged annual zonal wind stress τ for the Pacific and Atlantic Oceans; the black curve is the curl or torque of this wind stress, curl τ. By definition, curl τ is at a maximum at those latitudes where the wind stress curve shows the greatest change with latitude, e.g. at about 55° N where the wind stress changes from easterly to westerly, and at about 30° N where it changes from westerly to easterly (cf. Figure 4.10(a)).

(b) The circulation pattern that Munk calculated using the values of curl τ shown in (a). The volume transport between adjacent solid lines is $10^7 m^3 s^{-1}$. The greatest meridional flow occurs where curl τ is at a maximum, i.e. at about 55°, 30°, and 10° N; at these latitudes, flow is either southwards or northwards, rather than eastwards or westwards.

both friction and the wind. As far as friction was concerned, he considered not simply friction with coastal boundaries but friction associated with both lateral and vertical current shear; in this way, he included the effect of eddy viscosity in both the horizontal and the vertical dimensions (i.e. both A_h and A_z). Instead of using the torque (curl) of the hypothetical sinusoidally varying wind, he used the torque that would result from the east–west component of the averaged winds of the *real* Pacific and Atlantic (Figure 4.14(a)).

Figure 4.14(b) shows the pattern of ocean circulation that Munk calculated. If you compare this pattern with Figure 3.1 you will see that it bears a strong resemblance to the actual circulation in the Pacific and Atlantic, so much so that it is possible to identify major circulatory features, as indicated on the right-hand side of the diagram.

Thus, in many ways, Munk's model of ocean circulation reproduces the circulation patterns seen in the real ocean. What this means is that the oceanic process that he and others considered important—namely flow of water into different latitudes (i.e. regions in which the Coriolis force is different), and horizontal and vertical frictional forces—really are important in determining the large-scale current patterns that result from the wind in the real oceans.

Models of ocean circulation are being improved and refined all the time. Nevertheless, the basic 'tools' which oceanographers use to construct models of ocean circulation remain the same: they are the 'equations of motion'. The aim of Section 4.2.3 is to explain what these equations are and to convey something of how they are used. We have simplified the mathematical notation, but the general principles outlined form the basis for solving even very complicated dynamic problems.

4.2.3 THE EQUATIONS OF MOTION

It may surprise you to know that until relatively recently, calculations that led to advances in the understanding of ocean circulation were made without the help of computers. The diagrams in Figures 4.10, 4.12 and 4.14 were drawn by solving the appropriate mathematical equations and plotting the results graphically.

The equations that physical oceanographers need to solve in order to be able to describe the dynamics of the ocean are known as the **equations of motion**. These are simply Newton's Second Law of Motion:

$$\text{force} = \text{mass} \times \text{acceleration} \tag{4.2a}$$

applied to a fluid moving over the surface of the Earth. Equation 4.2a can be rearranged to give:

$$\text{acceleration} = \frac{\text{force}}{\text{mass}} \tag{4.2b}$$

In dealing with moving fluids, it is convenient to consider the forces acting 'per unit volume'. The mass of unit volume of fluid is given numerically by its density, ρ. We can therefore write equation 4.2b as:

$$\text{acceleration} = \frac{\text{force per unit volume}}{\text{density}} = \frac{1}{\rho} \times \text{force per unit volume} \tag{4.2c}$$

In order to determine the characteristics of flow in three dimensions, we must apply equation 4.2c in three directions at right angles to one another. The x, y, z coordinate system used by convention in oceanography is shown in Figure 4.15. The velocity in each of the x-, y- and z-directions is u, v and w, respectively, and so the corresponding accelerations (rates of change of velocity with time) are du/dt, dv/dt and dw/dt.

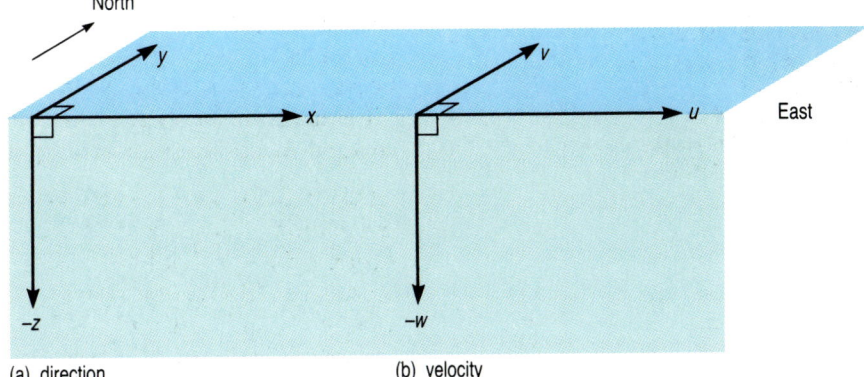

Figure 4.15 (a) The coordinate system commonly used in oceanography. The x-axis is positive eastwards and negative westwards; the y-axis is positive northwards and negative southwards; and the z-axis is positive upwards and negative downwards.

(b) The components of the current velocity in the x-, y-, and $-z$-directions are, by convention, u, v, and $-w$, respectively.

Before moving on, consider for a moment what forces might lead to acceleration in the horizontal x- and/or y-directions, and therefore need to be included in the equations of motion for those directions.

We hope you came up with the Coriolis force and the horizontal pressure gradient force, as well as perhaps wind stress and other frictional forces. The equation of motion appropriate to flow in the *x*-direction (i.e. easterly or westerly flow) may therefore be written:

$$\frac{du}{dt} = \frac{1}{\rho}\left(\begin{array}{l}\text{horizontal}\\ \text{pressure}\\ \text{gradient force in}\\ \text{the } x\text{-direction}\end{array} + \begin{array}{l}\text{Coriolis force}\\ \text{resulting in}\\ \text{motion in the}\\ x\text{-direction}\end{array} + \begin{array}{l}\text{other forces}\\ \text{related to}\\ \text{motion in the}\\ x\text{-direction}\end{array}\right)$$

For flow in the *y*-direction (i.e. northwards or southwards flow), the left-hand side of the equation is d*v*/d*t* and the right-hand side is identical to that above, except that 'in the *x*-direction' is replaced in each case by 'in the *y*-direction'.

In mathematical terms, the equations of motion for flow in the *x*- and *y*-directions may therefore be written:

$$\underset{\text{acceleration}}{\frac{du}{dt}} = \underset{\substack{\text{pressure}\\ \text{gradient}\\ \text{force}\\ \downarrow}}{\frac{1}{\rho}\left(-\frac{dp}{dx}\right.} + \underset{\substack{\text{Coriolis}\\ \text{force}\\ \downarrow}}{\rho f v} + \underset{\substack{\text{contributions}\\ \text{from other}\\ \text{forces}\\ \downarrow}}{\left.F_x\right)} \quad (4.3\text{a})$$

$$\frac{dv}{dt} = \frac{1}{\rho}\left(-\frac{dp}{dy}\right. - \rho f u + \left.F_y\right) \quad (4.3\text{b})$$

F_x and F_y may include wind stress, friction, or tidal forcing, depending upon the problem being investigated and the simplifications that can be made.

Note that the mathematical expressions used here have all been used already. The expressions for the horizontal pressure gradient forces are the same expressions as used in Chapter 3, and they have minus signs because the flow resulting from a horizontal pressure gradient is in the direction of decreasing pressure.

The expression for the Coriolis force has also been used already, although here *mfu* and *mfv* have been replaced by $\rho f u$ and $\rho f v$.

Why does the equation for flow in the x-direction have $\rho f v$ as the Coriolis term rather than $\rho f u$, and vice versa for flow in the y-direction?

The reason is, of course, that the Coriolis force acts at right angles to the current, so the component of the Coriolis force acting in the *x*-direction is proportional to the velocity in the *y*-direction and vice versa. (The minus sign before $\rho f u$ is a consequence of the coordinate system in Figure 4.15; in the Northern Hemisphere, for example, the Coriolis force acting on water flowing in the positive *x*-direction (i.e. towards the east) is in the *negative y*-direction (i.e. towards the south).)

As stressed already, the easiest situations to consider are those in which the ocean has reached an equilibrium or steady state, in which all the forces acting on it are in balance. In such situations, there is no acceleration and d*u*/d*t* and d*v*/d*t* in equation 4.3 are zero. This means that the equations become very much easier to solve, especially if they can be simplified further.

The easiest way to simplify the equations is to assume that one or other of the forces concerned may be ignored altogether. For example, in his work on wind-driven currents Ekman assumed that the ocean was homogeneous and that the sea-surface was horizontal. There were therefore no horizontal pressure gradients and dp/dx and dp/dy were zero.

Another way of keeping the equations simple is to express the contributions to F_x and F_y simply. For example, we may decide to assume that friction is directly proportional to current speed, in which case it can be written as Au and Bv, where A and B are constants. This is the approach that Stommel adopted in calculating the flow patterns in Figure 4.12, which clearly showed that the intensification of western boundary currents can be explained by the variation of the Coriolis parameter with latitude.

The examples of the work done by Ekman and Stommel have been quoted to illustrate a particular point. It is this: breakthroughs in understanding of dynamic situations come about as a result of someone realizing what the most important variable(s) might be and then simplifying the situation under consideration (and hence the equations of motion) so that the effect of the particular aspect being investigated can be seen more clearly.

So far, we have not considered the vertical or z-direction. The main forces that must be considered here are the pressure gradient force in the vertical direction and, of course, the force due to gravity, written ρg rather than mg because it is the weight per unit volume that we are interested in. The equation of motion for the z-direction may therefore be written as:

$$\frac{dw}{dt} = \frac{1}{\rho}\left(-\frac{dp}{dz} - \rho g + F_z\right) \qquad (4.3c)$$

Except beneath surface waves, vertical accelerations in the ocean are generally very small, so for many purposes, dw/dt may be assumed to be zero.

QUESTION 4.8 Rewrite equation 4.3c assuming that dw/dt = 0, and that there are no forces acting vertically other than the weight of the water and the vertical pressure gradient force. Do you recognize the equation you have written?

Thus, when there are no vertical accelerations (other than g) the equation of motion appropriate to the z-direction becomes simply the hydrostatic equation (equation 3.8). Indeed, for large-scale motion, the hydrostatic equation is a very good approximation to the 'vertical' equation of motion.

At this point we should mention the **principle of continuity**, another important tool in dealing with moving water. The principle of continuity expresses the fact that mass must be conserved, i.e. the mass of water flowing into a given space must equal the mass of water flowing out of that space. As seawater is virtually incompressible, continuity of mass is effectively continuity of volume. It follows that if the dimensions of a parcel of water change in a particular direction (e.g. the y-direction in Figure 4.16), they must also change in one or both of the other directions.

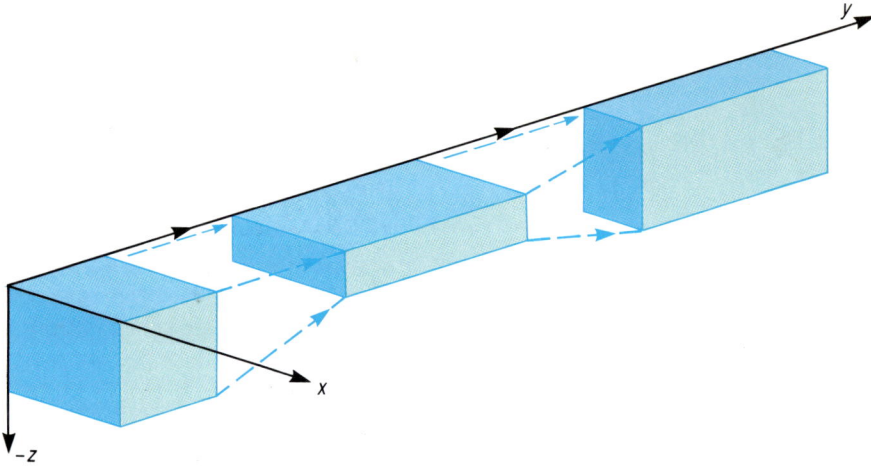

Figure 4.16 Continuity of volume during flow (here in the y-direction). *Dimensions* change but volume remains the same. Note that the parcel of water may change its shape as a result of spatial constraints (e.g. bottom topography) or because of changes in the current speed; e.g. a parcel of water entering a region of increased speed will become elongated in the direction of flow.

Thus, on passing through narrow straits, a wide shallow parcel of water will become elongated and, if possible, deeper; alternatively, if a flow of water is obstructed it may 'pile up', causing a local rise in sea-level.

As the amount of water flowing into a space must actually equal the amount of water flowing out of it *per unit time*, the *rate* of flow, i.e. velocity, is also important in continuity considerations. For example, a broad, shallow current entering narrow straits will become faster as well as perhaps becoming deeper.

QUESTION 4.9 Explain briefly how the flow pattern of the subtropical gyres exemplifies the principle of continuity.

The mathematical equation used to express the principle of continuity is:

$$\frac{du}{dx} + \frac{dv}{dy} + \frac{dw}{dz} = 0 \tag{4.3d}$$

which simply means that any change in the rate of flow in (say) the *x*-direction must be compensated for by a change in the rate of flow in the *y*- and/or *z*-direction(s). This continuity equation is used in conjunction with the equations of motion and provides extra constraints, enabling the equations to be solved for whatever dynamic situation is being investigated.

4.3 MODERN OBSERVATIONS OF THE NORTH ATLANTIC GYRE

In the South Atlantic Ocean the shape of the Brazilian coastline diverts a large proportion of the water carried by the South Equatorial Current so that it crosses the Equator as the Guyana Current and joins the North Equatorial Current (Figures 3.1 and 4.17(a)). Much of this water enters the Caribbean Sea through the passages between the islands of the Lesser Antilles, and continues through the Yucatan Channel into the Gulf of Mexico; the remainder continues to flow north-westwards in the Antillean Current outside the chain of islands. Most of the flow entering the Gulf of Mexico takes a direct path from the Yucatan Channel to the Straits of Florida, although some takes part in the circulation within the Gulf itself (Figure 4.17(a)).

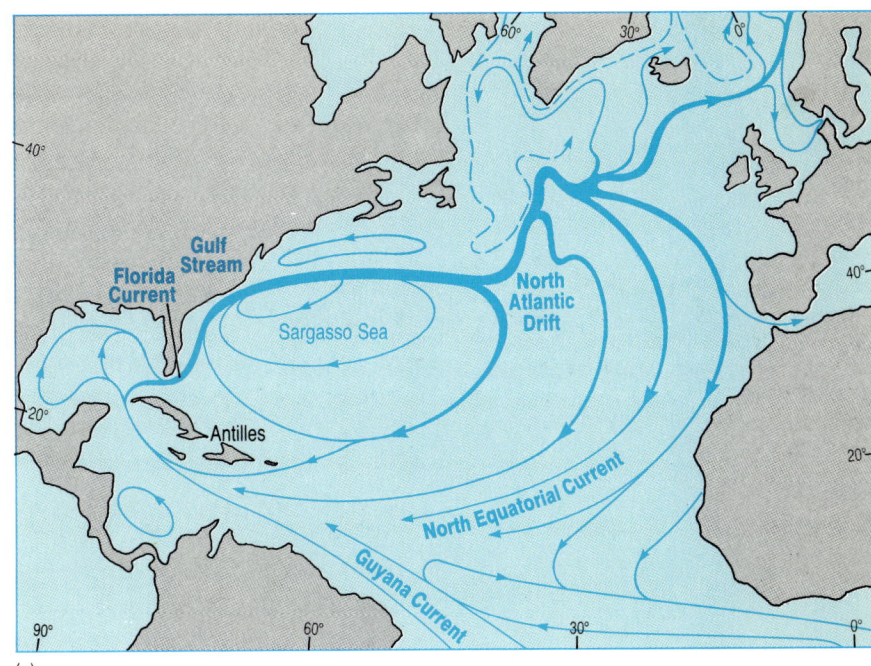

Figure 4.17 (a) The Gulf Stream in relation to the surface circulation of the Atlantic; the Stream consists of water from the equatorial current system (much of which comes *via* the Caribbean/Gulf of Mexico) and water that has recirculated in the North Atlantic subtropical gyre. The broken lines represent cool currents.

(b) Map to show the sea-floor topography off the east coast of the United States, and geographical locations of places mentioned in the text.

4.3.1 THE GULF STREAM SYSTEM

The Gulf Stream proper may be considered to extend from the Straits of Florida to the Grand Banks off Newfoundland. It does, however, have two fairly distinct sections, upstream and downstream of Cape Hatteras.

Between the Straits of Florida and Cape Hatteras, the current flows along the Blake Plateau (Figure 4.17(b)), following the continental slope, so that its depth is limited to about 800m. In this region, the flow remains narrow and well defined. Its temperature and salinity characteristics show that it is supplemented by water from two sources:

(i) the Antillean Current, which includes deep water from the South Atlantic that has been unable to enter the Caribbean Sea owing to the restricted depth of the passages between the Lesser Antilles; and

(ii) water that has recirculated in the Sargasso Sea.

Because of these contributions, particularly the recirculatory flow, the volume transport of the Gulf Stream increases as it flows northwards (*cf.* Question 4.1). Measurements made in 1976 indicate that the average transport of the Stream off Florida is about $30 \times 10^6 m^3 s^{-1}$; by the time it reaches the latitude of Cape Hatteras this has been increased to $85 \times 10^6 m^3 s^{-1}$. The maximum transport of about $150 \times 10^6 m^3 s^{-1}$ is reached at about 65° W, after which it begins to decrease again.

As the Gulf Stream continues beyond Cape Hatteras, it leaves the continental slope and moves into considerably deeper water (4000–5000m). While the current was following the continental slope, any fluctuation in its course had been limited and meanders had not exceeded about 55km in amplitude. Beyond Cape Hatteras there are no topographic constraints, and meanders with amplitudes in excess of 350km are common. These meanders often give rise to 'Gulf Stream rings' or eddies, which will be discussed in Section 4.3.5. Downstream from Cape Hatteras the flow also becomes more complex: it typically consists of a series of narrow vertical filaments with counter-currents (i.e. contrary flow) and eddies in between.

By the time it has reached the Grand Banks off Newfoundland (Figure 4.17(b)), the Gulf Stream has broadened considerably and become more diffuse. Beyond this area it is more correctly referred to as the **North Atlantic Drift** (or North Atlantic Current). Much of the water in the North Atlantic Drift turns south-eastwards to contribute to the Canary Current and to circulate again in the subtropical gyre (Figures 3.1 and 4.17(a)); other flows continue north-eastwards between Britain and Iceland.

4.3.2 GEOSTROPHIC FLOW IN THE GULF STREAM

Figure 4.18(a) and (b) show the distributions of temperature and salinity for a section across the narrowest part of the Straits of Florida, based on measurements made in 1878 and 1914. Figure 4.18(c) shows the velocity of the geostrophic current, as calculated from these T and S data. Figure 4.18(d) shows the distribution of current velocity based on *direct* measurements made by Pillsbury in the 1890s (Section 4.1.1); the correspondence between Figure 4.18(c) and (d) is remarkable. This single example of agreement between observed current speeds and the values calculated using the geostrophic method (Section 3.3.3) did much to increase confidence in the use of geostrophic calculations.

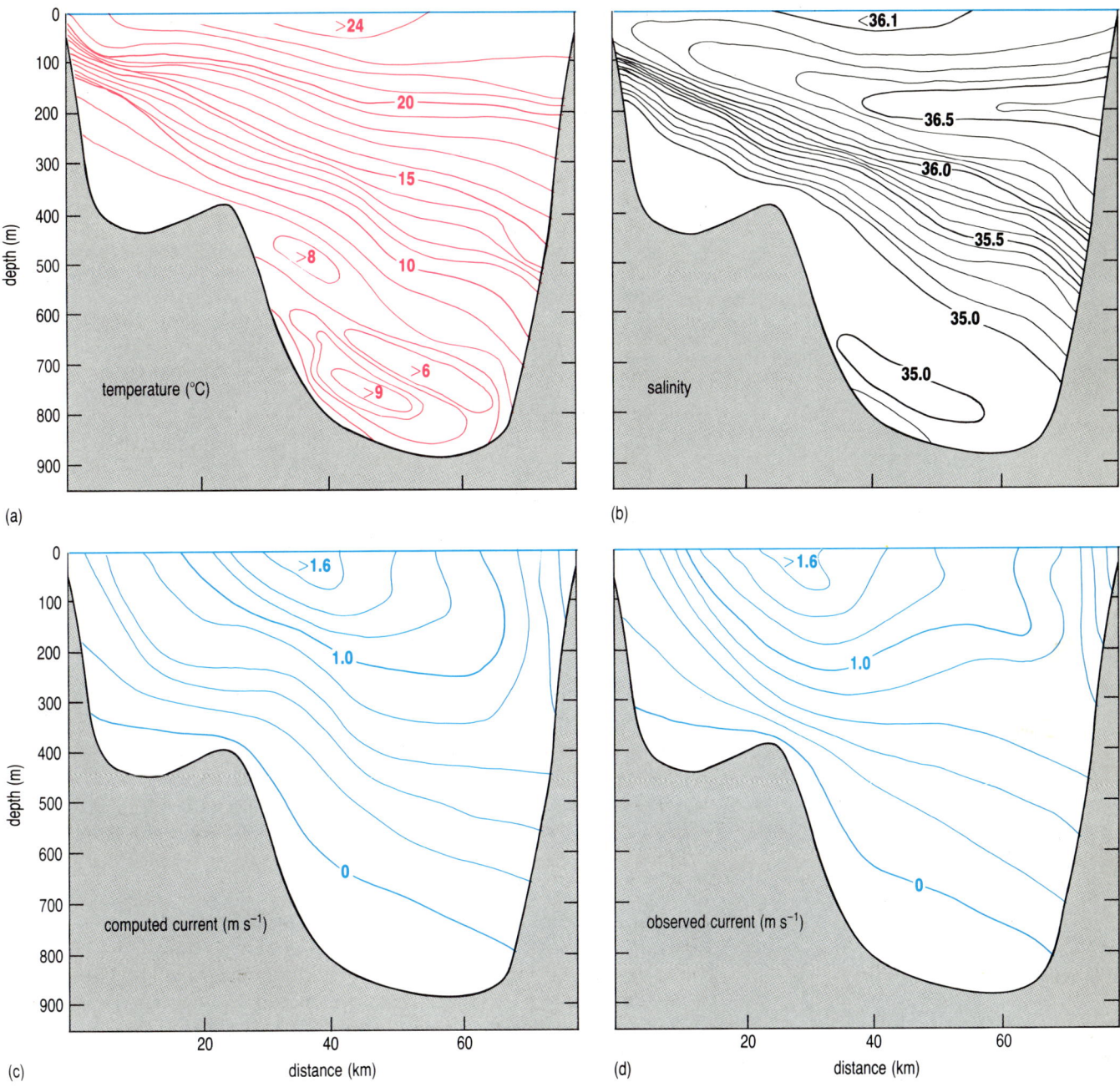

Figure 4.18 The distribution of (a) temperature (in °C) and (b) salinity, plotted from measurements made in 1878 and 1914, for an east–west section in the narrowest part of the Straits of Florida, between Fowey Rocks, a short distance south of Miami, and Gun Cay, south of Bimini Islands (see Figure 4.17(b)).
(c) The velocity distribution as calculated from the temperature and salinity data, on the assumption that the current has decreased to zero by the depth of the 0-contour; (d) based on direct measurements by Pillsbury.

QUESTION 4.10 (a) Do the sections in Figure 4.18 indicate that flow through the Florida Straits is barotropic or baroclinic?

(b) In which direction does the sea-surface slope therefore? Does this mean that flow is 'into the page' or 'out of the page' in Figure 4.18(c) and (d)? In other words, which side of the section is the Florida side and which the Bimini side?

(c) Does the sea-surface simply slope up from one side to the other, or does it have a more complex shape?

Calculations of the sea-surface slope indicate that the sea-level is about 45 cm lower on the Florida side of the Straits than the Bimini side.

Figure 4.19 shows the geostrophic velocity as calculated from *T* and *S* measurements across a section out from Cape Hatteras. The section illustrates how the current is no longer a relatively coherent flow, but instead consists of vertical filaments separated by counter-currents (shaded blue).

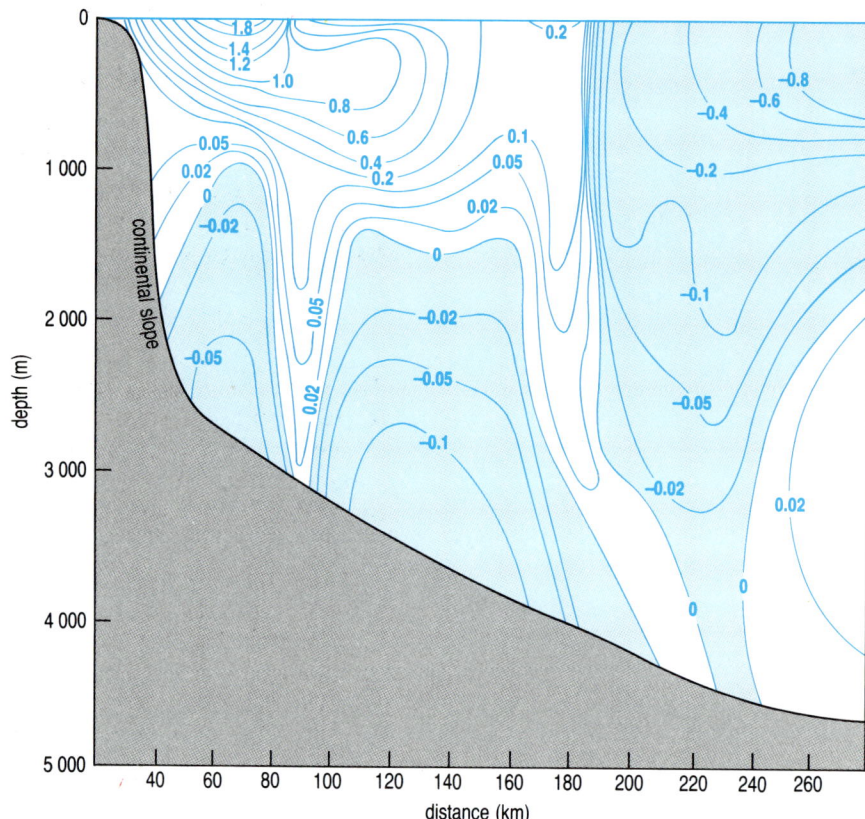

Figure 4.19 Geostrophic current velocity (in m s^{-1}) in the Gulf Stream off Cape Hatteras; the shaded areas (with negative velocity values) represent south-westerly flow. The horizontal scale is smaller than that in Figure 4.18.

In the shaded regions, does the horizontal pressure gradient force act towards the east or towards the west?

Towards the east. In the Gulf Stream, the flow is to the north (or strictly, north-east), and the Coriolis force acting to the right of the flow, i.e. towards the east, balances the horizontal pressure gradient acting to the left (west). The density distribution is such that at the depth corresponding to the zero-velocity contour, the horizontal pressure gradient has become zero because the effects of the sea-surface slope and the density distribution have balanced out with depth. Below the zero-velocity contour, the horizontal pressure gradient reverses. Flow is to the south (or strictly south-west), and the horizontal pressure gradient force to the left of the flow (i.e. to the east) is balanced by the Coriolis force to the right of the flow (i.e. to the west).

The velocity section shown in Figure 4.19 is one of a number that show a deep counter-current flowing south-westwards beneath the Gulf Stream. However, until relatively recently many oceanographers found the idea of a significant current close to the deep sea-bed, at depths of 3000–5000m, hard to believe. The determination of distributions of temperature, salinity and velocity is fraught with difficulty, especially if the T, S and direct current measurements have been made far apart. The sceptics could reasonably argue that other interpretations of the data, not involving counter-currents, were equally valid.

In 1965, Stommel developed a theory of the thermohaline circulation that supported the idea of such equatorward deep currents. However, it was freely drifting floats that in the 1950s and 1960s first provided direct evidence of Gulf Stream counter-currents. We will consider these direct current measurements in Section 4.3.3.

First, however, look at Figure 4.20 (overleaf) which shows T and S sections across the Gulf Stream between Chesapeake Bay and Bermuda, i.e. downstream of Cape Hatteras. The horizontal scale is much smaller than those of Figures 4.18 and 4.19, and the oceanographic stations were too far apart to allow any filaments or counter-currents to be resolved.

Where is the Gulf Stream on Figure 4.20?

The Gulf Stream is the region where isotherms and isohalines are close together and steeply sloping; its 'warm core' (20–22°C) may be clearly seen extending to depths of 200–300m. Note that the Sargasso Sea water is both warmer and more saline than the water on the coastal side of the Stream, which is influenced by cool freshwater input from land. The salinity distribution *on its own* would therefore cause the density contours, the isopycnals, to slope in the opposite direction to the isotherms and isohalines. However, as is usual in the oceans, the temperature distribution has by far the greater effect on the density distribution and isopycnals. Figure 4.20 thus illustrates the idea of the Gulf Stream as a narrow region of fast-flowing water associated with a frontal boundary between two different water masses and with large lateral variations in density (*cf.* Section 3.4).

4.3.3 DIRECT CURRENT MEASUREMENTS

As mentioned above, the first direct evidence for counter-currents within the Gulf Stream came from the motion of freely drifting floats; you have already encountered the track of a float caught in a deep southward-flowing Gulf Stream counter-current in Figure 3.29. Each float consists of an aluminium tube, about 6m long, which contains a battery and an acoustic transmitter, and a precisely determined amount of ballast. At the surface, the float is negatively buoyant and so initially it sinks; however, as it is less compressible than seawater its density increases more slowly with increasing pressure than does the density of the seawater. Eventually, at a density level that can be predetermined by adjusting the ballast, the float has the *same* density as the surrounding seawater. It is then neutrally buoyant and stops sinking; thereafter, it drifts with the water that surrounds it.

Neutrally buoyant floats are also known as Swallow floats, after John Swallow, the British oceanographer who invented them. They are cheap to make—the prototypes were constructed out of lengths of scaffolding—but in terms of ship time they can be very expensive. It is not easy for a ship to track more than a few floats at once, even though each has its own distinctive acoustic signal; and whilst tracking is in progress other activities are hampered. The original floats were designed to sink to about

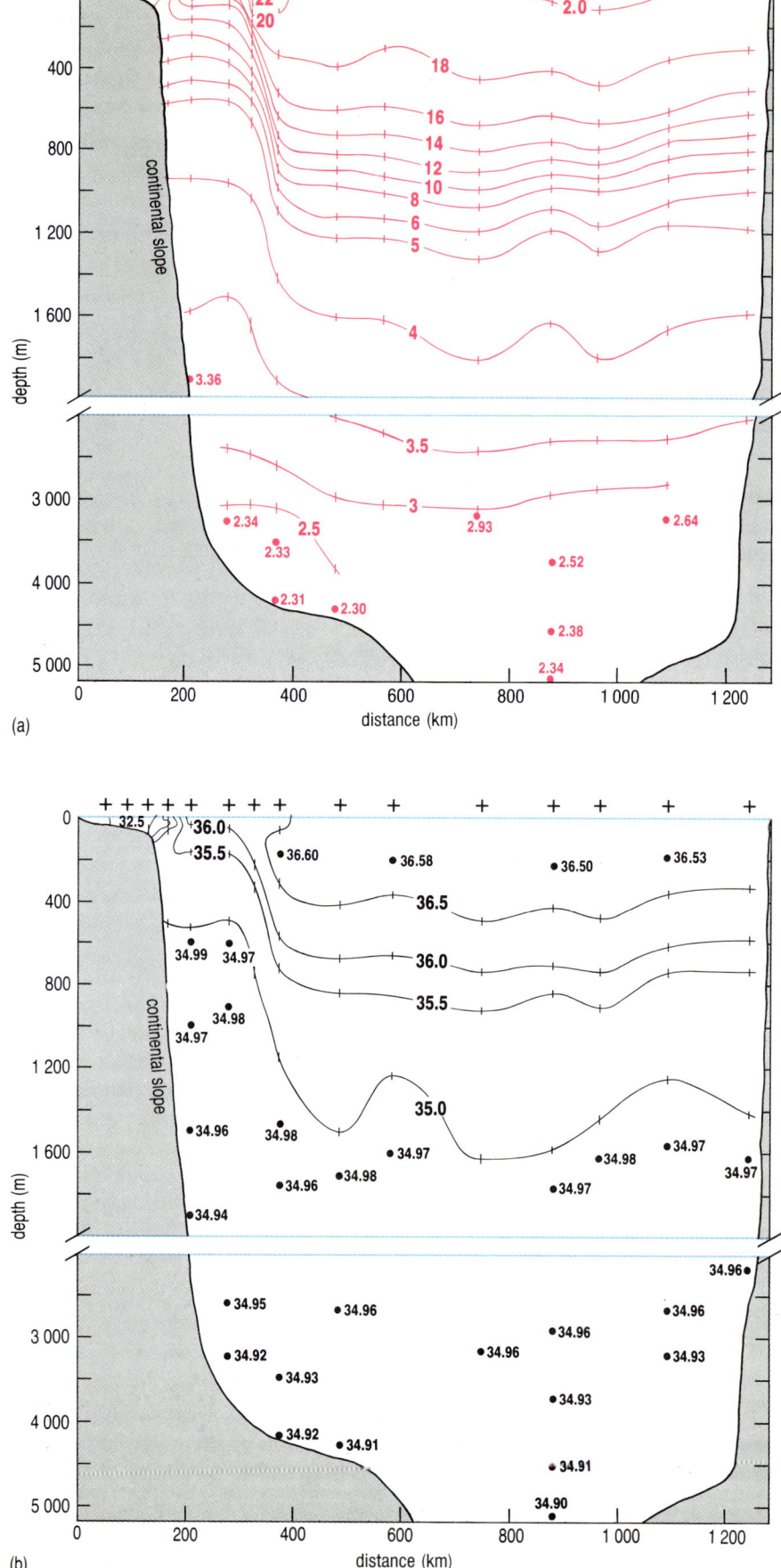

Figure 4.20 (a) Temperature (in °C) and (b) salinity sections across the Gulf Stream between Chesapeake Bay and Bermuda, based on measurements made between 17 and 23 April, 1932. Remember that these cross-sections, like those in Figures 4.18 and 4.19, would have been plotted using temperature and salinity measurements of water collected at a number of oceanographic stations, and at specific depths: the contours are interpolations based on the spot measurements. (Note that as it was expected that there would be more variability in the western part of the section, the oceanographic stations (shown as crosses) were closer together.)

1000 m, but in the 1970s neutrally buoyant floats were deployed in the sound channel between about 1000 and 2000 m. Such floats are called **Sofar** (*SO*und *F*ixing *A*nd *R*anging) floats, and large numbers of them can be continuously tracked over ranges of 1000 km or more, from receiving stations on the shore or the sea-bed.

Most neutrally buoyant floats only provide information about the horizontal flow at a given density level, and provide no information about the vertical component of current velocity. However, vertical flow has been investigated (in, for example, the Mediterranean) using floats with fins so arranged that the float rotates at an angular velocity proportional to the vertical current velocity.

Surface buoys that could be tracked visually or by radar (Figure 4.21) were already in use before neutrally buoyant floats were developed. However, drifting buoys only came into their own relatively recently with the advent of electronic acoustic and radio tracking systems, and satellite navigation. Surface buoys are inevitably affected by the wind. In order to reduce its effect, they generally have most of their volume below water and in addition have a parachute, or some other type of drogue, at a depth of about 20–30 m.

Sofar floats and surface buoys may, theoretically, be tracked until their batteries run down, which may be as long as several years. Other simpler types of drifters are not tracked at all. The current flow is simply deduced from the time and place at which they are found, either at sea or on the shore. In the nineteenth and early twentieth centuries the 'drift bottle' was commonly used. Today, a variety of cheap 'drifters' are used (plastic cards, for instance), mainly for small-scale local studies in estuaries, coastal waters, or semi-enclosed seas such as the North Sea and the Irish

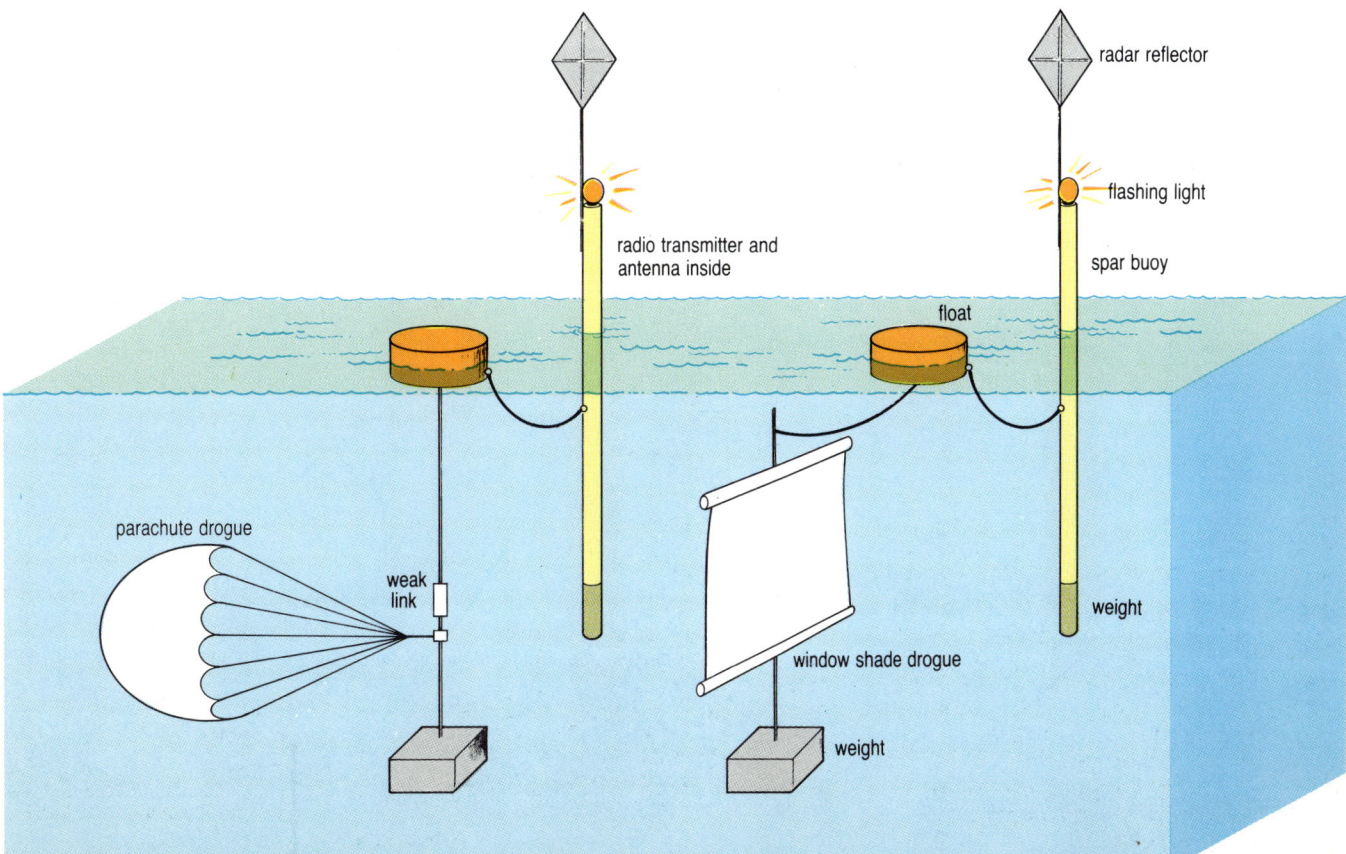

Figure 4.21 An example of a freely drifting surface buoy; two different types of drogues, to reduce the effect of the wind, are shown.

Sea. Each drifter carries a message asking the finder to inform a central agency of the position and time at which it was found, so that its trajectory may be deduced. Of the large number of drifters released in any one experiment, only a small proportion will be returned, and of those that are returned some may have been washed ashore many weeks earlier.

Interpretation of results obtained using these simple drifters is difficult. Drifters moving near the surface may be directly affected by the wind; others, designed to be just negatively buoyant at the sea-bed so that they trail along the bottom, are affected by friction with the sea-bed (as is the layer of water in which they move) and so their movement cannot be taken to represent that of the main body of water. The effects on the floats' trajectories of tidal flows and other fluctuating currents are also hard to assess.

In tracking or deducing the paths of drifting objects, we are effectively following the path taken by a parcel of water as it moves relative to the Earth; the velocity of the parcel of water may be calculated from its change of position with time. Such methods of current measurement are described as **Lagrangian**. Despite the disadvantages described above, much of our knowledge about oceanic circulation patterns and current velocities has been obtained by Lagrangian methods.

Alternatively, currents may be measured by **Eulerian** methods, in which the measuring instrument is held in a fixed position relative to the Earth, and flow of current past that point is measured.

The essential problem with Eulerian methods of current measurement is that of keeping the measuring instrument fixed in relation to the Earth.

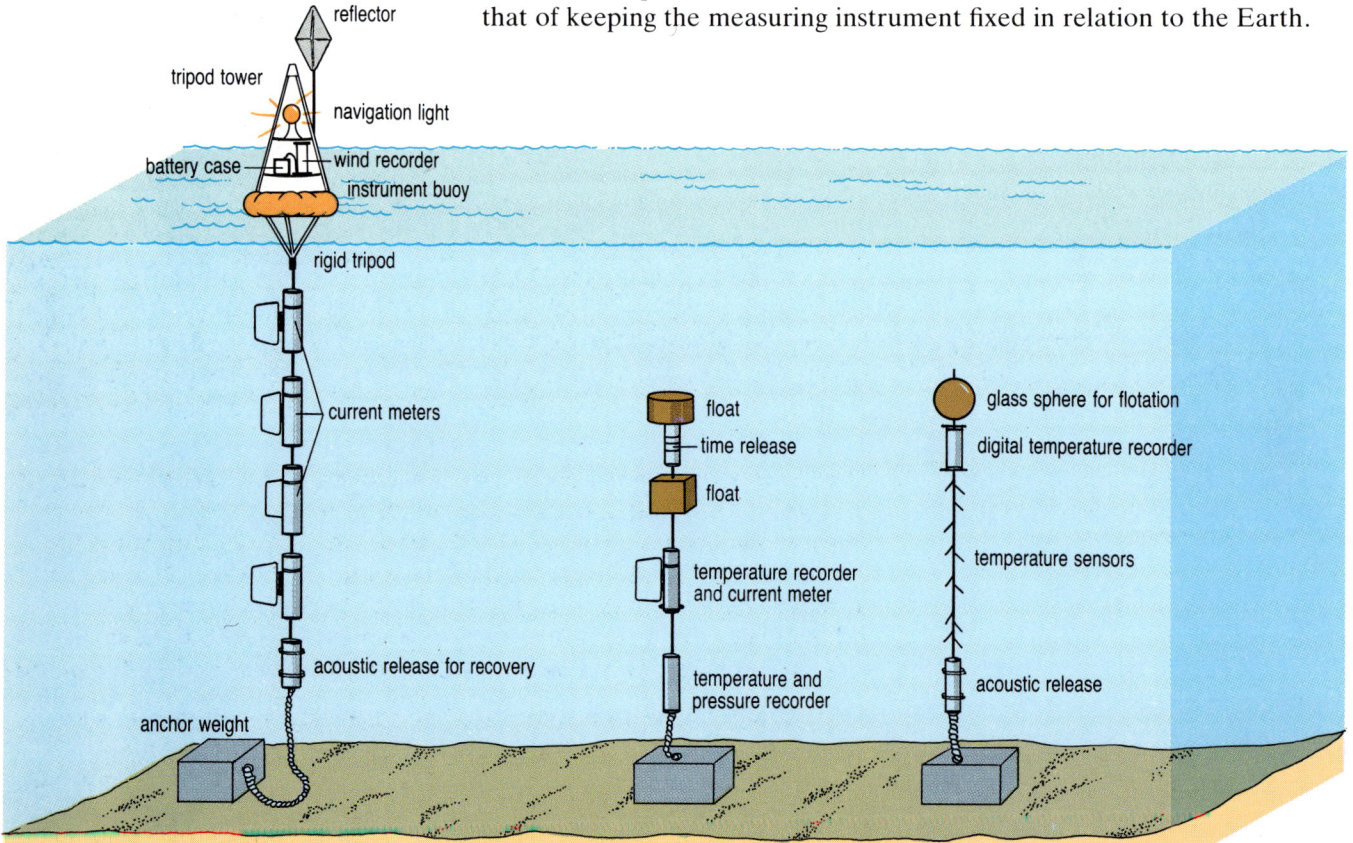

Figure 4.22 Some examples of moored current meters (not to scale), showing how the mooring can be used to accommodate equipment for measuring other variables of interest, such as wind speed, water temperature, etc.

(a) SURFACE BUOY WITH SUBSURFACE SENSORS

(b) SUBSURFACE BUOY SYSTEMS

Nowadays, this is commonly done by anchoring a subsurface buoy to the sea-bed and securing one or more current meters to the mooring cable (Figure 4.22(b)). Current meters *can* be suspended from a securely anchored surface buoy (Figure 4.22(a)), but such moorings are affected by surface wave motions which are transmitted to the meter; they are also particularly vulnerable to damage or displacement by shipping or fishing trawls. Current meters may also be deployed directly from anchored ships.

The most common type of current meter is that in which current flow causes a propeller to rotate, the speed of rotation being proportional to the current speed. Some current meters are designed to align themselves in the direction of flow so that the propeller rotates about an axis parallel to the current direction (Figure 4.23(a)). The Savonius rotor, which consists of two hollow S-shaped rotors mounted about a vertical axis, is very sensitive and responds even to very weak currents (Figure 4.23(b)). However, it only rotates in one direction, so that rapidly reversing flows caused, for example, by surface waves, all contribute to the total number of rotations recorded, whereas in the other type of meter the propeller rotates in both directions so that the effects of rapidly reversing flows cancel out.

(a)

(b)

Figure 4.23 (a) A typical propeller-type current meter. (b) A Savonius rotor. The meters are oriented with respect to the current by their large vanes.

In most modern current meters, current direction (measured by means of a magnetic compass) and speed are internally recorded on solid-state loggers. The record is retrieved at the end of the experiment, when the moorings are released from the anchor (perhaps by means of an acoustic signal) and the current meter bobs up to the surface. There have been developments in the transmission of current data to shore stations *via* satellite, so that they can be analysed immediately and used for forecasting.

When moored current meters were first used, their recovery rate was often as low as 50%. Technological advances have permitted the development of mooring systems that can be deployed in almost any part of the ocean for a considerable period. Such systems can withstand the effects of corrosion, strong currents and even fish, which sometimes bite through nylon mooring lines.

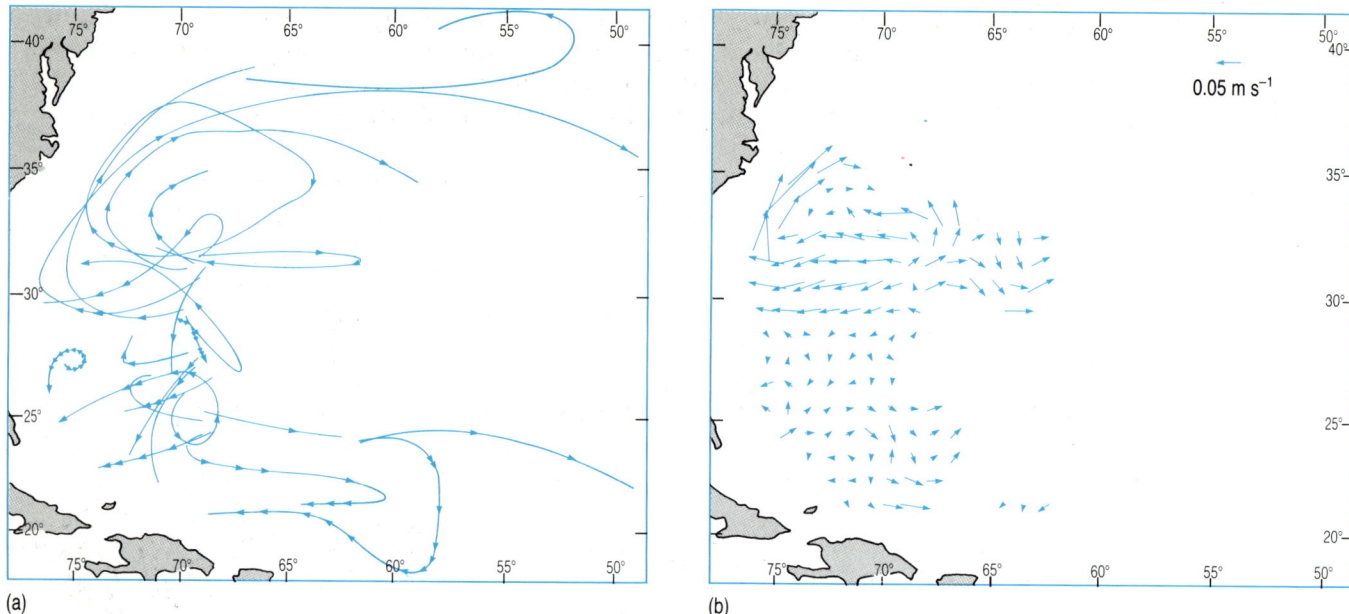

Figure 4.24 (a) The paths taken by a number of Sofar floats at a depth of 700 m in the western North Atlantic. The arrows are 100 days apart; note the high velocity of the floats caught in the Gulf Stream.

(b) The same region as in (a), showing the mean current flow as it would have been measured by fixed current meters.

Figure 4.24 shows the paths taken by a number of Sofar floats in the region of the Gulf Stream and, for comparison, the mean current flow as it *would* have been measured by means of moored current meters (although, in fact, such measurements were not made).

QUESTION 4.11 By comparing Figure 4.24(a) and (b), can you suggest some advantages and disadvantages of the Eulerian and Lagrangian approaches to current measurement?

In addition to the points covered in Question 4.11, you should also note that data obtained by Eulerian methods—i.e. variations in velocity at fixed, known points—are more easily used in theoretical studies (Sections 4.2.2 and 4.2.3) than are variations in velocity 'following the flow', which are obtained by Lagrangian methods.

4.3.4 MAPPING THE GULF STREAM USING WATER CHARACTERISTICS

As discussed in Section 4.1.1, seafarers have long been aware of the high temperatures associated with the Gulf Stream. They have also noted that the edge of the Gulf Stream is often marked by accumulations of *Sargassum* weed*, and that the waters of the Stream are a clear blue, contrasting strongly with the relatively murky waters between the Stream and the coast.

Since Franklin's time, there have been various surveys of the temperature distribution in the region of the Gulf Stream. These have greatly added to knowledge of the flow: for example, it was through measurement of surface-water temperatures that the north-easterly extension of the Gulf Stream towards Britain and Scandinavia was discovered by Captain Strickland in 1802.

Sargassum weed is a seaweed which, when torn loose from coastal areas, continues to grow and multiply vegetatively; it is kept afloat by air-bladders.

Figure 4.25 Three interpretations of temperature data collected in August 1953. The tracks along which measurements were made are shown as red lines. Interpretation (a) shows a single, simple stream, (b) a double stream with some branching.

Nevertheless, measurements made directly from ships are time-consuming and therefore expensive, and of necessity are relatively few in number and widely spaced. That this can lead to difficulties in interpreting the results is graphically illustrated by Figure 4.25(a)–(c).

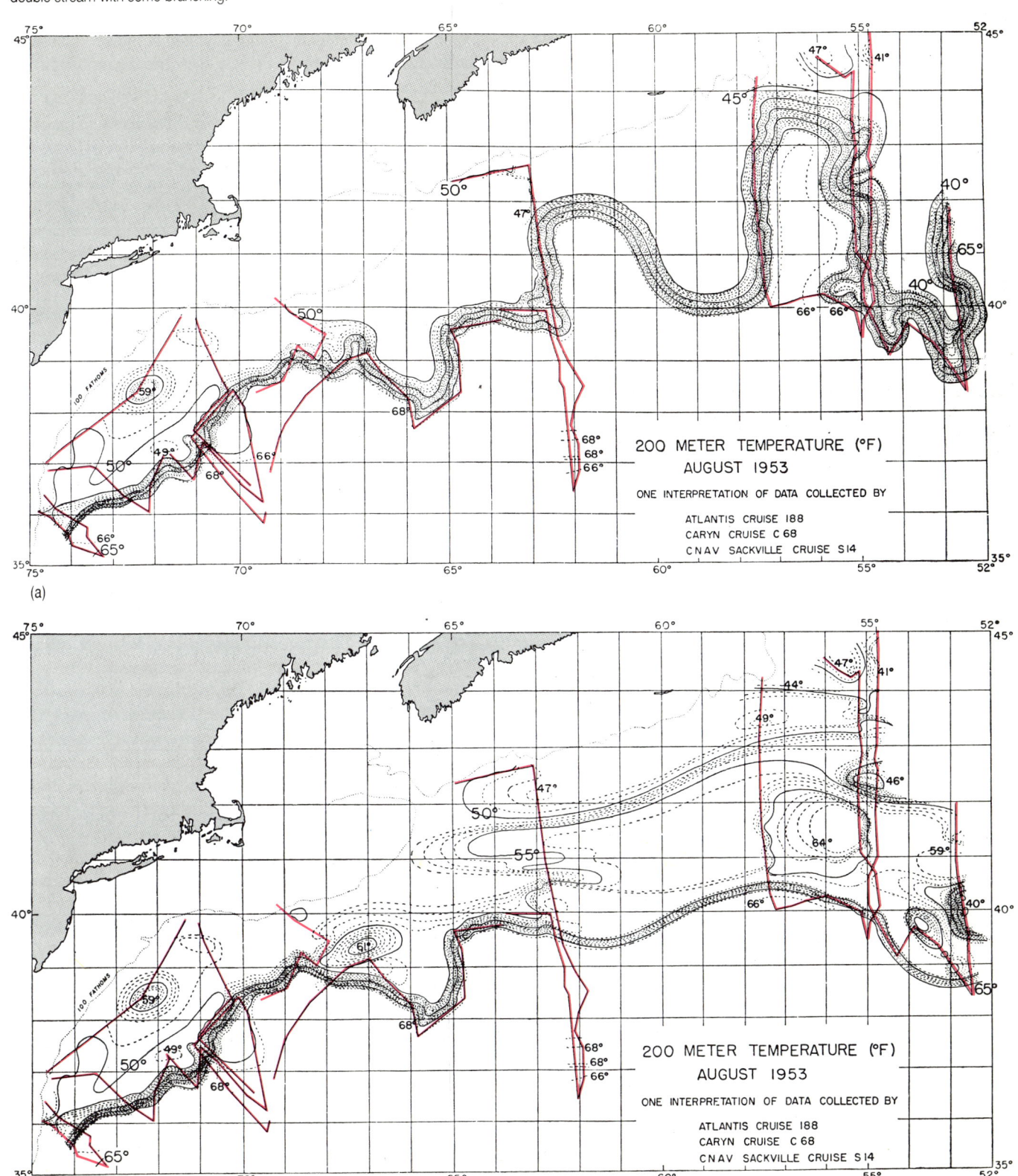

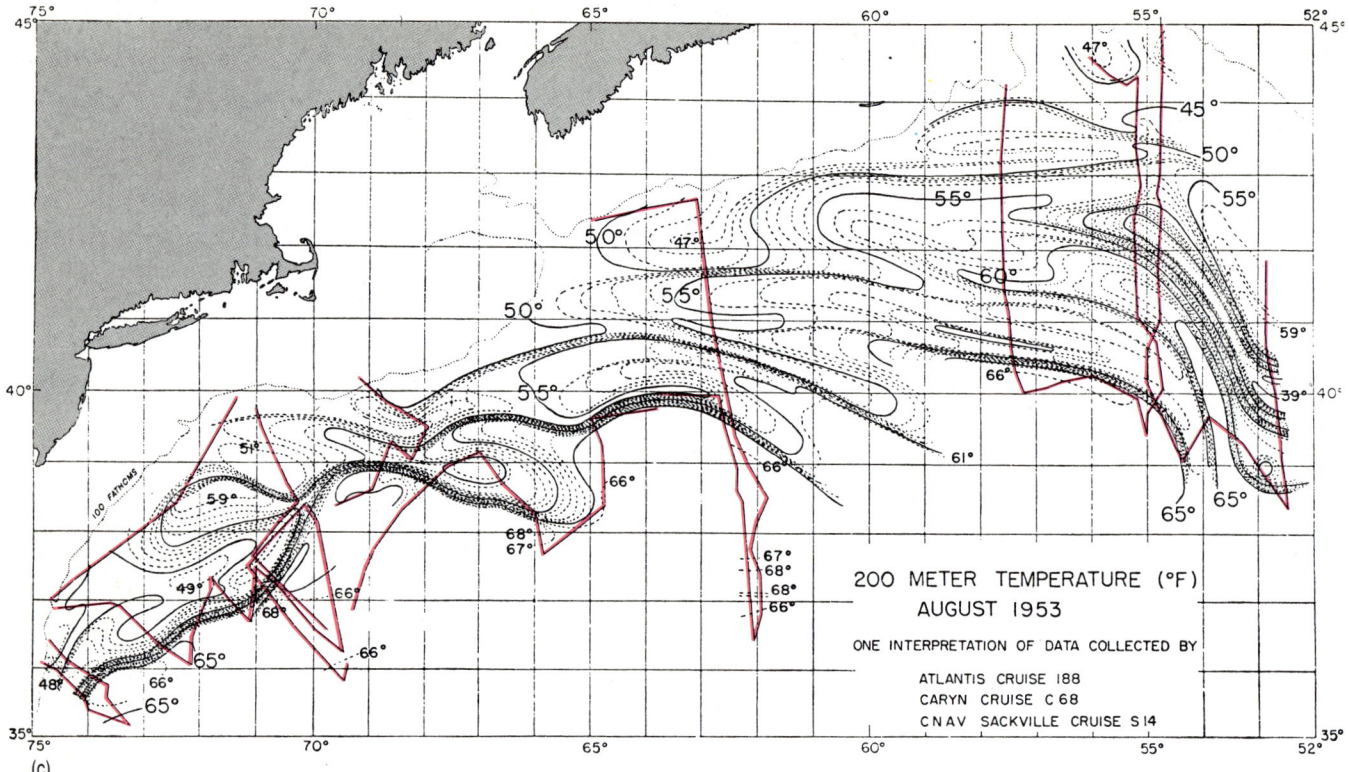

Figure 4.25 (c) This third interpretation of the temperature data shows the Gulf Stream as a series of disconnected fragments.

These maps show the results of a temperature survey published in 1955, and the important point to note is that *all three have been drawn using the same data*, collected while the ship moved along the tracks indicated by the red lines.

In addition to difficulties of interpretation, scientists using observations made from ships have to cope with the problem that measurements are made over periods of perhaps weeks and it is therefore impossible to obtain a 'snapshot' of the flow at any one time. This is not the case with satellite measurements, which enable us to see complex spatial variations of surface waters over a wide area effectively instantaneously.

Figure 4.26(a) and (b) show the distributions of sea-surface temperature and phytoplankton off the eastern coast of North America, as measured by the satellite-borne Coastal Zone Color Scanner (CZCS).

QUESTION 4.12 In Figure 4.26(a), the blue end of the colour range represents the coldest water and the orange-red the warmest water; the green region corresponds to cool water over the continental shelf and slope. Bearing this in mind, and referring to Figure 3.1 if necessary, can you identify:

(a) the water in the Labrador Current;

(b) the Sargasso Sea water and water flowing in the Gulf Stream?

4.3.5 GULF STREAM 'RINGS'

In Section 3.5.2, we discussed the generation of mesoscale eddies at regions of strong density contrast and velocity shear, such as the Gulf Stream. You will probably have noticed that the satellite images in Figure 4.26 show clearly the existence of such an eddy, on the landward side of

Figure 4.26 Distributions of (a) sea-surface temperature and (b) phytoplankton pigments, off the eastern coast of the United States; the data were collected on June 14, 1979, by the Coastal Zone Color Scanner (CZCS) on the Nimbus-7 satellite. The image in (a) is based on measurements of infrared radiation; the warmest water (shown red) is about 25 °C and the coldest (shown dark blue) is about 6 °C. The brown colour is the land and the white streaks are cloud (which often limits the usefulness of such images). The image in (b) is the same as that shown on the cover of this Volume. The highest concentrations of phytoplankton pigment are shown in brown; intermediate concentrations in red, yellow and green; and lowest levels in blue. Concentrations have been deduced from the relative absorption and reflection of red and green light by organic pigments.

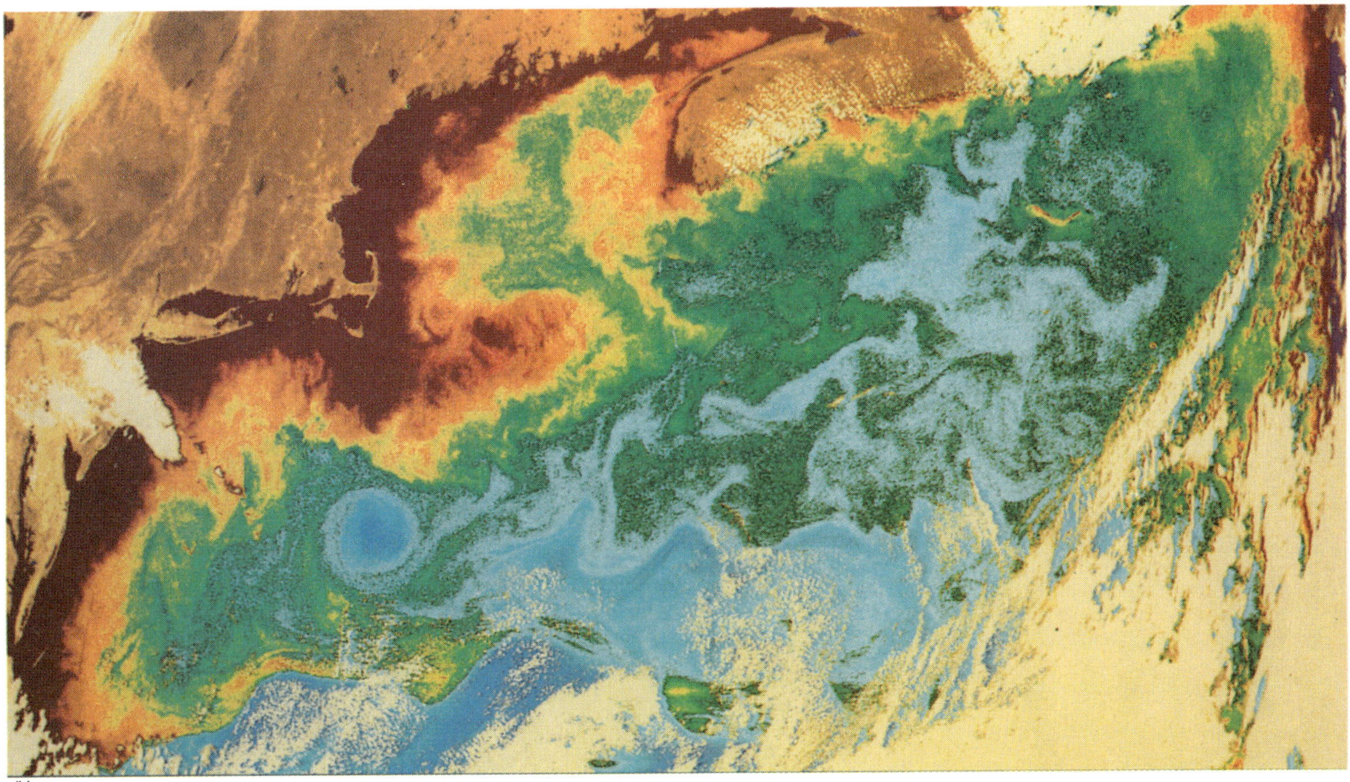

the Stream. Gulf Stream eddies form from meanders which break off to form independent circulatory systems, sometimes described as 'rings'. Figure 4.27(a) shows the evolution over the course of about a month of eddies on both the landward side and Sargasso Sea side of the Stream. Gulf Stream eddies are often described as 'cold-core' or 'warm-core' eddies.

QUESTION 4.13 Given how the eddies form (Figure 4.27(a)), would you expect the continental-margin side of the Gulf Stream to be characterized by warm-core or cold-core eddies? Is this borne out by Figure 4.26(a)?

Figure 4.26(a) and (b) show vividly the role that eddies play in transferring water properties across frontal boundaries. Together, the two images show how the formation of a warm-core eddy results in warm, relatively unproductive Sargasso Sea water being transferred across the Gulf Stream into the cool, productive waters over the continental margin.

The temperature section in Figure 4.27(b) shows that, like the western boundary current from which they form, the eddies extend to significant depths. Cold-core eddies may extend to the sea-floor at a depth of

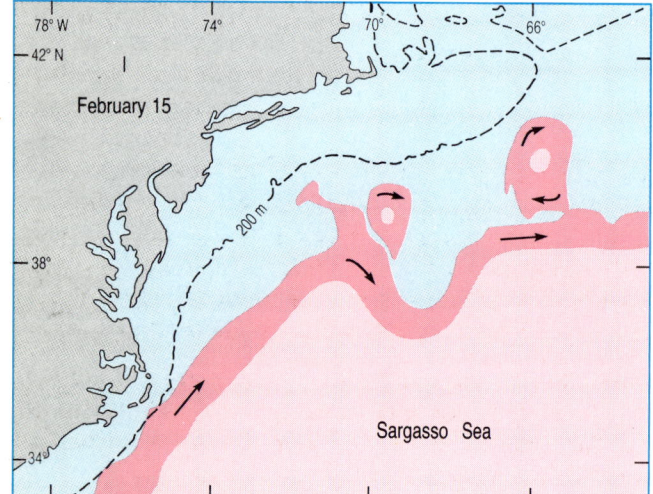
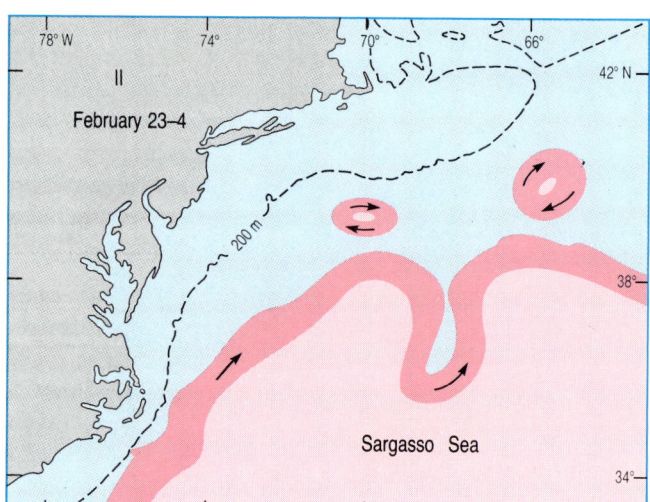
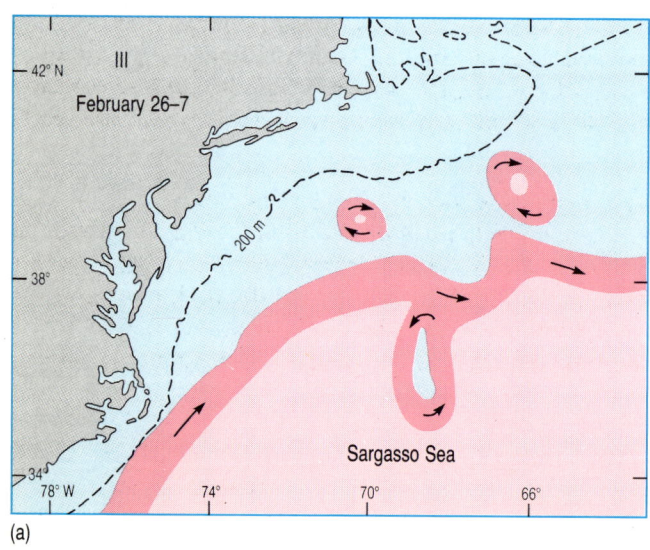
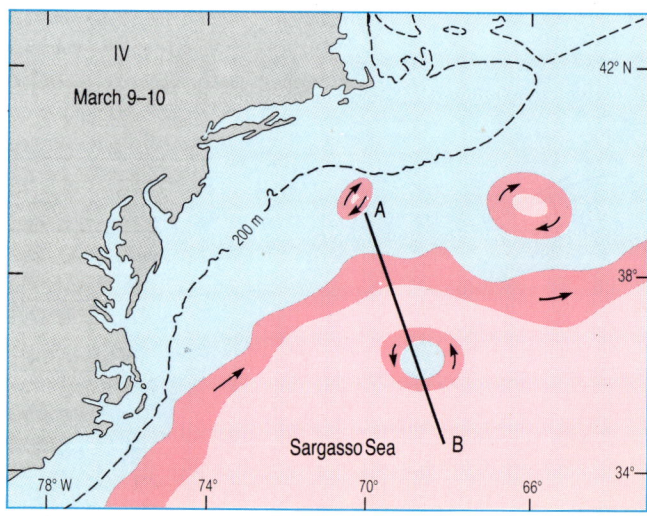

Figure 4.27 (a) The evolution of Gulf Stream eddies, as deduced from infrared satellite images made in February–March 1977. The warm Sargasso Sea water is shown as light pink, the cool continental shelf water as blue and the Gulf Stream as darker pink.

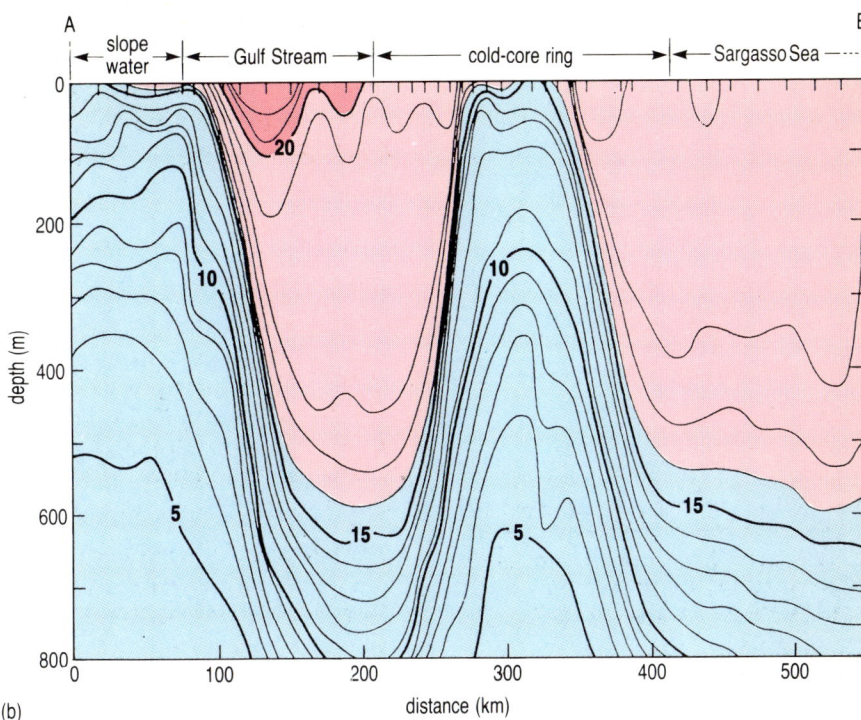

Figure 4.27 (b) Temperature section along the line shown in (iv), showing that the eddies extend to significant depths.

4000–5000 m; warm-core eddies are somewhat shallower but are nonetheless deep enough to impinge on the continental slope and rise when, after forming, they drift erratically towards the south-west. They often last until they are entrained back into the large-scale north-easterly flow of the Gulf Stream, and their lifetimes may be anything from a few months to a year; cold-core eddies generally survive somewhat longer.

Gulf Stream eddies are not only deep, they also extend over large areas. A newly formed cold-core eddy typically has a diameter of 150–300 km; a warm-core eddy has a diameter of about 100–200 km. During their lives, the eddies continually exchange energy, heat, water, nutrients and organisms with their surroundings; they also greatly affect the local exchange of heat and water between the ocean and the overlying atmosphere. Some idea of the influence of the eddies on the North Atlantic as a whole may be gauged from the fact that at any one time as much as 15% of the area of the Sargasso Sea may be covered by cold-core eddies, and as much as 40% of the continental shelf water by warm-core eddies.

Eddy generation may also be important in transferring water characteristics *between* oceans. Eddies similar to Gulf Stream rings have been observed forming from the Agulhas Current (see Figure 3.1) which, with a volume transport of about $130 \times 10^6 \, m^3 s^{-1}$, is the major western boundary current in the Southern Hemisphere. These eddies may be an important agent in the transfer of water between the Indian and Atlantic Oceans.

Returning for a moment to the Gulf Stream rings in Figure 4.27, what can you say about the directions of rotation of cold-core and warm-core eddies?

Cold-core eddies are always cyclonic (anticlockwise in the Northern Hemisphere) and warm-core eddies are always anticyclonic; this is true of all mesoscale eddies, not just Gulf Stream rings. You have already encountered this idea in Question 3.10: the 'highs' on Figure 3.30 correspond to eddies with warm central regions and the lows to eddies with cold central regions.

All Gulf Stream eddies, whether warm-core or cold-core, contain a ring of Gulf Stream water. Rotational velocities are highest in this ring—as much as $1.5–2.0 \text{ms}^{-1}$—and decrease both towards the centre of the eddy and towards the outer 'rim'. Such information has been obtained largely through direct current measurements. Satellite images like those in Figure 4.26 are extremely useful for showing horizontal variations in water properties, and they reveal flow patterns whose complexity could not have been fully appreciated through traditional oceanographic techniques, but they cannot provide information about current velocity. A remote-sensing technique that *can* provide information about current velocity is satellite altimetry, which you have already encountered in Section 3.3.4. Figure 3.21 showed a map of the *mean* sea-surface topography, compiled using data from a large number of passes; the sea-surface topography recorded during *one* pass provides a near-instantaneous picture of the sea-surface and so may be used to deduce the velocities of surface currents, if geostrophic equilibrium is assumed. Figure 4.28 shows how the shape of the sea-surface along a south-east–north-west satellite track in the region of the Gulf Stream changed over the course of 21 days. The Gulf Stream itself shows up clearly, as does a cold-core ring, which moves away to the side of the satellite track during the period in question.

Satellite altimetry is very exciting to physical oceanographers as it enables them to *see* the dynamic topography of the sea-surface (Section 3.3.4). Also, comparison of directly measured current velocities with values calculated from the observed sea-surface slopes should enable the depths at which geostrophic currents become zero (i.e. the depths at which isobaric surfaces become horizontal) to be accurately determined.

This concludes our survey of recent measurements and observations of the Gulf Stream—an example of an intense western boundary current. It is likely that as more becomes known about the other western boundary currents, they will be seen to share many of the characteristics of the Gulf Stream. Before moving on to look at the equatorward-flowing eastern limbs of the subtropical gyres—the eastern boundary currents—we will briefly mention some methods of current determination that have not been discussed so far.

4.3.6 OTHER METHODS OF CURRENT MEASUREMENT

The methods of current measurement discussed in Section 4.3.3 all depend on motion caused by the current itself—either the motion of floats drifting with the current, or the rotatory motion of propellers within fixed current meters. An alternative approach to current measurement is to use the various ways in which moving water interacts with electromagnetic and acoustic waves.

There are two ways in which sound may be used in the oceans to determine current velocity. The first method involves measurement of the

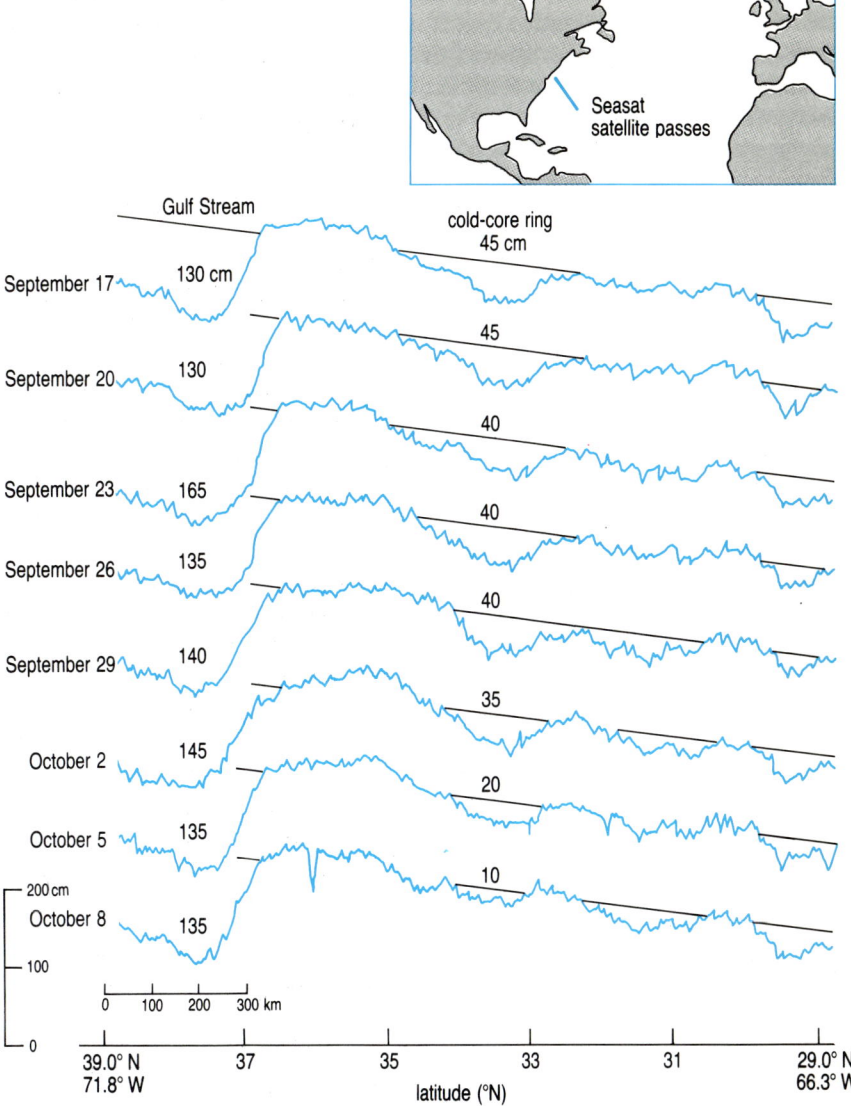

Figure 4.28 Variation in the height of the sea-surface along a satellite track (see inset map) in the western North Atlantic. The measurements were made by the SEASAT radar altimeter from 17 September to 8 October 1978. The black line represents the local height of the marine geoid, and the distances in centimetres are departures from this level.

time taken for sound to travel between two hydrophones a fixed distance apart. The speed of sound is affected by currents because the seawater which is being alternately compressed and rarefied by the passage of the sound wave is itself moving (see Figure 4.29).

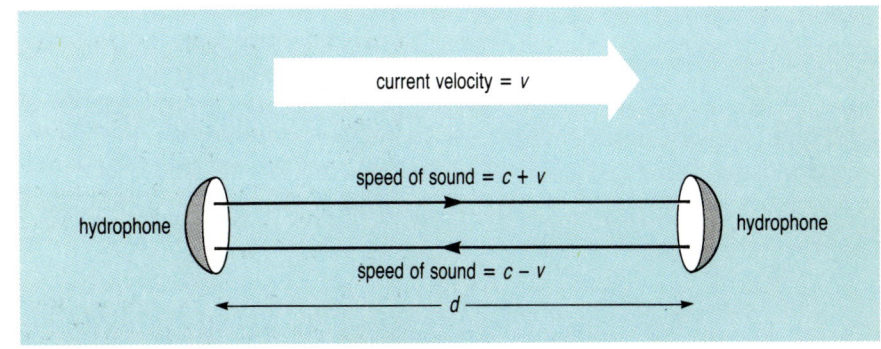

Figure 4.29 One of the acoustic methods for determining current velocity. The current velocity v can be calculated from the times taken for sound to travel between two hydrophones positioned distance d apart. The travel times are $d/(c+v)$ and $d/(c-v)$, and as d is known, v can be determined even if the local value of c is not known.

Another way in which sound may be used to measure current velocity involves exploiting the Doppler effect whereby the measured frequency of vibration is affected by relative movement between the source of the vibration and the point of measurement. Two narrow-beam echo-sounders are aimed at a particular volume of water and the shift in frequency between the sound waves emitted by the hydrophones and those back-scattered by particles in the water is measured. This Doppler shift in frequency is proportional to the current speed, which can therefore be determined. This technique, and a similar technique which employs lasers (light) instead of sound waves, is particularly useful for measuring the short-period fluctuations in velocity associated with turbulence. Thanks to rapid technological advances, since the late 1980s it has become routine practice for ship-mounted acoustic Doppler current profilers to be used for continuous measurement of current velocities to depths of several hundred metres, while the ship is actually in motion (see Figure 5.4(c) and (d)).

The Doppler shift may be used in a different way to measure surface currents. The technique known as Ocean Surface Current Radar, or OSCR, exploits the fact that radio waves reflected from surface waves having half the radio wavelength, and travelling either directly towards or away from the radio transmission station, interfere constructively to produce a large signal, analogous to the Bragg reflections observed in crystallography. At any one time and location, waves at the sea-surface have a range of wavelengths which includes very small waves of the wavelengths that interact with radio waves. The Doppler frequency spectrum of the reflected radio signals therefore includes two 'Bragg lines', which in the absence of a current are symmetrical about zero Doppler shift. When the surface water on which the waves are propagating is moving, carrying the small waves with it, the frequency of the reflected radio waves is further shifted and the two Bragg peaks are no longer symmetrically disposed about zero Doppler shift. The magnitude of the current is derived by measuring the displacement of the Bragg lines from their 'zero current' positions.

The two techniques just described, which use the phenomenon of the Doppler shift of frequencies, are relatively recent innovations in the measurement of current velocity. A more traditional method, first suggested by Faraday in 1832, uses the fact that an electromotive force (e.m.f.) is induced in a conductor moving across a magnetic field. Seawater is a conductor, and as it flows across the lines of the Earth's magnetic field, an e.m.f. is generated in it. The strength of this e.m.f. is proportional to the current speed, the width of the current and the strength of the Earth's magnetic field in a vertical direction. It is measured by means of two electrodes, one either side of the current, connected by an insulating cable. In practice, the value measured is always less than would be expected from the simple theory, because the sea-bed conducts electricity. A correction can be made for this if a number of direct current measurements are made while the electromagnetic system is in operation. This method is most successful when used in narrow confined currents, for example in the Straits of Dover or the Gulf Stream off Florida.

We have described only a few of the ways in which currents can be measured, but the selection given here should give you some idea of the wide variety of techniques available.

4.4 COASTAL UPWELLING IN EASTERN BOUNDARY CURRENTS

Eastern boundary currents are less spectacular than western boundary currents and have received less attention. Nevertheless, the upwelling which is often associated with them has biological implications of considerable ecological and economic importance.

In Chapter 3, you saw how the average motion of the wind-driven layer is 90° *cum sole* of the wind direction (Figure 3.6(b)), and hence how cyclonic winds lead to a surface divergence and upwelling in mid-ocean (Figure 3.23(a) and (b)). Coastal upwelling is the result of a divergence of surface water away from the coastal boundary.

In which direction must the wind blow in order to cause most upwelling?

Because Ekman transport is 90° *cum sole* of the wind direction, the most offshore transport of surface water, and the most upwelling, occurs in response to *longshore equatorward* winds, *not* offshore winds.

The situation is complicated by the fact that the offshore movement of surface water leads to a lowering of sea-level towards the coast. This results in a landwards horizontal pressure gradient, which in turn generates geostrophic flow towards the Equator. The equatorward geostrophic current, combined with the offshore wind-driven current, results in a surface current directed offshore and towards the Equator. How this arises is explained in detail in Figure 4.30 and its caption (overleaf), which you should study carefully.

The ecological importance of upwelling lies in the fact that phytoplankton in the sunlit waters need nutrients in order to grow, and so where nutrient levels are low, primary productivity is also low. In regions of upwelling, the supply of nutrients in surface waters is continually being replenished by water from greater depths, and so the production of phytoplankton (and hence of higher trophic levels) is high. By contrast, regions of convergence—the Sargasso Sea for instance—are characterized by sinking (Figure 3.23(c) and (d)), so that surface waters are almost permanently depleted in nutrients and productivity is low. Furthermore, in regions of convergence and sinking the thermocline is depressed (Figure 3.23(d)) so that turbulence in the wind-mixed surface layer may lead to phytoplankton being carried down below the photic zone where they cannot grow; in regions of upwelling the thermocline is raised (Figure 3.23(b) and Figure 4.30(d)) so the depth to which phytoplankton can be carried by turbulence is restricted, and the plant cells can therefore remain longer in the photic zone.

Figure 4.31 illustrates the marked effect that wind has on upwelling, and hence primary productivity, in the surface waters of the Canary Current off north-west Africa.

QUESTION 4.14 Upwelling off the coast of north-west Africa occurs in response to the North-East Trade Winds. By reference to Figure 2.3, explain why upwelling occurs all year round north of Cape Blanc, but varies seasonally south of Cape Blanc, as shown by the difference between Figure 4.31(a) and (b).

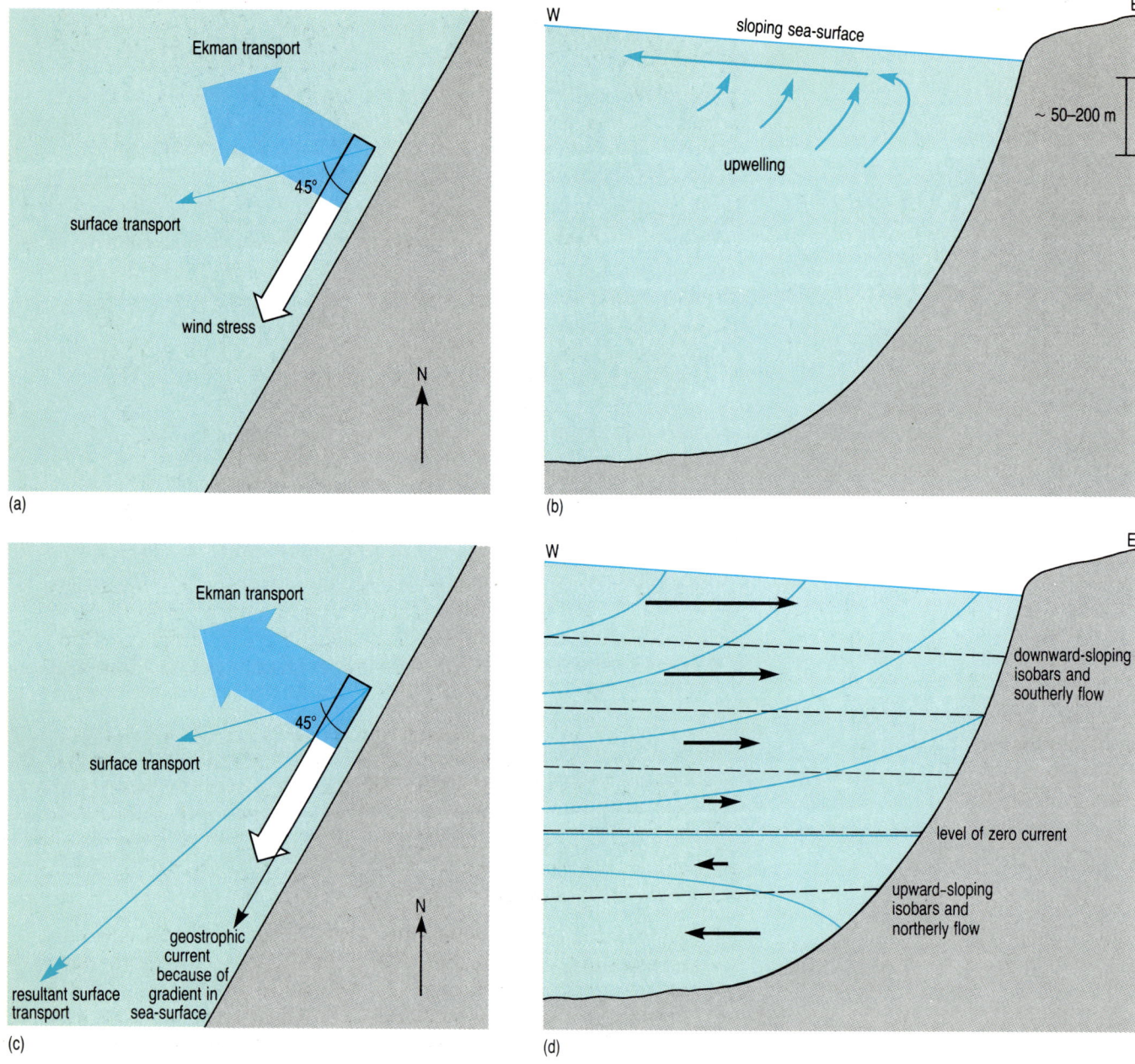

Figure 4.30 Diagrams (*not to scale*) to illustrate the essentials of coastal upwelling (here shown for the Northern Hemisphere).

(a) Initial stage: wind stress along the shore causes surface transport 45° to the right of the wind, and Ekman transport (average motion in the wind-driven layer) 90° to the right of the wind (*cf.* Figure 3.6(b)).

(b) Cross-section to illustrate the effect of conditions in (a): the divergence of surface waters away from the land leads to their replacement by upwelled subsurface water, and to a lowering of sea-level towards the coast.

(c) As a result of the sloping sea-surface, there is a horizontal pressure gradient directed towards the land (black arrows in (d)) and a geostrophic current develops 90° to the right of this pressure gradient. This 'slope' currrent flows along the coast and towards the Equator. The resultant surface transport, i.e. the transport caused by the combination of the surface transport at 45° to the wind stress and the 'slope' current, still has an offshore component, so upwelling continues.

(d) Cross-section to illustrate the variation with depth of density (the blue lines are isopycnals) and pressure (the black lines are isobars and the horizontal arrows represent the direction and relative strength of the horizontal pressure gradient force). Isopycnals slope up towards the shore as cooler, denser water wells up to replace warmer, less dense surface waters. The shoreward slope of the isobars decreases progressively with depth until they become horizontal; at this depth the horizontal pressure gradient force is zero, and so the velocity of the geostrophic current is also zero. At greater depths, isobars (and isopycnals) slope *down* towards the coast indicating the existence of a *northerly* flow; a deep counter-current is a common feature of upwelling systems.

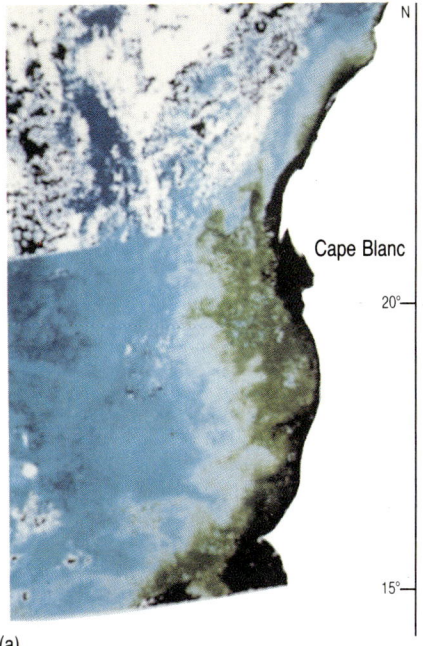 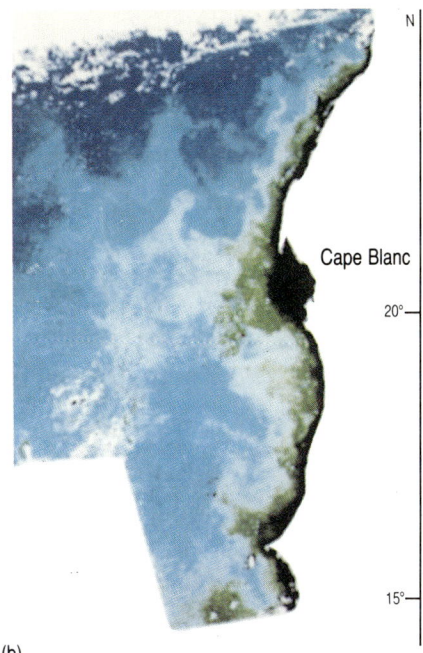 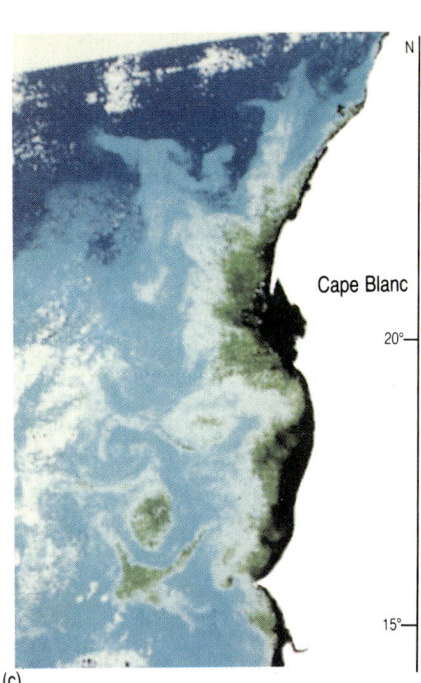

Figure 4.31 CZCS images showing the concentration of chlorophyll *a* pigment in surface waters off the coast of north-west Africa, in each case averaged over about a month: (a) March–April 1983; (b) November 1982; (c) November 1983. In these images, clouds and land are shown as white and turbid coastal waters are black. The colours represent pigment concentration according to a logarithmic scale: dark blue is smallest concentration; dark green largest.

Upwelling along the western coasts of continents thus varies seasonally according to the strength of the Trade Winds and the position of the Intertropical Convergence Zone. These same factors may also lead to differences between one year and another; compare Figure 4.31(b) and (c), for the month of November in 1982 and 1983.

Coastal upwelling is most marked in the Trade Wind zones, but it can occur wherever winds cause offshore movement of water. Nevertheless, it is difficult to investigate the process of upwelling directly because it occurs episodically, and because the *average* speeds of upward motion are very low—of the order of 1–2 metres per day. Indirect methods must be used and, as with indirect methods of measuring currents, these may be based either on the *causes* or the *effects* of upwelling.

The cause of coastal upwelling is the offshore movement of water in response to wind stress. Since the upwelled water rises to replace that moved offshore by the wind, the rate at which water upwells is the *same* as that with which it moves offshore. Hence, the rate of upwelling may be calculated using equation 3.4, which tells us that it must be directly proportional to the wind stress and inversely proportional to the sine of the latitude. This method of calculating the rate of upwelling gives reasonable results only if the assumption that a steady state has been attained is valid, *and* there is adequate information about local winds. However, you should bear in mind that the average *speed* of the wind is not the best indicator of the amount of upwelling it will induce, because the wind *stress* is proportional to something like the square of the wind speed (equation 3.1). Thus, occasional strong winds have a disproportionately large influence on water movement, and upwelling rates fluctuate greatly in response to fairly small changes in wind speed.

Although there are chemical and biological indicators of upwelling (Figure 4.31), in practice it is the physical characteristic of temperature that is most often used to identify and investigate regions of upwelling.

Water that upwells to the surface comes from subsurface layers, at most only a few hundred metres deep but nevertheless significantly colder than that at the surface (Figure 4.32). Relatively cold surface water does not always imply upwelling, however; it may simply result from advection of water from colder regions, by currents or in mesoscale eddies. When *subsurface* measurements are available, the clearest indication of coastal upwelling is the upward slope of isotherms (and associated isopycnals) towards the coast (Figure 4.30(d)).

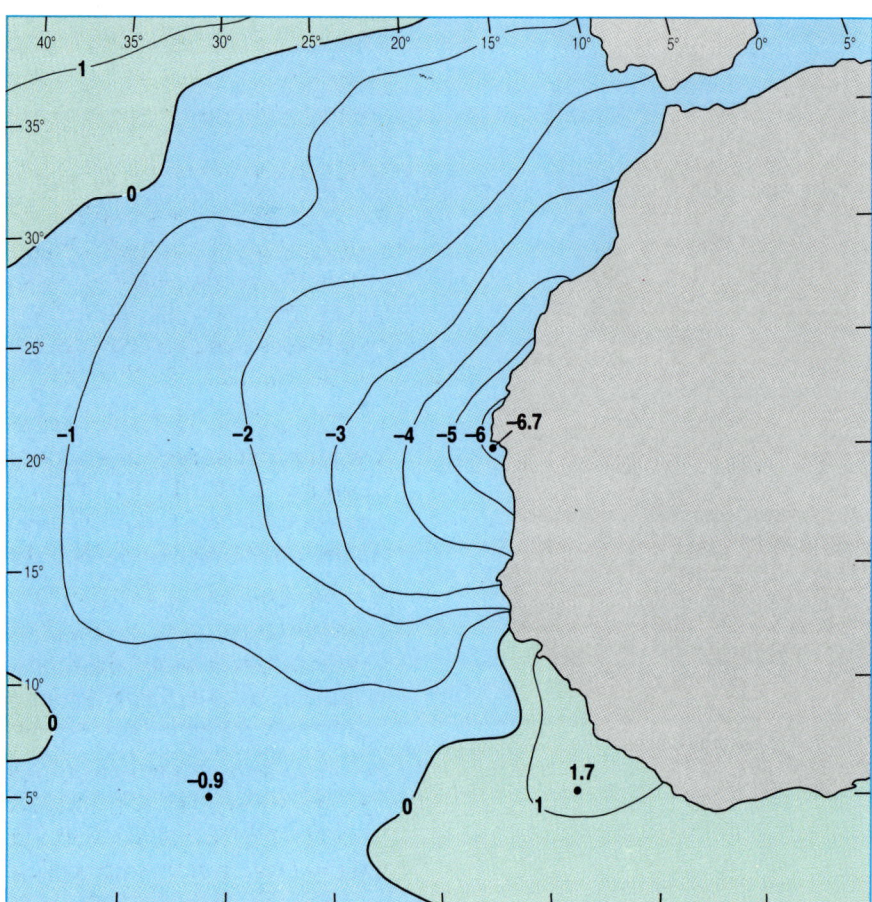

Figure 4.32 Mean anomaly in the sea-surface temperature off north-west Africa for April. This large negative anomaly (i.e. region of large differences between actual surface temperatures and average surface temperatures for these latitudes) is mainly attributable to upwelling (*cf.* Figure 4.31).

This discussion of coastal upwelling in eastern boundary currents concludes our survey of observations and ideas about the North Atlantic subtropical gyre. Before leaving the topic of upwelling, however, we should emphasize that upwelling does not only occur in response to longshore or cyclonic winds. It may also occur, on a local scale, as a result of subsurface currents being deflected by bottom topography. In Chapter 5, you will see how the pattern of winds in the region of the Equator leads to some of the most extensive areas of upwelling in the world.

4.5 SUMMARY OF CHAPTER 4

1   The subtropical gyres are characterized by intense western boundary currents and diffuse eastern boundary currents. In the North Atlantic, the western boundary current is the Gulf Stream, and the eastern boundary current is the Canary Current.

2   Exploration of the east coast of America, followed by colonization, trading and whaling, led to the western North Atlantic in general, and the Gulf Stream in particular, being charted earlier than most other areas of ocean. Two of the most notable charts were made by De Brahm and by Franklin and Folger (in the late eighteenth century) and Maury (in the mid-nineteenth century); Maury was also the first to encourage systematic collection and recording of oceanographic and meteorological data.

3   The Gulf Stream consists of water that has come from equatorial regions (largely via the Gulf of Mexico) and water that has recirculated within the subtropical gyre. The low-latitude origin of much of the water means that the Gulf Stream has warm surface waters, although the warm 'core' becomes progressively eroded by mixing with adjacent waters as the Stream flows north-east.

4   The prevailing Trade Winds cause sea-levels to be higher in the western part of the Atlantic basin than in the eastern part, and the resulting 'head' of water in the Gulf of Mexico provides a horizontal pressure gradient acting downstream. The flow leaving the Straits of Florida therefore has some of the characteristics of a jet.

5   The Gulf Stream follows the continental slope as far as Cape Hatteras where it moves into deeper water and has an increasing tendency to form eddies and meanders; the flow also becomes more filamentous, with cold counter-currents. Beyond the Grand Banks, the current becomes even more diffuse and is generally known as the North Atlantic Drift.

6   Flow in the Gulf Stream is in approximate geostrophic equilibrium, and the strong lateral gradients in temperature and salinity mean that the flow is baroclinic. Confidence in practical application of the geostrophic method was greatly increased when it was successfully used to calculate geostrophic current velocities in the Straits of Florida.

7   The fast, deep currents in the Gulf Stream are associated with the steep downward slope of the isotherms and isopycnals towards the Sargasso Sea. The Gulf Stream may be regarded as a ribbon of high-velocity water forming a front between the warm Sargasso Sea water and the cool waters over the continental margin. This frontal region with strong velocity shear is subject to baroclinic instabilities and, especially downstream of Cape Hatteras, to mesoscale eddy production. Eddies that are formed from meanders with an anticyclonic tendency are known as warm-core eddies, and those formed from meanders with a cyclonic tendency are called cold-core eddies. These eddies extend to significant depths.

8   One of the tools for studying fluid flow is the concept of vorticity, or the tendency to rotate. All objects on the surface of the Earth of necessity share the component of the Earth's rotation appropriate to the latitude; this is known as planetary vorticity and, because vorticity is defined as $2 \times$ angular velocity, is equal to $2\Omega \sin \phi$ and given the same symbol as

the Coriolis parameter, f. Rotatory motion *relative* to the Earth is known as relative vorticity, ζ. The vorticity of a fluid parcel relative to fixed space—its absolute vorticity—is given by $f + \zeta$. In the absence of external influences, potential vorticity $(f + \zeta)/D$ (where D is the depth of the water parcel) remains constant. Away from coastal waters and other regions of strong velocity shear, f is much greater than ζ. By convention, an anticlockwise rotatory tendency is described as positive, and a clockwise one as negative (regardless of hemisphere).

9 Early theories about oceanic circulation were restricted because the effects of the Earth's rotation—the Coriolis force and hence geostrophic currents—were not appreciated. In the present century, ideas about large-scale ocean circulation have developed dramatically. Stommel demonstrated that the intensification of the western boundary currents of subtropical gyres is a consequence of the increase in the Coriolis parameter with latitude, and that western intensification can be explained in terms of vorticity balance.

10 Sverdrup showed that when horizontal pressure gradient forces, caused by sea-surface slopes, are taken into account, the total wind-driven, meridional (north–south) flow is proportional to the torque, or curl, of the wind stress. Sverdrup's ideas were extended by Munk who used real wind data (rather than a sinusoidally varying wind distribution, as used by Stommel and Sverdrup) and allowed for frictional forces resulting from turbulent mixing in both vertical and horizontal directions. The circulation pattern he derived bears a close resemblance to that of the real oceans.

11 The equations of motion—i.e. the mathematical equations used to investigate water movements in the oceans—are simply Newton's Second Law, force = mass × acceleration, applied in each of the three coordinate directions. They are most easily solved by considering equilibrium flow, in which there is no acceleration. When this is done for the equation of motion in a vertical direction, it becomes the hydrostatic equation.

12 Another important principle governing flow in the oceans is the principle of continuity, which expresses the fact that the mass (and, because water is virtually incompressible, volume) of water moving into a region per unit time must equal that leaving it per unit time.

13 Direct methods of measuring currents divide into two categories. Lagrangian methods provide information about circulation *patterns* and involve tracking the motion of an object on the assumption that its motion represents that of the water that surrounds it; they include the use of Swallow/Sofar floats, drifting surface buoys and cheap 'disposable' drifters. Eulerian methods are those in which the measuring instrument is held in a fixed position, and the current flow past that point is measured; the instrument most commonly used is a rotary current meter kept in a fixed position by means of a mooring. Increasingly, currents are being measured using acoustic or electromagnetic methods.

14 Circulation patterns can also be inferred from the distribution of properties of the water, such as temperature or chlorophyll content. Satellite images can provide near-instantaneous pictures of the sea-surface, and avoid the problems that arise from interpolating between widely spaced measurements, perhaps also taken over a long period of time.

15   Satellite radar altimetry may be used to determine variations in the height of the sea-surface (i.e. departures from the geoid), and hence effectively provide a direct measure of the dynamic topography of the sea-surface.

16   The eastern boundary currents of the subtropical gyres are associated with coastal upwelling which occurs in response to equatorward longshore winds. Areas of divergence and upwelling are characterized by cooler than normal surface waters, a raised thermocline and, because nutrients are continually being supplied to the photic zone, high productivity of phytoplankton, the basis of marine life. By contrast, areas of convergence and sinking, such as the regions of the subtropical gyres, are the oceanic equivalent of barren deserts.

Now try the following questions to consolidate your understanding of this Chapter.

QUESTION 4.15   Both Franklin and Rennell identified the Gulf Stream to be what we know today as a geostrophic current. True or false?

QUESTION 4.16   What is meant by the last line of the quotation at the beginning of Section 4.2.1: '. . . . and so on to viscosity'?

QUESTION 4.17   Bearing in mind that the Gulf Stream is an intense frontal region, can you suggest why it is often delineated by accumulations of seaweed?

QUESTION 4.18   It is thought that the meanders and eddies that characterize the Gulf Stream downstream of Cape Hatteras may be related to a chain of seamounts that extends out from the continental shelf (Figure 4.17(b)). Bearing in mind information given in Section 4.2.1, can you suggest why changes in water depth over the seamounts might lead to meanders and eddies?

QUESTION 4.19   We have said that flow in the Gulf Stream is in approximate geostrophic equilibrium. Why would it not be strictly valid to regard the meandering flow in the eddy-generating region downstream of Cape Hatteras as in geostrophic equilibrium?

QUESTION 4.20   Which of the following statements (a–d) concerning Lagrangian and Eulerian methods of current measurement are true, and which are false?

(a)   Average current velocities determined by Lagrangian methods are likely to be *minimum* estimates.

(b)   Eulerian methods require the cooperation of people other than those involved in the research programme, whereas Lagrangian methods do not.

(c)   The two methods of current measurement that make use of Doppler shift in frequency are both Lagrangian methods.

(d)   Calculation of surface currents using ship's drift (dead-reckoning) is an Eulerian technique.

CHAPTER 5 — OTHER MAJOR CURRENT SYSTEMS

In the previous Chapter we considered the subtropical gyres; here, we turn our attention towards both higher and lower latitudes. We will begin with the equatorial current systems, which consist largely of the low-latitude limbs of the subtropical gyres, under the influence of the Trade Winds.

5.1 EQUATORIAL CURRENT SYSTEMS

The first thing to notice about the equatorial current systems (Figure 3.1) is that they are not symmetrically disposed with respect to the Equator, but are shifted several degrees of latitude to the north of it.

By reference to Figure 2.3, can you suggest why this is?

As discussed in Section 2.1, the Intertropical Convergence Zone, the zone along which the Trade Winds of the two hemispheres meet, is distorted towards land in the hemisphere experiencing summer. There is a much greater proportion of land in the Northern Hemisphere than in the Southern and so the mean position of the ITCZ is shifted north of the Equator, most notably so in the Pacific.

In order to understand current flow in low latitudes, it is important to remember the following:

1 The Coriolis force is zero on the Equator.

2 Nevertheless, by latitudes of about 0.5° it is large enough to have a significant effect.

Figure 5.1 shows schematically the winds, and surface currents and Ekman transports that result, in the region of the Equator. These simplified pictures represent what actually happens, to varying extents, in each of the three oceans. Look first at Figure 5.1(a). The short blue arrows represent the Ekman transport—the *net* transport in the wind-driven layer; these transports are to the left of the wind in the Southern Hemisphere, to the right of the wind in the Northern Hemisphere, and *in the direction of the wind on the Equator*. The South-East Trade Winds blow *across* the Equator, with the result that there is a *divergence* to the south of the Equator; this is known as the **Equatorial Divergence.**

Remember that the ITCZ is characterized by light winds (the Doldrums) and so there is no significant Ekman transport in the region *between* the two Trade Wind zones. Water moving across the Equator in response to the South-East Trade Winds therefore converges with water in this region (at about 4° N). On the other hand, there is a *divergence* at about 10° N as a result of water moving *away* from the Doldrum region in response to the North-East Trade Winds.

Figure 5.1(b) is a highly exaggerated diagrammatic north–south section across the Equator, showing the vertical and meridional circulation in the mixed surface layer above the thermocline/pycnocline. There is little horizontal variation in density in the upper part of this layer, particularly in the top few tens of metres, and surfaces of constant pressure are more or less parallel to surfaces of constant density, and both are parallel to the sea-surface; in other words, conditions may be regarded as barotropic. By the depth of the thermocline, however, there are significant lateral

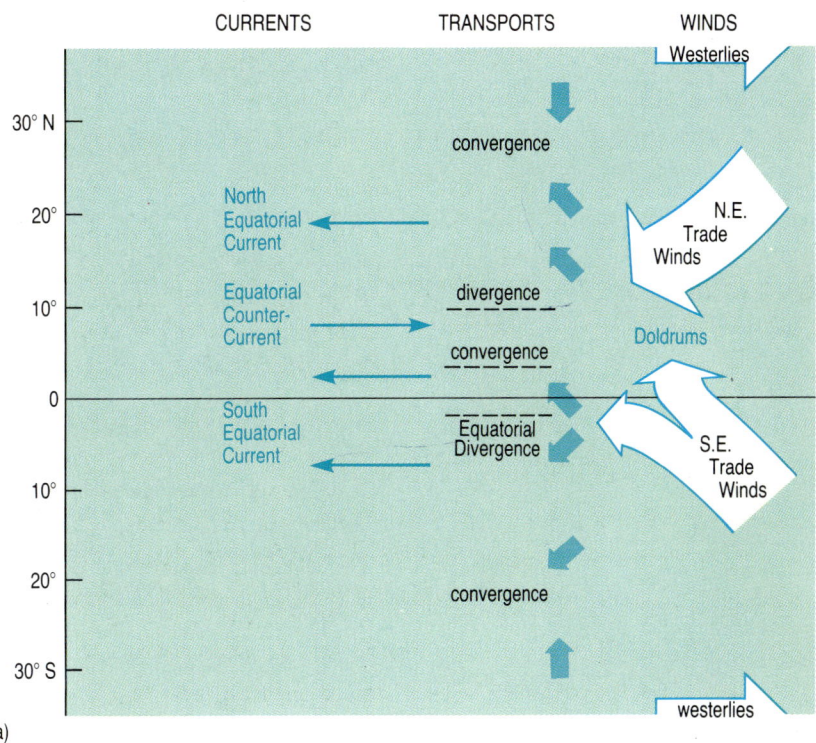

(a)

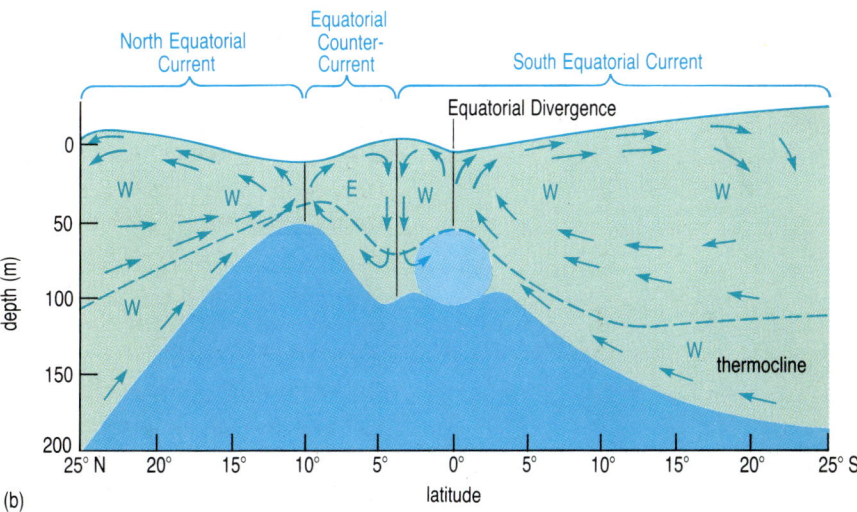

(b)

Figure 5.1(a) The relationships between the wind direction, the surface current and the Ekman transport (the total transport in the wind-driven layer, shown by short blue arrows), in equatorial latitudes. Note the Doldrum belt between about 5° and 10° N.

(b) North–south section diagrammatically showing the vertical and meridional circulation in equatorial latitudes, and the shape of the sea-surface and thermocline. Regions of eastward and westward flow are indicated by the letters E and W. The darker blue region (in which the geostrophic current is assumed to be zero, cf. Figure 3.19(a)) is the deep water below the thermocline. The blue oval at about 100 m depth at the Equator represents the Equatorial Undercurrent discussed in Section 5.1.1. (Note that the vertical scale is greatly exaggerated.)

variations in density; the slopes in the thermocline are therefore contrary to those of the sea-surface, and conditions have become baroclinic.

Figure 5.1(b) is most easily interpreted in terms of easterly and westerly geostrophic flow if we apply the simplified picture illustrated in Figure 3.19(a), and assume that all of the water above the bottom of the thermocline behaves as one homogeneous layer (shown lighter blue). This is a justified assumption in low latitudes, because here the density gradient, or **pycnocline**, between the mixed surface layer and deeper, colder waters is relatively sharp (*cf.* Figure 3.18).

QUESTION 5.1 Bearing this in mind, explain how the three regions of westward flow in Figure 5.1(b)—i.e. (i) the **North Equatorial Current** between 25° and 10° N, (ii) the **South Equatorial Current** between about 4° N and about 0°, and (iii) the South Equatorial Current between 0° and about 20° S—may each be explained in terms of a geostrophic balance between horizontal pressure gradient forces and the Coriolis force.

Westward flow in surface waters is also a direct result of the Trade Winds. These blow at about 45° to the Equator and surface flow in the South and North Equatorial Currents is 45° *cum sole* of this, i.e. to the west in both cases.

The easterly **Equatorial Counter-Current** between about 4° and 10° N may be explained as follows. The overall effect of the Trade Winds is to drive water towards the west, but the flow is blocked by the land masses along the western boundaries. As a result, in equatorial regions the sea-surface slopes up towards the west, causing an eastward horizontal pressure gradient force. Because winds are light in the Doldrums, water is able to flow 'down' the horizontal pressure gradient in a current that is contrary (i.e. 'counter') to the prevailing wind direction. Furthermore, even this close to the Equator, there is some deflection by the Coriolis force. The deflection is to the right and therefore towards the Equator, and so contributes to the convergence at about 4° N (along with the transport across the Equator resulting from the South-East Trades). The sea-surface therefore slopes up from about 10° N to about 4° N, giving rise to a northward horizontal pressure gradient force which drives a geostrophic current towards the east.

This general pattern of westward-flowing North and South Equatorial Currents and an eastward-flowing Counter-Current may be observed in all three oceans. The South Equatorial Current is, on average, the strongest. Often in the Atlantic and occasionally in the Pacific, it shows two cores of maximum velocity, one at latitude about 2° N and the other at about 3–5° S, as shown schematically in Figure 5.1(a). The Equatorial Counter-Current is best developed in the Pacific, where it reaches its maximum speed between 5° and 10° N, some distance below the surface. The Equatorial Counter-Current in the Atlantic is present throughout the year only in the eastern part of the ocean, between 5° and 10° N, where it is known as the Guinea Current. In both the Pacific and the Atlantic, there is also a weak South Equatorial Counter-Current which can be distinguished in the western and central ocean between about 5° and 10° S (not shown in Figure 5.1).

The equatorial current system of the Atlantic does not have the generally simple east–west flows of the Pacific. This is partly because of the relative narrowness of the ocean basin (it is about one-third of the width of the Pacific) combined with the influence of the shape of the African and American coastlines on the direction of current flow.

QUESTION 5.2 There is another reason, also connected with the shape of the continental masses, why surface currents in the tropical Atlantic are not simply east–west. Can you say what it is by looking at Figure 2.3?

It is important to remember that there will be many differences between the generalized picture of equatorial currents that we have been discussing, and the pattern that may exist in the oceans at any one time. Changes in the extent, speed and even the direction of the currents occur in response to changes in the overlying wind pattern, especially changes in the position of the ITCZ. This is particularly true in the Indian Ocean and the western Pacific Ocean, where seasonally reversing winds known as Monsoons have dramatic effects on current speeds and directions (these will be discussed in Section 5.2).

5.1.1 THE EQUATORIAL UNDERCURRENT

A major feature of equatorial circulation, shown on Figure 5.1(b), is the **Equatorial Undercurrent**. Such an Undercurrent occurs in all three oceans, although it is only a seasonal feature in the Indian Ocean.

Equatorial Undercurrents flow from west to east, below the direct influence of the wind, yet they *are* wind driven. How can this be?

The effect of the wind is transmitted downwards to deeper layers via turbulence (eddy viscosity) and is mainly confined to the mixed surface layer above the thermocline/pycnocline (Section 3.1.1). In low latitudes, the mixed surface layer is thin—as shown in Figure 5.1(b), it is only 50–100m thick in the vicinity of the Equator. This is because there is no winter cooling of surface waters which, in higher latitudes, causes them to become denser, destabilizing the upper part of the water column, so that it may more easily be mixed by wind and waves. In addition, as mentioned earlier, the pycnocline is a *sharper* boundary in low latitudes than in other regions; this is partly because surface heating is most intense there.

Bearing in mind information given in Figure 2.2(b) and Section 2.3.1, can you give the other reason for the low density of surface waters in low latitudes?

The cumulonimbus activity associated with the ITCZ means that the equatorial zone is characterized by high precipitation, and the heavy rainfall significantly lowers the density of surface waters. Indeed, salinity has a greater control over the density distribution in the equatorial zone than in most other regions of the ocean.

Now study Figure 5.2 which shows a schematic east–west section across the equatorial ocean. The Trade Winds blow from east to west and drive a westward flow in a fairly well-defined mixed surface layer. As a result of this westward transport (and despite some return flow in the Equatorial

Counter-Current) water piles up against the western boundary of the ocean; there is therefore a sea-surface slope up towards the west, and a concomitant adjustment in the thermocline so that it slopes *down* to the west. Because the mixed layer is thin, the horizontal pressure gradient which results from the sea-surface slope extends to greater depths than the effect of the wind. Hence, although current flow 'down' the horizontal pressure gradient is opposed by the wind-driven current in the mixed surface layer, this is not the case *below* the mixed layer. There is therefore an eastward jet-like current *in the thermocline*. This is the Equatorial Undercurrent.

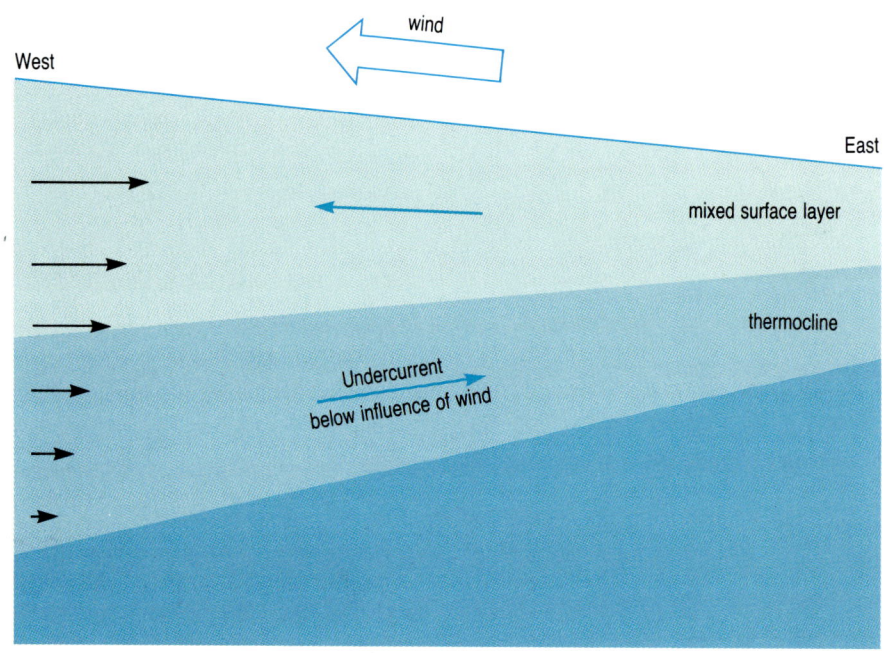

Figure 5.2 Schematic east–west cross-section across the equatorial ocean, showing how the slope in the sea-surface resulting from the Trade Winds, and the resulting horizontal pressure gradient, leads to the generation of the Equatorial Undercurrent. The lengths of the arrows indicate the relative magnitude of the zonal horizontal pressure gradient. The vertical scale is greatly exaggerated.

The most powerful Equatorial Undercurrent is that in the Pacific (see Figure 5.3). Its existence had been suspected as early as 1886, but it was not investigated properly until 1951 when it was fortuitously rediscovered by Townsend Cromwell, a young oceanographer who was leading an expedition to study tuna fish. It is therefore often referred to as the Cromwell Current.

Figure 5.3 shows the relationship between the mean wind stress and the resulting slopes in the sea-surface and thermocline in the Pacific (*cf.* the schematic diagram in Figure 5.2). In Figure 5.3(c), the position of the core of the Undercurrent (the region of highest velocity) is indicated by the blue crosses; note that in the eastern ocean, where the thermocline is especially shallow, and upwelling brings cooler water to the surface, the flow in the Undercurrent may extend to the surface. The current is very fast, especially in the eastern Pacific; velocities in the core may reach $1.0-1.5\,\mathrm{m\,s^{-1}}$.

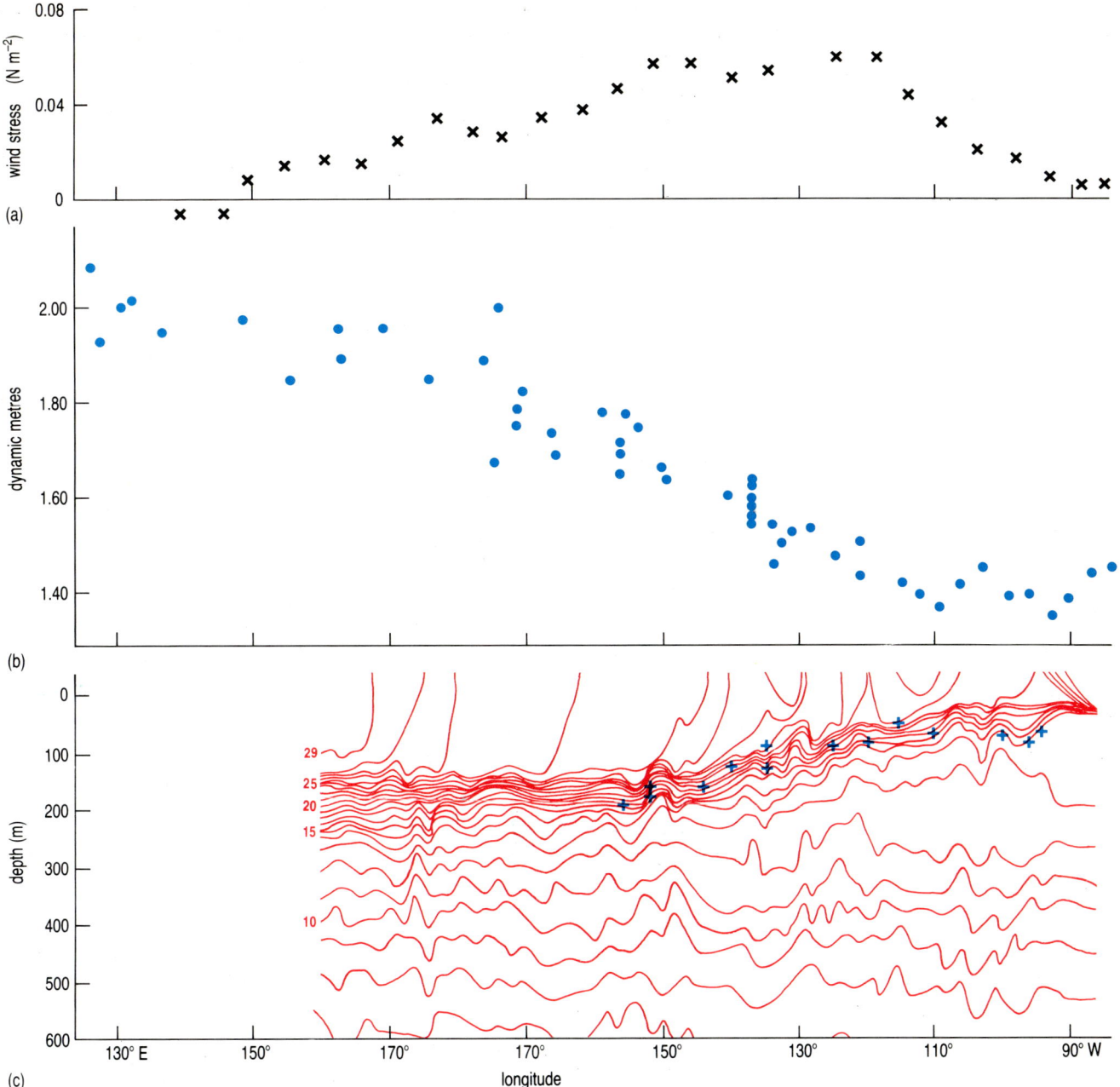

Figure 5.3(a) Mean eastward wind stress along the Equator in the Pacific, and (b) the dynamic height of the sea-surface, assuming no horizontal pressure gradient at 1 000 m; remember that dynamic metres are numerically very similar to geometric metres.
(c) Vertical distribution of temperature (°C) along the Equator in the Pacific. The blue crosses indicate the position of the core of the Cromwell Current.

In the context of its contribution to ocean circulation, however, it is not the speed of a current so much as its volume transport that is important. We may make an estimate of the volume transport of the Equatorial Undercurrent in the Pacific by reference to Figure 5.4(a), a vertical cross-section showing the distribution of velocity in the vicinity of the Equator. Note that the vertical scale is greatly exaggerated so that the actual shape of the Undercurrent—that of a horizontal 'ribbon' of water about 1500 times wider than it is thick—is not immediately apparent.

QUESTION 5.3 Use Figure 5.4(a) to estimate (i) the cross-sectional area and (ii) the average velocity of the Cromwell Current. Hence calculate its volume transport in $m^3 s^{-1}$. (*Note*: 1° of latitude is about 110 km.)

Recent measurements indicate that the average volume transport in the Cromwell Current is about $40 \times 10^6 m^3 s^{-1}$. This average conceals wide seasonal variations and in the northern summer the volume transport may be as high as $70 \times 10^6 m^3 s^{-1}$.

In the Atlantic, the transport in the Equatorial Undercurrent is about one-third of that in the Cromwell Current. The Atlantic Equatorial Undercurrent was first observed in 1886, by Buchanan, one of the scientists who had earlier worked on the *Challenger* oceanographic expedition. It was then forgotten about and was rediscovered in 1959; it is sometimes called the Lomonosov Current, after the vessel from which it was observed. This rediscovery was not accidental. Oceanographers were keen to know whether the Pacific Equatorial Undercurrent had a counterpart in the Atlantic; if none could be found, that would indicate that the Cromwell Current was caused by some peculiarity of the Pacific; if one *was* found, it would seem much more likely that an Undercurrent was

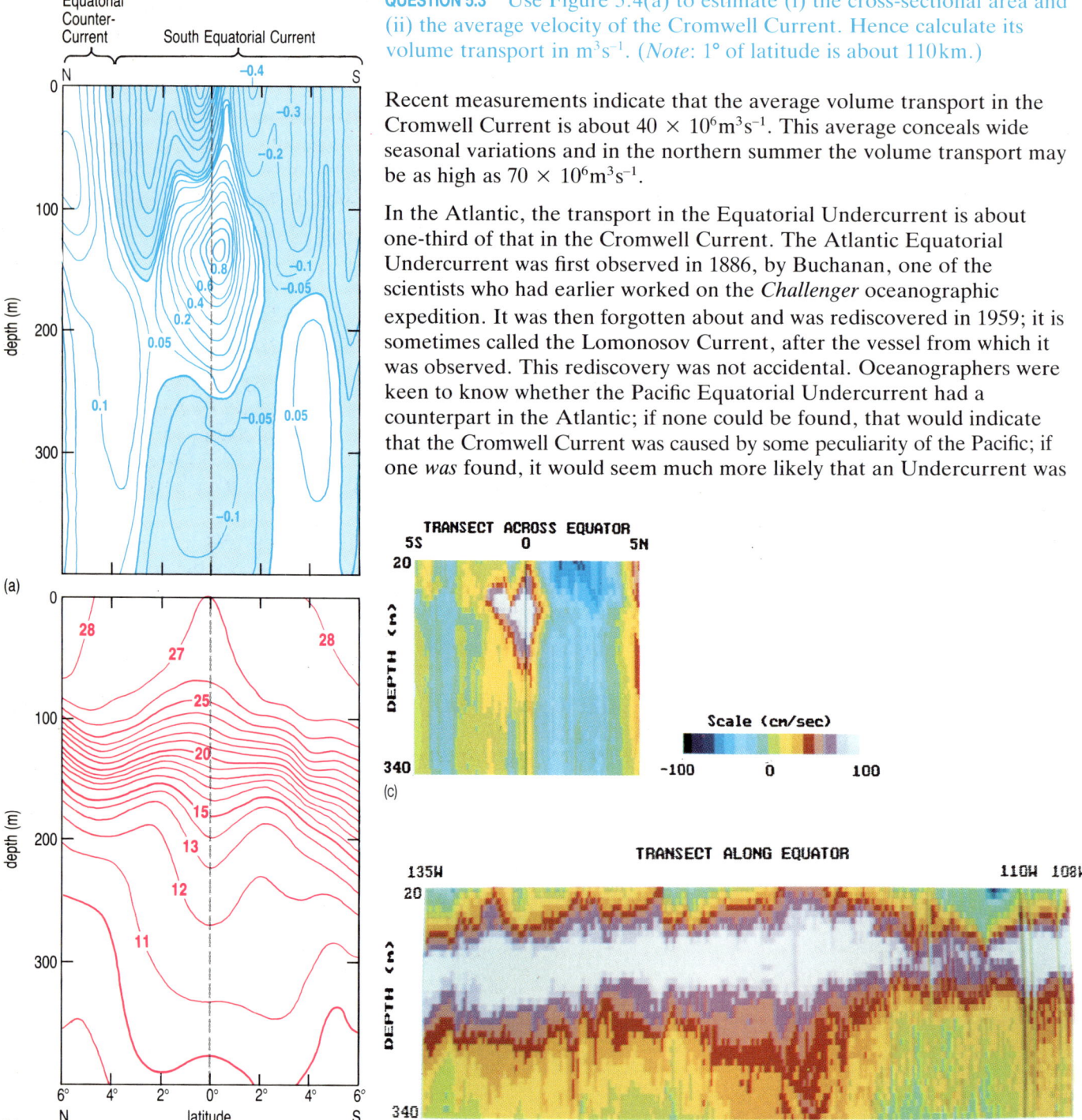

Figure 5.4 *Left:* Section across the equatorial Pacific along 150°W between about 6°N and 6°S, showing the distribution of (a) velocity (ms^{-1}; the shaded region corresponds to westward flow) and (b) temperature (°C). Note that the core of the Cromwell Current is in the thermocline (cf. Figure 5.3(b)) and that in the region of the core the isotherms are spread apart and the thermocline is weakened. The vertical scale is greatly exaggerated. *Above:* sections (c) across the Cromwell Current at 110°W, and (d) along it, between 135° and 108°W. These sections were made using an acoustic Doppler current profiler which works on the principle outlined in Section 4.3.6.

a consquence of some fundamental aspect of equatorial circulation. It is now known that an Equatorial Undercurrent is an intrinsic component of equatorial circulation, although the explanation of its generation given above probably represents only part of the story.

Another aspect of the Equatorial Undercurrent that may be explained in a number of ways is the fact that it is generally aligned along the Equator, although the Trade Winds that drive it blow over a fairly wide latitude band. One of the reasons for this is that, on both sides of the Equator, the geostrophic flow resulting from the wind-induced eastward horizontal pressure gradient (Figure 5.2) is *towards* the Equator; once on the Equator the water can flow directly 'down' the pressure gradient.

Nevertheless, Equatorial Undercurrents are *not* always centred on the Equator—their cores have been observed to be as much as 100km away. This may be caused by northerly or southerly winds, or by the presence of islands. However, because the current is flowing east, when it strays from the Equator it is always deflected back by the Coriolis force: if it strays to the north, the Coriolis force acts to turn it to the south; if it strays to the south, the Coriolis force acts to turn it to the north. This is in contrast to the action of the Coriolis force on westward-flowing currents, which are always deflected polewards (*cf.* Figure 1.2(b)).

Figure 5.4(c) and (d) show transverse and longitudinal sections across the Cromwell Current. They show vividly how modern technology may be used to reveal complex current structures.

It is important to remember that the cross-section across the equatorial ocean shown in Figure 5.1(b) is very schematic: at any one time and place the actual pattern may look very different. Before leaving this discussion of the main equatorial currents, look at Figure 5.5, a temperature section across the eastern Atlantic made during August and September 1963. For convenience, the 17°C isotherm may be taken to represent the bottom of the thermocline.

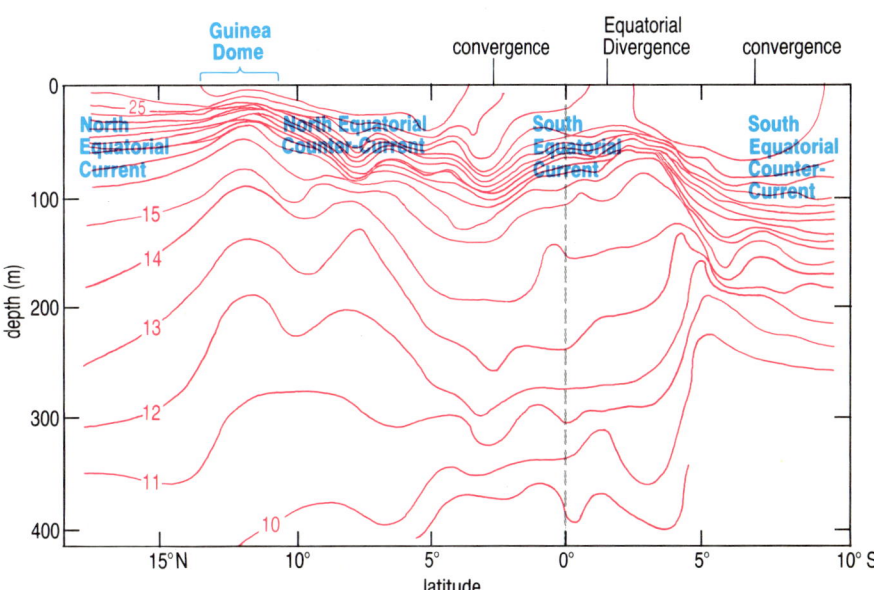

Figure 5.5 The vertical temperature distribution (in °C) across the Equator in the Atlantic at 20° W, plotted from measurements made in August and September, 1963 from the research vessel *John Pillsbury*.

Compare Figure 5.5 with Figure 5.1(b), and try to identify the North Equatorial Current and the South Equatorial Current.

The two regions of westward flow that make up the South Equatorial Current may be seen in the region above the thermocline between about 8° S and the Equatorial Divergence (thermocline sloping up towards the north), and between the Equatorial Divergence and about 3° N (thermocline sloping down towards the north). The change in the direction of slope of the thermocline seems to occur a few degrees to the south of the Equator; this probably means that the South Equatorial Current has moved position with respect to the Equator but geostrophic equilibrium has not yet been re-established. The region of westward flow corresponding to the North Equatorial Current may be seen between about 12° and 18° N (thermocline sloping downwards to the north).

Note that two counter-currents are visible: a North Equatorial Counter-Current in about the position shown in Figure 5.1(b) (thermocline sloping down towards the south) and a South Equatorial Counter-Current (thermocline sloping up to the south) south of about 8° S.

Can an Equatorial Undercurrent be discerned?

No obvious spreading of the isotherms can be seen in the upper part of the thermocline in the region of the Equator, but the spreading of the 13–17 °C isotherms is probably related to mixing associated with an Equatorial Undercurrent. Such well-mixed regions where temperature and density vary little with depth are sometimes referred to as thermostads or **pycnostads**. A similar thermostad/pycnostad may be seen on Figure 5.4(b).

Figure 5.6 repeats Figure 5.5 but here the various regions of current flow have been labelled. Do not worry if you found it difficult to interpret Figure 5.5; it was included both to illustrate the extent to which diagrams such as Figure 5.1 are simplifications of the complex distributions observed in real temperature sections, and to give you a chance to look at some real data.

A feature of Figure 5.6 that is particularly interesting is the region of 'doming' isotherms at about 12° N. This is known as the 'Guinea Dome', and will be discussed in the next Section.

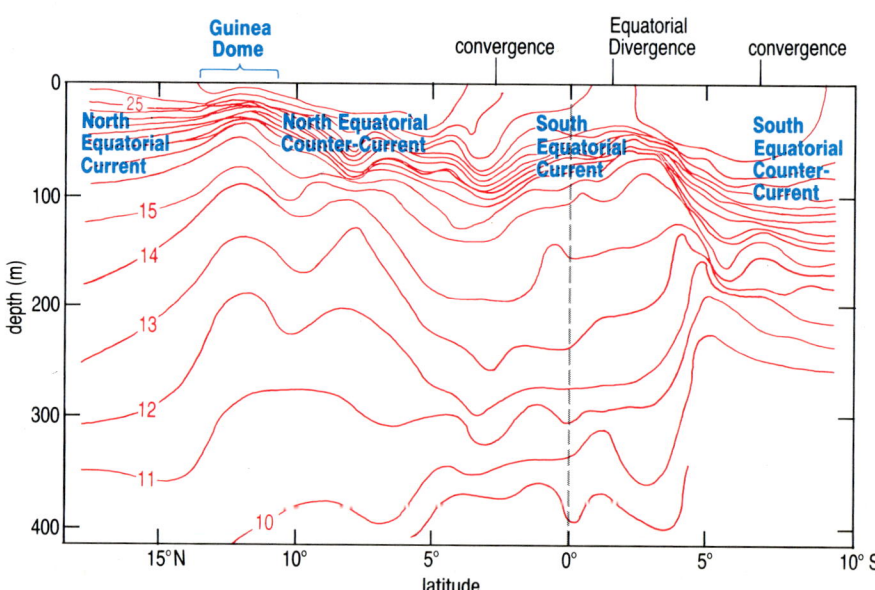

Figure 5.6 As for Figure 5.5, but with the regions corresponding to the various currents identified. Note that both a North and South Equatorial Counter-Current are visible. The doming of isotherms at about 12° N is the Guinea Dome which will be discussed in Section 5.1.2.

5.1.2 UPWELLING IN LOW LATITUDES

Figure 5.7 is a schematic map showing the different upwelling regions of the eastern tropical Atlantic. A similar map could be drawn for the eastern Pacific, but the pattern in the Indian Ocean is somewhat different, mainly because of its monsoonal wind system.

QUESTION 5.4 (a) From what you have read previously (especially in connection with Question 4.14), can you suggest why coastal upwelling occurs all year round to the north of Cape Blanc and to the south of Cape Frio, but only seasonally between these latitudes?

(b) By reference to Figure 5.1, explain why there is upwelling in the region indicated by the blue shading just to the south of the Equator.

Mid-ocean equatorial upwelling, like coastal upwelling, is subject to seasonal variations. The main period of mid-ocean upwelling in the eastern equatorial Atlantic is from July to September. This upwelling maximum is *partly* the result of increased divergence of surface water in response to the seasonal increase in the strength of the South-East Trade Winds; however, the main reason is probably that this increase in wind strength increases the flow in the South Equatorial Current, so that the north–south slope in the thermocline becomes so steep that it intersects

Figure 5.7 The positions of the various upwelling areas in the tropical Atlantic Ocean, in relation to the eastwards flow in the equatorial current system (the westward-flowing North and South Equatorial Currents have been omitted for the sake of clarity). The upper part of the diagram may be compared with Figure 4.31(c).

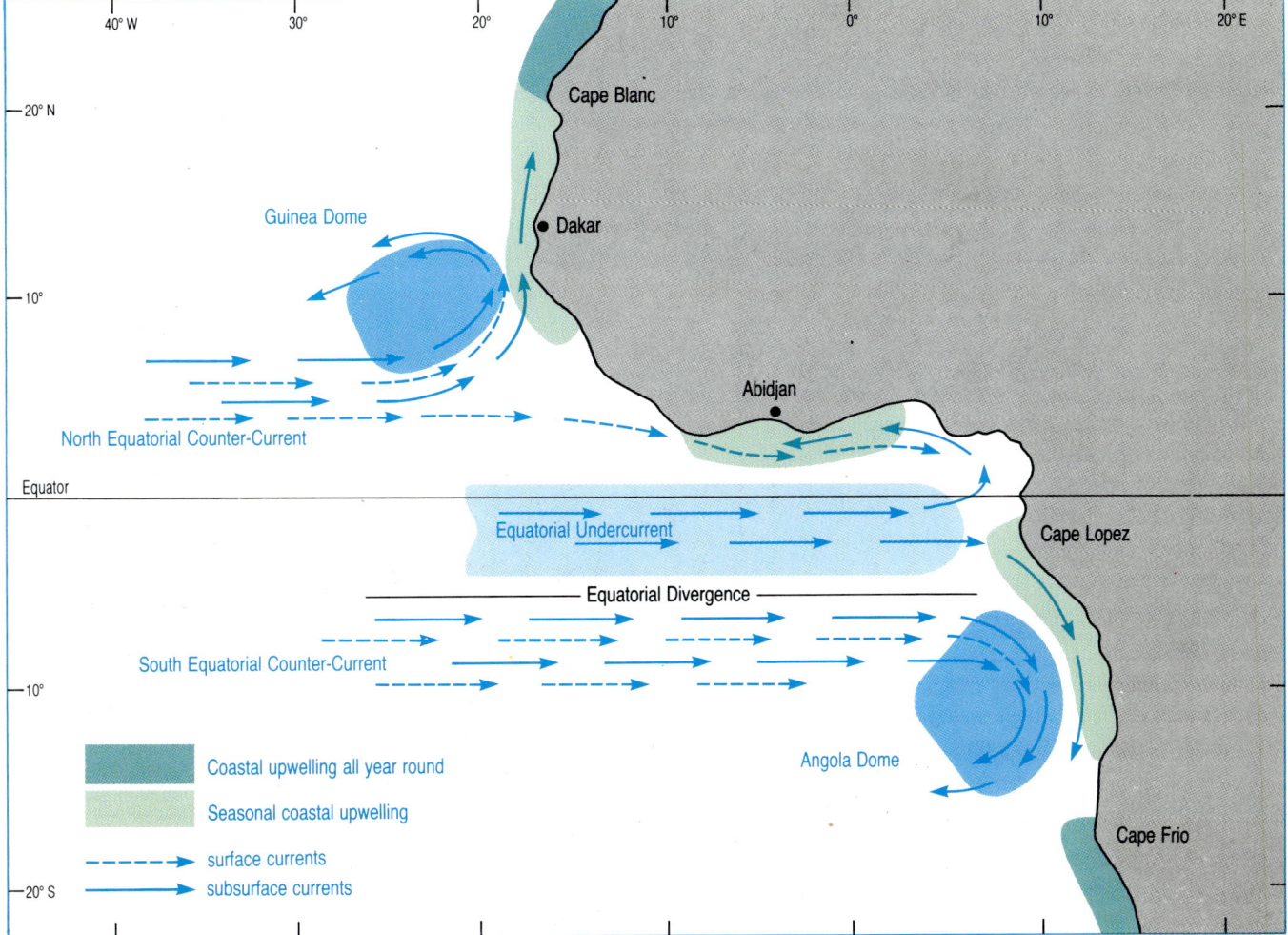

the surface (*cf*. Figure 5.1(b)), allowing deeper water to come to the surface.

Remember also that in low latitudes the thermocline slopes down towards the western coastal boundary, in response to the westward wind stress (Figure 5.2). Because the mixed surface layer is therefore thinner in the eastern ocean, not only is it more easily affected by the wind, but also sub-thermocline water may more readily be brought up to the surface. This is the main reason why upwelling is a feature of the *eastern* equatorial oceans in particular.

In certain localities of the eastern tropical Atlantic and Pacific, the isotherms bow up into a dome-like shape. Figure 5.7 shows the position of the Guinea Dome (also visible on Figure 5.6) and the Angola Dome; a similar dome in the tropical Pacific is named the Costa Rica dome. These *thermal domes* are not well understood, and seem to have a number of causes. Their existence is certainly linked with the flow patterns of the equatorial current systems: Figure 5.7 shows how the positions of the Guinea Dome and the Angola Dome relate to flow in the North and South Equatorial Counter-Currents, while Figure 5.6 shows how the Guinea Dome corresponds to the northern edge of the region of sloping isotherms associated with the North Equatorial Counter-Current.

The degree of 'doming' varies seasonally. During the northern summer when the ITCZ is in its most northerly position, the Guinea Dome protrudes into the thermocline, but the Angola Dome is weak. In the southern summer, the situation is reversed: the Guinea Dome no longer distorts the thermocline, and the Angola Dome is at its most prominent (although it is never as strong a feature as the Guinea Dome).

QUESTION 5.5 Winds are generally light and variable in the vicinity of the ITCZ. In some places, however, they do significantly affect surface waters, particularly in regions where the atmospheric pressure is especially low.

(a)   Will winds be cyclonic or anticyclonic around such regions?

(b)   How does this help to explain the fact that the Guinea Dome 'protrudes' into surface waters when the ITCZ is in its most northerly position?

As in other areas of upwelling, productivity of phytoplankton and hence of other organisms is enhanced in the surface waters above thermal domes. For this reason, the position of the Guinea Dome may be clearly seen on Figure 4.31(c), even though the image was made late on in the season.

To round off this Section, look at Figure 5.8 which shows the variation in the depth of the 21°C isotherm (which corresponds roughly to the middle of the thermocline) and, for comparison, the distribution of sea-surface temperature, in the tropical Atlantic, both at the same time of year.

QUESTION 5.6 Does Figure 5.8 show the situation in the northern summer (July to September) or the southern summer (January to March)?

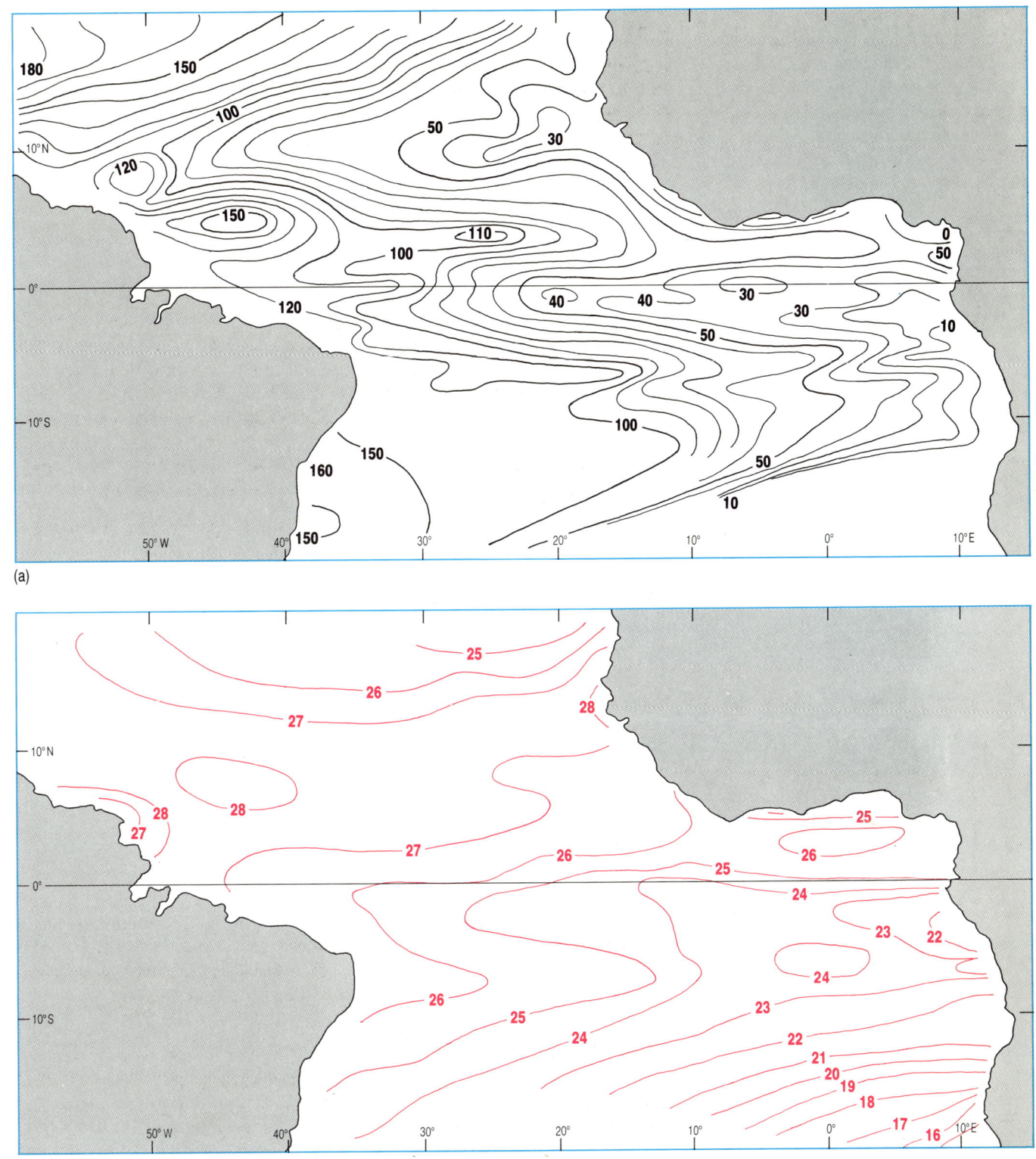

Figure 5.8 For use with Question 5.6.
(a) The average depth (in metres) of the 21°C isotherm (taken as corresponding to the middle of the thermocline) in the tropical Atlantic at one season of the year.
(b) The mean sea-surface temperature (°C) in the tropical Atlantic at the same season of the year as (a).

5.2 MONSOONAL CIRCULATION

The word 'monsoon' is derived from an Arab word meaning 'winds that change seasonally'.

From Figure 2.3, which regions of the ocean are affected by monsoons?

The most immediately obvious region is the Indian Ocean, north of about 15° S; also affected by seasonally reversing winds, however, is the westernmost part of the Pacific Ocean, including the regions of the Malaysian and Indonesian archipelagos. Here, we will concentrate on the monsoonal reversals over the Indian Ocean.

5.2.1 MONSOON WINDS OVER THE INDIAN OCEAN

The seasonal changes in the winds over the Indian Ocean were mentioned in Section 2.1 and will now be discussed in a little more detail. In the northern winter (Figure 2.3(b)), the air over southern Asia is cooler and denser than air over the ocean, and so the surface atmospheric pressure is greater over the continent than over the ocean. The resulting pressure gradient leads to a northerly or north-easterly flow of air from Asia to south of the Equator. This flow of air is the North-East Monsoon. After crossing the Equator, the flow is turned to the left by the Coriolis force and converges with the South-East Trades at about 10–20° S. As the year progresses, increased heating weakens the high pressure over southern Asia. By the northern summer, a low has developed, so that from May/June to September a southerly or south-westerly wind blows across the region. This is the South-West Monsoon, the stronger of the two monsoons. The monsoonal nature of the winds may also be thought of as a manifestation of the seasonal change in the position of the ITCZ, from about 20° S in January to about 25° N, over Asia, in July (see also Figure 2.12).

During which season of the year, i.e. during which monsoon, does the wind pattern over the Indian Ocean most resemble that in the other two oceans?

Figure 2.3 shows that the wind pattern over the Indian Ocean most resembles that over the other two oceans in the southern summer/northern winter. At this time of year the North-East Monsoon blows across the ocean, and the wind field resembles that associated with the North-East Trade Winds.

The different types of weather characteristic of the two monsoons are well known. In January and February, during the North-East Monsoon, the winds bring dry cool air to India from the Asian land mass. During May and June, during the South-West Monsoon, the winds cross the Arabian Sea and bring humid maritime air to India. The moisture that provides the heavy monsoon rains is partly a direct result of evaporation from the warmed surface of the Arabian Sea, and partly the result of upward convection of warm moist air above the Arabian Sea which leads to the formation of cyclonic vortices which draw in *more* moisture-laden air from adjacent regions (*cf.* Section 2.3.1).

A particularly interesting aspect of the South-West Monsoon is the appearance over the western side of the ocean of an intense, southerly low-level atmospheric jet (Figure 5.9). This resembles an oceanic western boundary current, although here it is the high table lands of east Africa that form the western boundary.

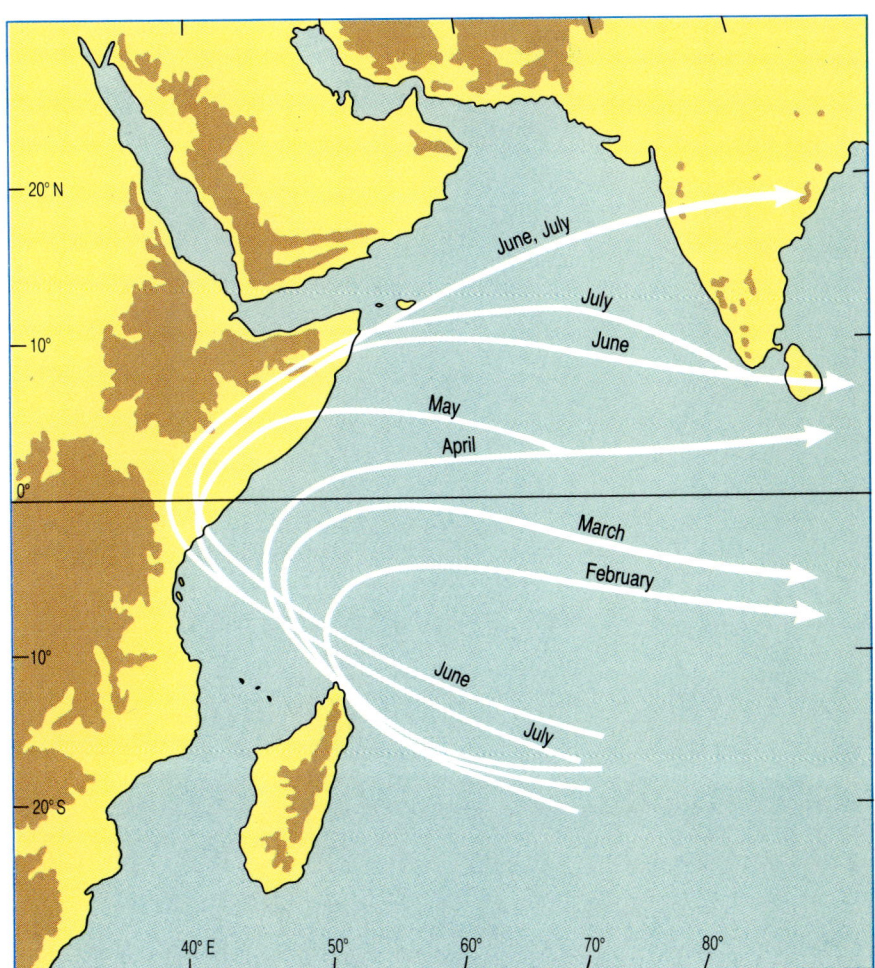

Figure 5.9 Monthly positions of the core of the low-level (1–2 km) atmospheric jet over the Indian Ocean. Brown shading represents land higher than 1 km above sea-level.

5.2.2 THE CURRENT SYSTEM OF THE INDIAN OCEAN

As you might expect, the surface circulation of the northern Indian Ocean changes seasonally, in response to the monsoons, but most resembles that of the other two oceans in the northern winter. At this time of year, both a North and a South Equatorial Current are present, as well as an Equatorial Counter-Current (Figure 5.10(a)). In the northern summer, by contrast, the flow in the North Equatorial Current reverses and combines with a weakened Equatorial Counter-Current to form the South-West Monsoon Current (Figure 5.10(b)). The South Equatorial Current is still present, although its flow is not as strong as during the North-East Monsoon.

The most spectacular seasonal change is the reversal of the Somali Current, off east Africa. This reversal has been known about for centuries: in the ninth century Ibn Khordazbeh observed that 'the sea flows during the summer months to the north-east', and 'during the winter

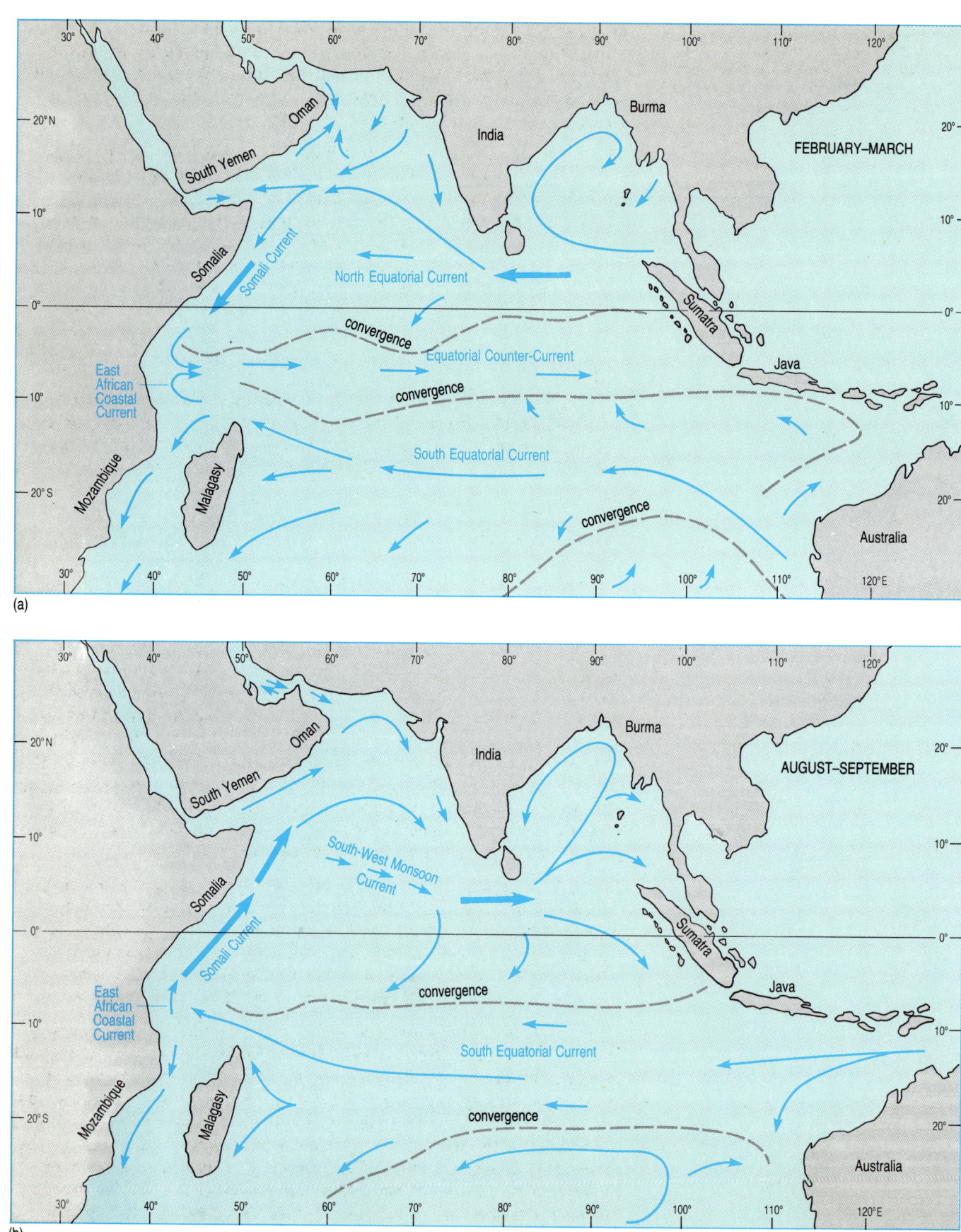

(a)

(b)

Figure 5.10 (Opposite) Surface currents in the Indian Ocean (a) during February and March (in the North-East Monsoon) and (b) during August and September (in the South-West Monsoon). The dashed lines are regions of convergence of surface water.

months to the south-west'. So, during the North-East Monsoon the Somali Current flows to the south-west, while during the South-West Monsoon it is a major western boundary current. At this time of year, its surface velocity may reach $3.7\,\text{m s}^{-1}$, and the volume transport in the upper 200 m of its flow is of the order of $60 \times 10^6\,\text{m}^3\text{s}^{-1}$.

The low-level atmospheric jet mentioned in the previous Section (Figure 5.9) is thought to play an important role in the generation of the intense coastal upwelling that occurs off Somalia during the South-West Monsoon. Upwelling also occurs at this time of year off Arabia, both along the coast where the north-easterly current diverges from it, and offshore in response to local cyclonic winds. The Somali and Arabian regions of upwelling are the most intense in the Indian Ocean.

Would you expect open-ocean upwelling to occur in the region of the Equator, as it does in the Pacific and Atlantic?

No, because these regions of upwelling occur as a result of the South-East Trade Winds blowing across the Equator and causing a surface divergence just to the south of it (Figure 5.1(a)). Figure 2.3 shows that these wind conditions are not typical of either season of the year.

Would you expect there to be an Equatorial Undercurrent in the Indian Ocean?

No, at least certainly not as a permanent feature. The existence of an Undercurrent depends on there being wind stress towards the west, to drive a surface current and hence cause a sea-surface slope up to the west; this would provide an eastward horizontal pressure gradient force to drive an eastward current below the wind-driven layer. In the equatorial Indian Ocean, the direction of the wind varies seasonally, but best fulfils the necessary conditions for an Undercurrent during the first few months of the year. The first direct observations of the Undercurrent were made during the International Indian Ocean Expedition (1959–65) when, at the end of the North-East Monsoon it was seen to be an ocean-wide feature; but this may not have been typical. In general, in the Indian Ocean the Undercurrent seems to be a stronger, more long-lived flow in the western part of the ocean than in the central or eastern parts.

Figure 5.10(a) and (b) showed the averaged surface current flow during the two monsoon seasons. *Between* the monsoons, there are westerly winds over the central equatorial ocean, and these drive an eastward surface jet along the Equator. In the relatively short inter-monsoon period, there is also an increase in the large number of cyclonic and anticyclonic eddies, which range in size from 100 to 1000 km across.

The complexity of the surface circulation of the Indian Ocean—which contrasts with the relative simplicity of the gyral systems of the Pacific and Atlantic Oceans—is a result of the frequency and rapidity with which the overlying wind system changes. Wind speeds and directions change so fast that there is not always time for the upper ocean to adjust so that it is in equilibrium with the wind—the ocean's response time, or its 'memory', is many times longer than that of the atmosphere. Indeed, one of the most interesting questions that can be asked about the ocean circulation is: How is it possible for the ocean to react as fast as it does? This question will be addressed in the next Section.

5.3 THE ROLE OF LONG WAVES IN OCEAN CIRCULATION

We have seen that the stress of the Trade Winds—particularly the South-East Trades—across the equatorial oceans causes the sea-surface to slope up towards the west, and the thermocline to slope down (Figures 5.2 and 5.3). The question we now wish to consider is: how fast do the slopes of the sea-surface and thermocline respond to seasonal changes in the wind?

In the Atlantic, the South-East Trades are at their weakest during March–April and at their strongest during August–September. Figure 5.11 shows how the slopes of the sea-surface and thermocline across the Atlantic basin vary over the course of the year.

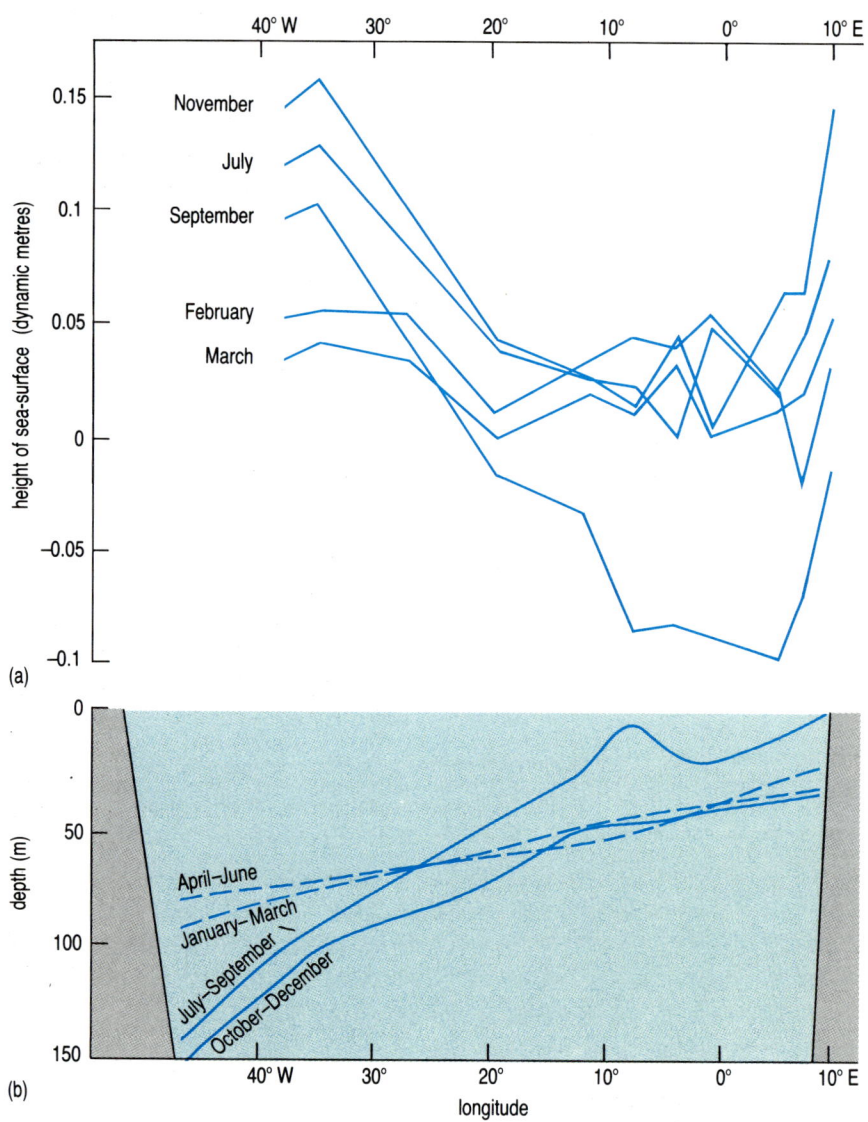

Figure 5.11 (a) Seasonal variations in the height of the sea-surface across the equatorial Atlantic. (The curves appear jagged because they are based on mean values at particular longitudes.)

(b) Seasonal variations in the depth of the 23 °C isotherm across the equatorial Atlantic.

Do the minimum and maximum slopes of the sea-surface and thermocline correspond to the periods of minimum and maximum wind stress, respectively?

Yes, they do, as far as the monthly mean values on Figure 5.11 allow us to tell. The important point about this is not the correlation itself but the fact that it indicates that the upper waters of the Atlantic Ocean *respond very quickly* to changes in the overlying wind field.

This fast response time cannot be explained simply in terms of water being transported across the equatorial Atlantic and 'piling up' at the western boundary: the upper ocean as a whole must somehow adjust to the overlying wind and to the fact that there is a boundary along the western side of the ocean. In other words, the ocean in the central and eastern parts of the basin must in some sense 'know about' or 'have felt' the western boundary. The way in which the mid-ocean receives information about the existence of a boundary is through perturbations or disturbances that travel through the ocean as pulses or waves. This is analogous to the way in which *we* receive information concerning the world about us—through light waves or sound waves.

These wave-like perturbations not only enable water in mid-ocean to respond to the existence of coastal boundaries, they also transmit the effects of changes in the overlying wind field from one region of ocean to another, much faster than would be possible simply through transportation of water in wind-driven currents. It is believed that the speed of the reversal of the Somali Current is one example of such an effect: as described in Section 5.2.2, the surface waters off Somalia flow fairly fast south-westwards during the North-East Monsoon, but can nevertheless be moving *very* fast in the opposite direction only a few months later. This dramatic 'switch' is thought to be possible because the upper ocean on the western side of the ocean 'feels' some effect of winds blowing in the central region, as well as being directly driven by local winds.

A number of different types of waves are found in the ocean. We are all familiar with the wind waves which occur at the surface of the ocean, in which particles of water are displaced from their 'normal' or equilibrium position in a vertical direction and return under the influence of gravity. The types of waves that are of interest so far as ocean circulation is concerned have wavelengths which may be anything from tens to thousands of kilometres, periods ranging from days up to months, or even years, and vertical displacements which vary from centimetres up to tens of metres. These vertical displacements may be more or less constant with depth (Figure 5.12(a)); alternatively, they may be greatest where there is a strong vertical density gradient—in the thermocline, for example (Figure 5.12(b)). In the first case, the vertical density distribution is not affected by the passage of the wave, and so if isobaric and isopycnic surfaces are initially parallel to one another, they will remain so (Figure 5.12(a)). For this reason, such waves are sometimes referred to as 'barotropic waves'; we will refer to them as 'surface waves' although it should be remembered that motion associated with them extends to significant depths, because their wavelengths are much greater than the depth of the ocean. In the second case, the vertical density distribution *is* affected by the passage of the waves, so that density surfaces are caused to intersect isobaric surfaces (Figure 5.12(b)); such waves are therefore referred to as 'baroclinic' waves. Baroclinic waves generally have much larger amplitudes than surface (barotropic) waves.

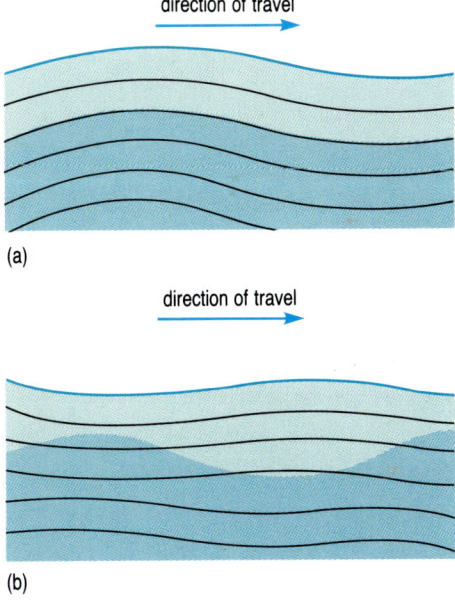

Figure 5.12 Examples of (a) a 'surface' long wave and (b) a long wave in the thermocline. In (a), the surface ocean as a whole moves up and down, and isobaric and isopycnic surfaces remain parallel. Such waves are therefore described as 'barotropic'. In (b), the passage of the wave changes the vertical density distribution, so that isopycnic surfaces are alternately compressed and separated. In addition, there are pressure variations over the surface of the density interface so that isobaric and isopycnic surfaces intersect; such waves are therefore described as 'baroclinic'.

Because these disturbances have very long wavelengths and periods, they are significantly affected by the Coriolis force. As a result, motion occurs in a *horizontal* as well as a vertical direction. This may be seen most clearly in the flow patterns associated with **Kelvin waves** and planetary or **Rossby waves**, the two classes of waves that are most important so far as ocean circulation is concerned.

5.3.1 OCEANIC WAVE GUIDES AND KELVIN WAVES

A general feature of all wave motions, through water or any other medium, is that where the physical characteristics of the medium change with position, waves seeking to cross the line of change may be reflected, or in some other way deflected, so that they become trapped within a **wave guide**. Common examples of wave guides employed in communications technology are optical fibres and coaxial cables, both of which are used to carry information along a specified path. An example of a wave guide in the ocean is the sound channel, a depth zone over which the velocity of sound in seawater is relatively low, and within which sound waves may be trapped by refraction. Sounds emitted in the sound channel—by, for example, Sofar floats—are transmitted over distances of thousands of kilometres with relatively little attenuation.

Now imagine a parcel of water in the Northern Hemisphere, moving northwards with a coastal boundary on its right. The Coriolis force acts to deflect the parcel to the right, but beyond a limited extent this is impossible because the coastal boundary is in the way. Water piles up against the boundary, giving rise to an offshore horizontal pressure gradient force; this keeps the parcel of water moving parallel to the coast, in a geostrophic current. Consequently, a coastal boundary constrains the way in which water can move in response to the forces acting on it. As a result, coasts may act as wave guides to the perturbations known as Kelvin waves (Figure 5.13(a)), which can travel as surface (barotropic) waves or as baroclinic waves.

In a Kelvin wave, perturbations of the sea-surface or of the thermocline propagate parallel to and close to the coast, as though unaffected by the Earth's rotation, because the Coriolis force directed towards the coast is opposed by a horizontal pressure gradient force that results from the slope of the sea-surface. Thus, a necessary condition for the propagation of Kelvin waves is that the horizontal pressure gradient force and Coriolis force act in opposition.

QUESTION 5.7 Given that necessary condition, is it possible for a Kelvin wave in the Northern Hemisphere to propagate with the coastal boundary on the left? Or for a Kelvin wave in the Southern Hemisphere to propagate with the coastal boundary on the right?

Kelvin waves are similar to surface wind waves in that the principal maintaining force is gravity; but particle movement within the wave is such that the amplitude of the vertical displacement is greatest at the coast and decreases exponentially away from it, so that at any point and any time the Coriolis force balances the pressure gradient due to the slope of the sea-surface or thermocline (Figure 5.13(b)).

A Kelvin wave may be regarded as being 'trapped' within a certain distance of the coast, because outside that distance its amplitude has

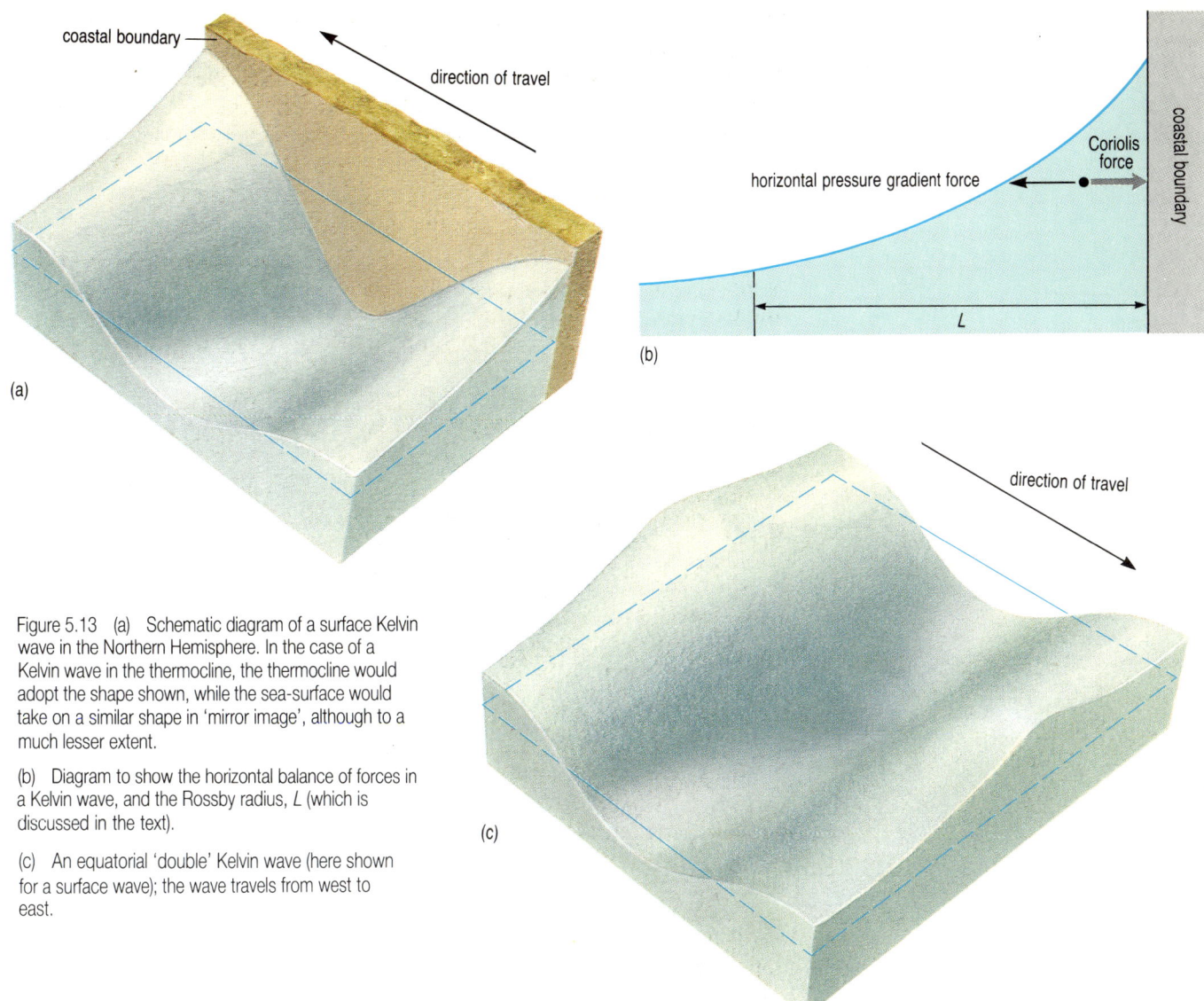

Figure 5.13 (a) Schematic diagram of a surface Kelvin wave in the Northern Hemisphere. In the case of a Kelvin wave in the thermocline, the thermocline would adopt the shape shown, while the sea-surface would take on a similar shape in 'mirror image', although to a much lesser extent.

(b) Diagram to show the horizontal balance of forces in a Kelvin wave, and the Rossby radius, L (which is discussed in the text).

(c) An equatorial 'double' Kelvin wave (here shown for a surface wave); the wave travels from west to east.

decayed away so as to be hardly discernible. This distance is known as the **Rossby radius of deformation** (L) and can be calculated from $L = c/f$ where f is the Coriolis parameter and c is the wave speed. For a surface Kelvin wave, c is of the order of 200 m s^{-1}; for a Kelvin wave with its maximum displacement in the thermocline, c is typically between 0.5 and 3 m s^{-1}. In mid-latitudes, the Rossby radius for a Kelvin wave in the thermocline is generally about 25 km.

QUESTION 5.8 What is the Rossby radius L for a Kelvin wave in the thermocline at 10°N, if its speed is 2 m s^{-1}? (You will need to use the fact that $\Omega = 7.29 \times 10^{-5}$ s^{-1}.)

Thus, in low latitudes, coastal Kelvin waves are not as closely trapped to the coast as they are in mid-latitudes.

The Rossby radius of deformation is not simply a measure of the degree to which Kelvin waves are trapped. More generally, it is the distance that a wave with speed c can travel in time $1/f$, and it therefore provides a guide to the distance that a wave (i.e. a disturbance) can travel before being significantly affected by the Coriolis force. As f is zero at the Equator and a maximum at the poles, so Rossby radii decrease from infinity at the Equator to a minimum at the poles. The tendency for disturbances in current patterns to take on a curved or gyral character therefore *increases* with increasing distance from the Equator.

Almost everybody has experienced the effects of a Kelvin wave at first hand. The twice-daily rise and fall of sea-level corresponding to high and low water occurs in the form of coastal Kelvin waves, which progress anticlockwise round ocean basins (i.e. with the coast on the right) in the **amphidromic systems** of the Northern Hemisphere, and clockwise round basins in the Southern Hemisphere.

We have seen that coastal Kelvin waves may travel along coastal boundaries because the Coriolis force cannot play the same part in the balance of forces as it usually does. Along the Equator, the Coriolis force actually *is* zero and a similar effect arises, leading to the existence of an **equatorial wave guide**. An equatorial Kelvin wave is like two parallel coastal Kelvin waves (one in each hemisphere) joining at the Equator, which they 'feel' as a boundary (Figure 5.13(c)). Like coastal Kelvin waves, equatorial Kelvin waves propagate with the 'boundary' on the right in the Northern Hemisphere and on the left in the Southern Hemisphere. As a result, Kelvin waves may only propagate *eastwards* along the equatorial wave guide.

Surface equatorial Kelvin waves travel very fast, at about $200\,\text{m s}^{-1}$. Their Rossby radius of deformation can be shown to be about 2000 km, and so they can hardly be regarded as 'trapped' at all. However, this is not the case for Kelvin waves in the thermocline. They are much slower, with c typically between 0.5 and $3.0\,\text{m s}^{-1}$, and have Rossby radii of 100 – 250 km. They may be detectable at the surface, as sea-level is slightly raised above regions where the thermocline is depressed, and slightly depressed above regions where the thermocline is raised.

Kelvin waves in the thermocline can have dramatic effects, particularly in low latitudes where the mixed surface layer is thin. They may be generated by an abrupt change in the overlying wind field, as occurs for instance in the western Atlantic when the ITCZ moves northwards over the region and it comes under the influence of the South-East Trades. This causes a disturbance in the upper ocean (*cf.* Figure 5.14(a)), which travels eastwards along the equatorial wave guide as a double Kelvin wave (this takes about 4–6 weeks) and, on reaching the coast, splits into two coastal Kelvin waves, each travelling away from the Equator (Figure 5.14(b)). In

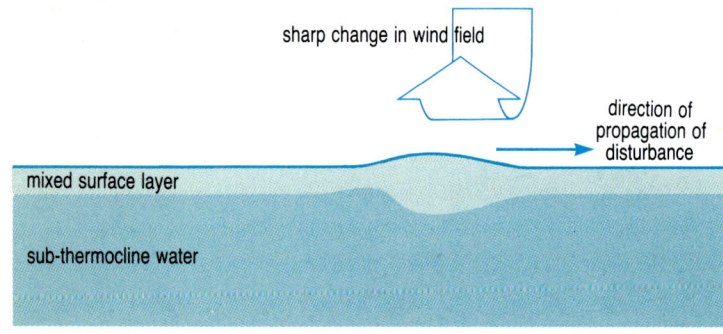
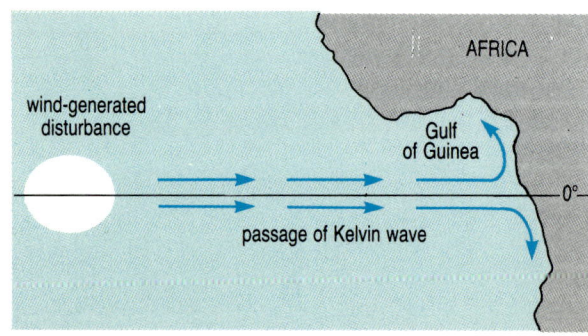

Figure 5.14 (a) Schematic diagram showing a disturbance of the upper ocean caused by an abrupt change in the overlying wind field.

(b) Such a disturbance may be generated in the western Atlantic and travel eastwards as an equatorial Kelvin wave; at the eastern boundary, this splits into two coastal Kelvin waves, which cause seasonal upwelling in the Gulf of Guinea (*cf.* Figure 5.7).

the region of the disturbance where the thermocline bulges upward, colder, deeper water comes nearer to the surface. By the time the wave reaches the coasts of Ghana and Ivory Coast (north of the Equator) and Gabon (south of the Equator), this *upwelling* of cold, sub-thermocline water is detectable at the surface; hence the regions of 'seasonal coastal upwelling' off Abidjan and south of Cape Lopez, shown in light-green on Figure 5.7.

Why is this 'pulse' not detected earlier, as it moves across the western and central equatorial Atlantic?

As discussed in Section 5.1.1, the equatorial thermocline slopes up from the west and is nearest to the surface at the eastern coast. The passage of the Kelvin wave only brings cold, sub-thermocline water to the surface where the depth of the thermocline is already fairly shallow.

Kelvin waves occur in the atmosphere as well as the ocean, as do Rossby or planetary waves.

5.3.2 ROSSBY WAVES

Rossby or planetary waves propagate zonally, along the Equator, but also along other lines of latitude. They arise from the need for potential vorticity to be conserved (Section 4.2.1). Imagine a parcel of water at some given latitude ϕ in the Northern Hemisphere and assume that, initially, it has no relative vorticity—i.e. no rotational movement in relation to the Earth, and no current shear. If the parcel of water is displaced polewards, it has entered a region of higher positive planetary vorticity, f (Figure 5.15(a)(i)). Its potential vorticity $(f + \zeta)/D$ must remain constant, and so to compensate for its gain in f the water parcel will acquire negative relative vorticity ζ and will tend to rotate clockwise, causing adjacent particles to move clockwise around it (Figure 5.15(a)(i)). If it moves southwards and overshoots its original latitude, its loss in planetary vorticity will be compensated for by a gain in positive relative vorticity, and it will circulate anticlockwise (Figure 5.15(a)(ii)). A water parcel may swing back and forth about its original latitude, alternately gaining planetary vorticity while losing relative vorticity, and losing planetary vorticity while gaining relative vorticity.

Now imagine a row of such water parcels in a current or airstream flowing along a line of latitude. If the flow is displaced polewards or equatorwards, horizontal oscillations of the kind described above may occur, so that the flow undulates about the original line of latitude (Figure 5.15(b)); these undulations are Rossby or planetary waves. In the ocean, the scale of such undulations is of the order of hundreds of kilometres; in the atmosphere, it varies from about 5000 to 20000 km.

From your reading of Chapter 2, can you give an example of Rossby waves in the atmosphere?

The large-scale undulations in the jet stream of the upper westerlies, shown in Figure 2.7, are atmospheric Rossby waves.

In both the atmosphere and the ocean, the overall effect of the clockwise and anticlockwise rotations associated with Rossby waves is to cause the wave-*form*—i.e. the undulations—to move *westwards* relative to the flow.

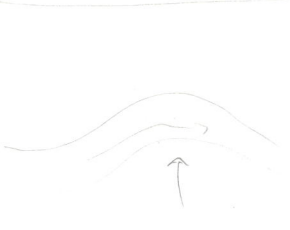

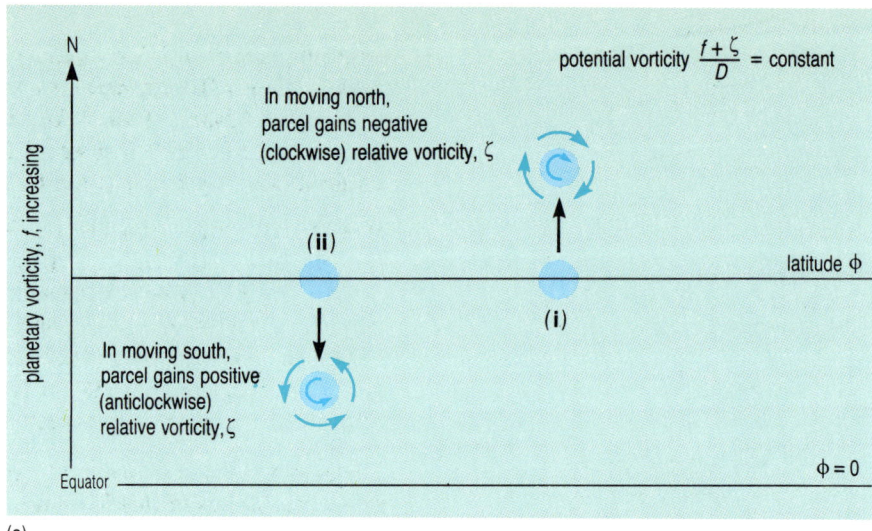

Figure 5.15 (a) Diagram to show how in a Rossby wave the need to conserve potential vorticity $(f + \zeta)/D$ leads to a parcel of water oscillating about a line of latitude ϕ while alternately gaining and losing relative vorticity ζ. For details, see text.

(b) The path taken by a current or airstream affected by a Rossby wave. Note that the flow pattern is characterized by anticyclonic and cyclonic eddies, and that the wave-form moves *westwards* relative to the current or airstream

This is true even if the flow is itself moving eastwards, as is the case with the jet stream in the upper westerlies. However, flow velocities in the atmosphere may reach 100 m s^{-1} and so Rossby waves in an airstream may move eastward *relative to the Earth*, while still moving westward relative to the airstream. If the eastward motion of air in the airstream is approximately equal to the westward motion of the wave-form, stationary Rossby waves result. In the ocean, flow velocities rarely reach one metre per second and so even in eastward-flowing currents, Rossby waves nearly always move westward relative to the Earth.

The way in which Kelvin and Rossby waves affect ocean circulation depends on the latitude. At middle and high latitudes, information about a change in the wind stress propagates mainly westwards, by means of Rossby waves, so the ocean near western boundaries is affected by events in mid-ocean to a much greater extent than the ocean near eastern boundaries. By contrast, at low latitudes information can travel westwards by Rossby waves *or* eastwards by Kelvin waves in the equatorial wave guide.

In addition, because of the equatorial wave guide, the upper ocean in low latitudes can respond to changing winds much *faster* than is possible away from the Equator. This is partly because the equatorial wave guide supports both Rossby *and* Kelvin waves, and partly because Rossby waves travel fastest there. For example, a Rossby wave can take as little as three months to travel west across the equatorial Pacific, whereas it could take ten years to cross the Pacific at 30°N or 30°S.

It is the equatorial wave guide that enables the Indian Ocean to respond so quickly to changes in the wind direction. When the winds over the Indian Ocean change from westerlies to north/north-easterlies, the upper ocean can 'rearrange itself', so that the sea-surface slopes up to the west instead of the east, in less than a week. Away from the equatorial wave guide, Rossby waves alone would take up to five years to establish such a balance across a basin the size of the Indian Ocean.

It would not be appropriate to go further into the details of either Rossby or Kelvin waves in this Volume; however, one of the most intriguing aspects of these waves is that equatorial Kelvin waves reaching an eastern boundary, in addition to splitting and travelling poleward along the coast, may be partially reflected as a Rossby wave. This can be seen in the computer-generated diagrams shown in Figure 5.16.

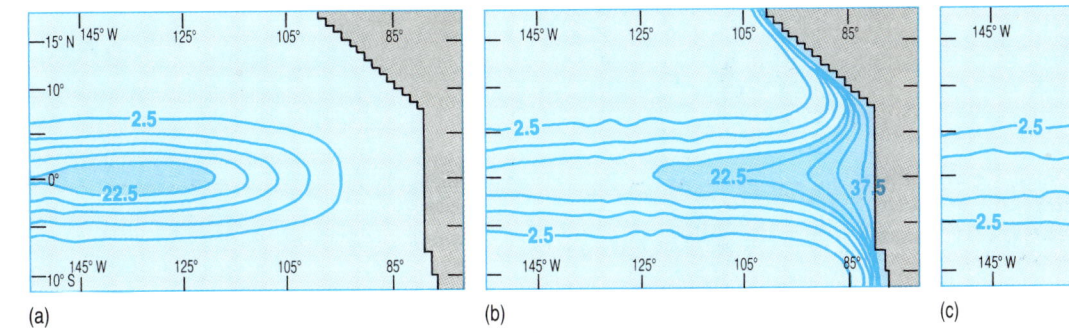
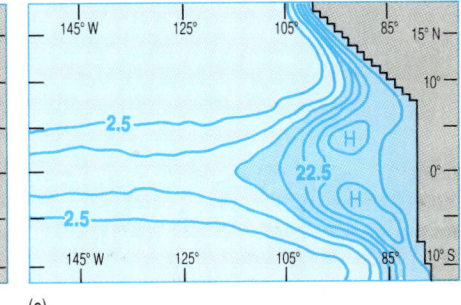

Figure 5.16 Computer-generated diagrams showing the progress, from mid-Pacific to the South American coast, of an internal equatorial Kelvin wave. The contour numbers may be regarded as either the depression of the thermocline in metres or the accompanying rise in sea-level in centimetres. The diagrams show the situation at successive monthly intervals. In (c), the equatorial Kelvin wave has split into two poleward-travelling coastal Kelvin waves. Note that the coastal boundary has the effect of considerably amplifying the disturbance. The equatorial Kelvin wave has also just been partially reflected as an equatorial Rossby wave, as can be seen by the circular contours which result from the rotatory motion associated with the wave. (Note that because the two eddies are on *either side of the Equator*, both are anticyclonic and lead to topographic highs (H), although the northerly one is clockwise and the southerly one anticlockwise (*cf.* Figure 5.15 (b).)

5.4 EL NIÑO

An El Niño event is a climatic fluctuation, centred in the Pacific, that occurs every 2–10 years. The most obvious sign that an El Niño event is underway is the appearance of unusually warm water off the coast of Ecuador and Peru. This generally occurs within a few months of Christmas, and the name El Niño—meaning 'the Christ Child'—was originally simply the local name for the seasonal increase in the temperature of coastal waters that occurs around Christmas time.

El Niño events are perturbations of the *ocean-atmosphere system*. It is not known whether the perturbations originate in the atmosphere or the ocean, but for convenience we will start by considering what happens in the atmosphere during an El Niño event.

The prevailing winds over the equatorial Pacific are the South-East Trades. Their strength depends on the difference in surface atmospheric pressure between the subtropical high pressure region in the eastern South Pacific where cool, dry air converges and subsides—and the low pressure region over Indonesia—where warm, moist air rises, producing cumulonimbus clouds and heavy rainfall (Figure 5.17). During an El Niño event, the Indonesian Low has anomalously high pressure (i.e. is a weaker low than usual) and moves eastwards into the central Pacific, while the South Pacific High becomes anomalously low. The South-East Trades weaken, often becoming westerly in the western Pacific.

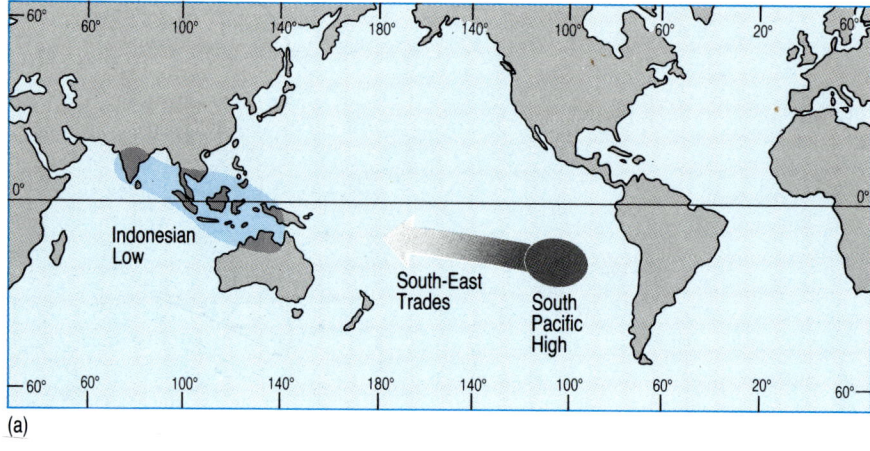

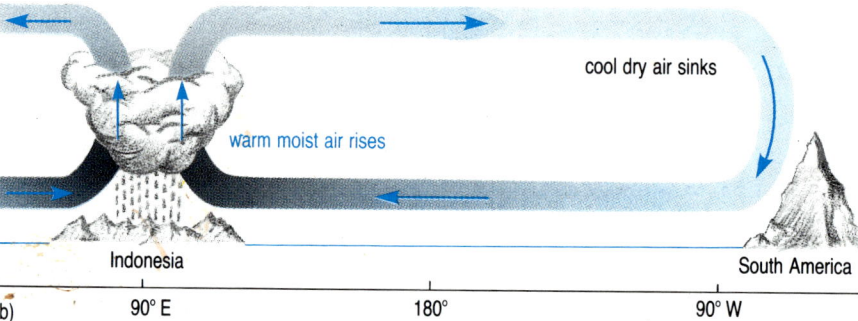

Figure 5.17 (a) Schematic map showing the positions of the Indonesian Low and the South Pacific High. (This map should be studied in conjunction with Figure 2.3, showing the global wind pattern.)

(b) The zonal atmospheric circulation between the Indonesian Low and the South Pacific High.

By reference to Figure 5.3, can you suggest what effect a relaxation of the South-East Trades will have on the upper waters of the equatorial Pacific?

The sea-surface slope will 'collapse', so that both it and the thermocline become near-horizontal, enabling a considerable volume of warm mixed-layer water to move eastwards across the ocean (Figure 5.18). In the western Pacific, the collapse in the Trade Winds occurs abruptly, and so the resulting change in the upper ocean—a depression in the thermocline accompanied by a slight rise in sea-level (*cf.* Figure 5.14(a))—propagates eastwards along the Equator as a pulse, or series of pulses, of Kelvin waves. At the eastern boundary, the equatorial Kelvin waves split into northward- and southward-travelling coastal Kelvin waves (*cf.* Figure 5.14(b)), as well as being partially reflected as Rossby waves (*cf.* Figure 5.16).

Figure 5.19 summarizes the main differences between normal conditions in the Pacific basin, and conditions during an El Niño event. As shown in Figure 5.19(a), the highest sea-surface temperatures, of 28–29°C, are normally found in the western ocean; during an El Niño event, this area

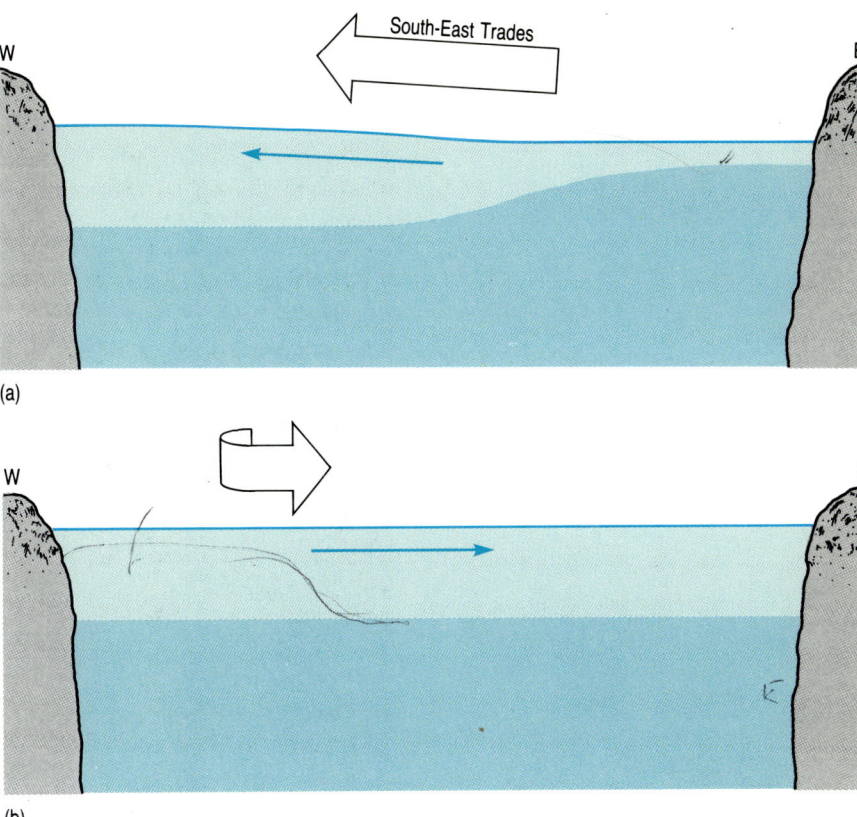

Figure 5.18 Cross-section along the Equator in the Pacific (a) in a normal year, and (b) during an El Niño event.

of exceptionally high sea-surface temperature moves into the central ocean. As mentioned above, the vigorous convection of moist air associated with the Indonesian Low also moves eastwards into the central ocean during an El Niño.

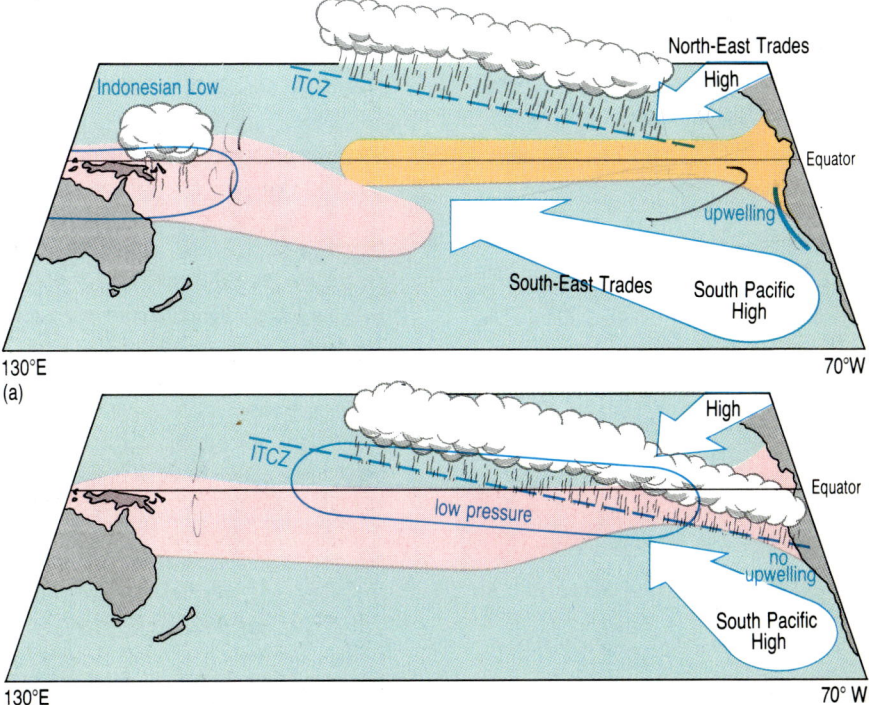

Figure 5.19 Schematic diagrams showing conditions in the Pacific (a) in a normal year, and (b) during an El Niño event. Note that although no El Niño can be labelled 'typical', the features shown in the diagram seem to occur in most El Niños. The red tone indicates regions where the sea-surface temperature is higher than about 28 °C; the orange tone in (a) indicates regions that are normally dry.

> In what way could this be the *result* of an eastward movement of exceptionally warm surface water, rather than the indirect cause of it?

A warm sea-surface leads to increased upward convection of moist air; as discussed in Section 2.3.1, the increase in convection is particularly marked when sea-surface temperatures exceed 28–29°C. The eastward movement across the Pacific of the Indonesian Low and the exceptionally warm surface water are therefore intimately linked. Thus, although the general eastward movement of warm mixed-layer water may be explained in terms of the slackening of the Trade Winds in response to a change in the strength and position of the Indonesian Low, this is clearly only part of the story. How and why El Niño events develop will only become clearer when the feedback loops between atmosphere and ocean are better understood.

It is important that El Niño events *are* better understood, because only then can they be accurately *predicted*. The anomalous climatic conditions associated with these events cause havoc over a large part of the globe. The best-known natural calamity characteristic of an El Niño event is probably the collapse of the anchoveta fishery off Peru, but its effects are considerably more wide-ranging than this. Regions of the Pacific basin that are normally wet experience drought, and regions that are normally arid experience destructive torrential rain. Cyclones are more frequent and occur eastward of their normal limit (Section 2.3.1); during the El Niño event of 1982–83, French Polynesia was struck by five cyclones although it is normally unaffected by them. Furthermore, there are indications that anomalous climatic conditions outside the Pacific basin—extremely cold winters in North America and Eurasia, for example—are also in some way linked with El Niño events.

5.5 CIRCULATION IN HIGH LATITUDES

Between about 50° and 70° of latitude, surface winds have a cyclonic tendency, associated with the subpolar low pressure regions (Figure 2.2). In northern high latitudes, where current flow is modified by the presence of land masses, cyclonic surface gyres result. The **subpolar gyre** in the North Pacific consists of the Oyashio and Alaska Current; gyral flow in the North Atlantic is disrupted by the Greenland land mass (Figure 3.1). In southern high latitudes, there are no land masses to impede flow and the westerly winds drive the (eastward-flowing) West Wind Drift or **Antarctic Circumpolar Current**, while the easterly winds blowing off the ice-covered continent drive the Antarctic Polar Current which flows west in a narrow zone around most of Antarctica.

5.5.1 THE ARCTIC SEA

The Arctic Sea is largely surrounded by continental masses; it is linked to the Pacific via the Bering Straits—which are only about 45 m deep—and to the Atlantic mainly via the Norwegian and Greenland Seas (Figure 5.20). Deep Atlantic water is prevented from entering these seas, and hence from entering the Arctic Sea, by shallow submarine plateaus which extend from Greenland to Scotland. The significance of this for the deep circulation of the ocean will become clear in Chapter 6.

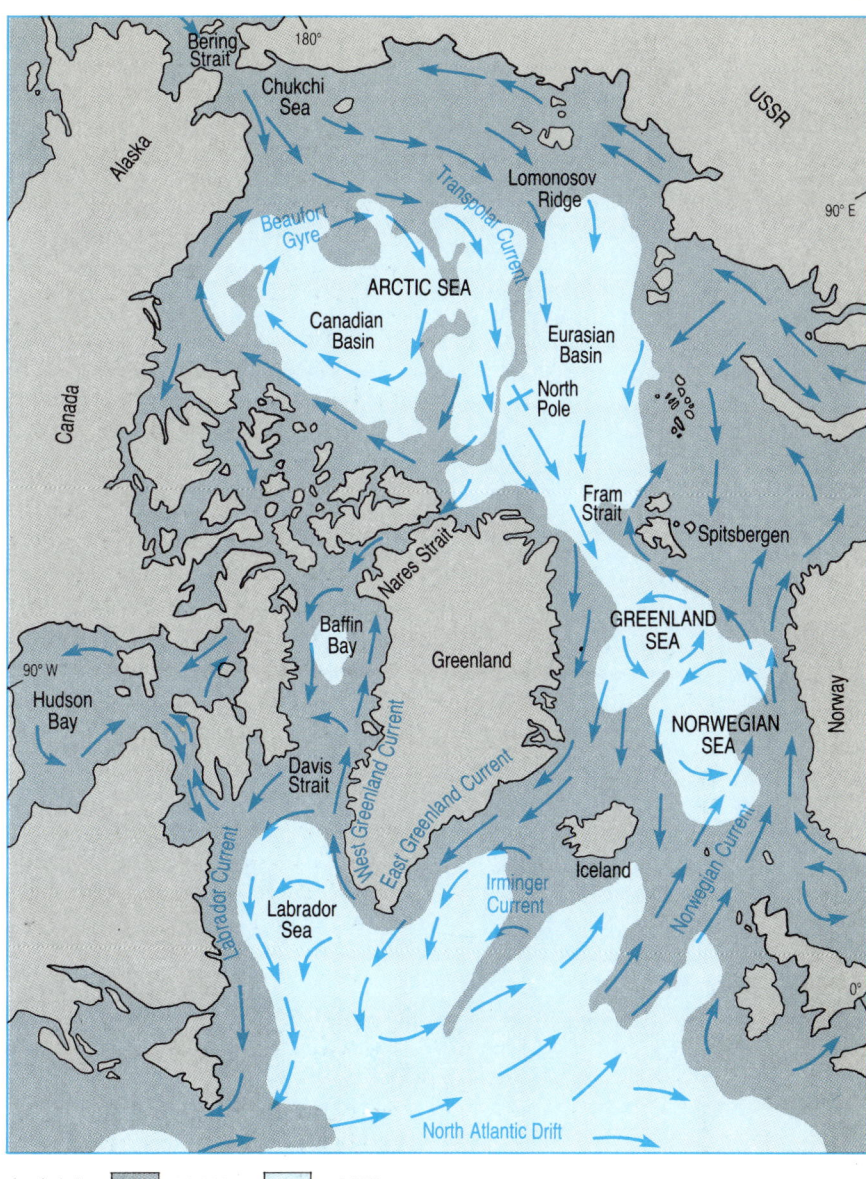

Figure 5.20 The bathymetry and surface currents of the Arctic Sea and the adjacent seas of the North Atlantic.

The enclosed nature of the Arctic region also greatly affects the nature of its ice cover. Figure 5.21 shows the seasonal variation in the ice cover of both the Arctic and the Antarctic.

Which of the two regions shows the lesser variability in ice cover?

The Arctic. Much of the Arctic ice remains 'locked' in the Sea all year round; only about 10% leaves annually, via the Fram Straits between Greenland and Spitsbergen (see Figure 5.20). The central part of the Arctic Sea is permanently covered with ice (Figure 5.21(a)), and most Arctic pack ice is several years old. This situation contrasts markedly with that in the Southern Ocean where the limits of ice cover shift over about 20° of latitude during the course of a year (Figure 5.21(c) and (d)), and most of the ice cover is re-formed annually.

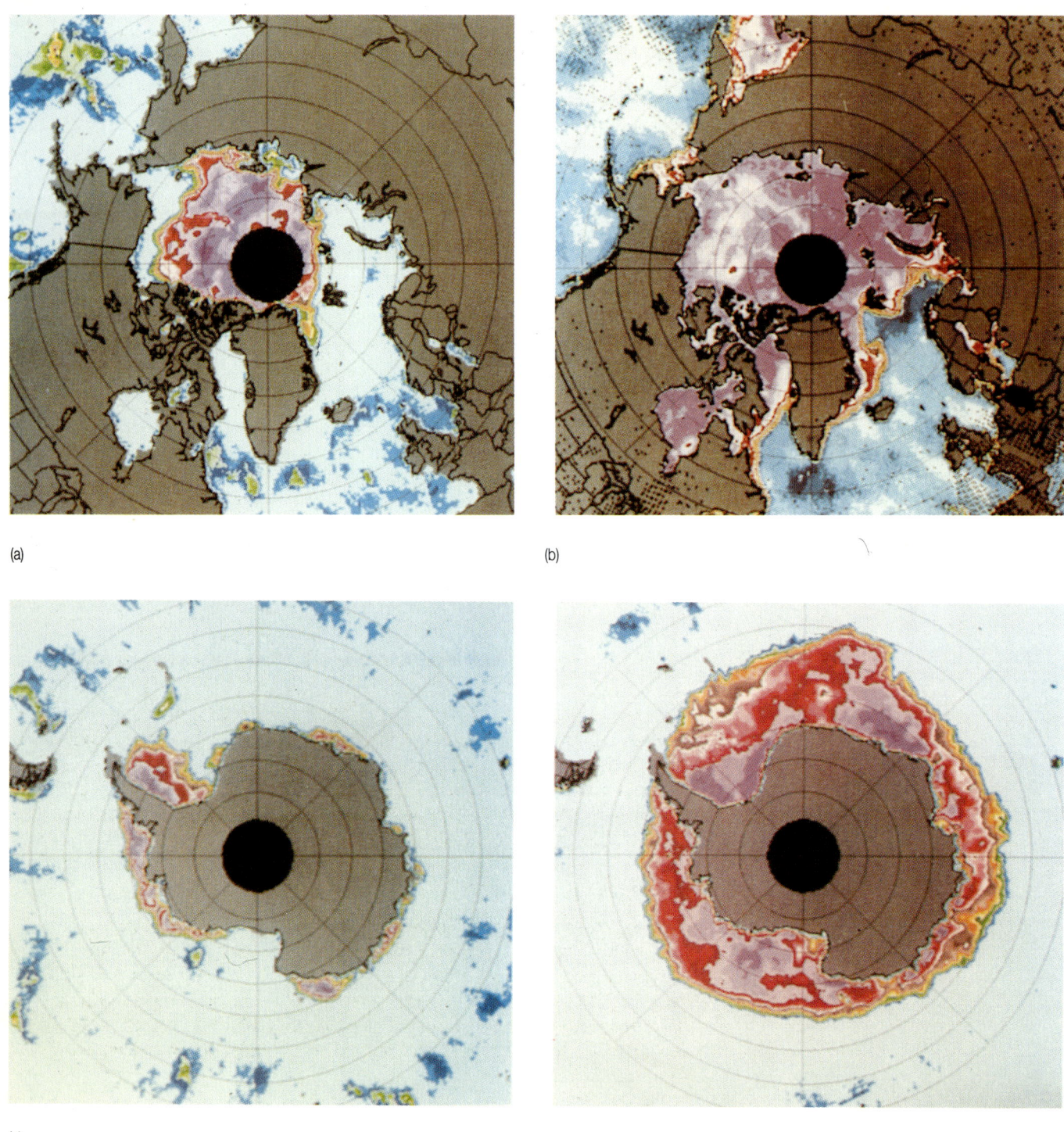

Figure 5.21 Seasonal changes in ice cover in northern and southern high latitudes, as determined using microwave measurements obtained from the *Nimbus* satellite programme during 1978–79. The different colours represent the percentage of sea-surface covered by ice: a pinky-red tone indicates 100% coverage, while a light blue tone represents a coverage of 20% or less.

(a) and (b) Near-minimum and near-maximum ice cover in the Arctic region (images are for 1–7 September and 3–7 February, 1979, respectively).

(c) and (d) Near-minimum and near-maximum ice cover in the Antarctic region (images are for 1–7 February, 1979, and November 1978, respectively).

Even in those parts of the Arctic Sea where the ice cover is permanent, the pack ice does not form a solid mass. Under the influence of winds and currents it continually cracks and shifts, so that layers of ice raft over one another; in addition, pressure ridges form, locally increasing the ice thickness from 3–4m to 40–50m.

If so much of the Arctic Sea is permanently covered by a thick layer of ice, how has the general surface circulation shown in Figure 5.20 been determined?

The general circulation in the upper layers of the Arctic Sea was initially deduced from the average motion of the ice, as revealed by the movement of ice-bound ships and camps on the ice; more recently, buoys fixed in the ice have been tracked by satellite. There are difficulties in separating pack-ice motion caused by local winds from that related to general current patterns, but it seems that the shorter-period fluctuations are largely related to local winds, and longer-period motion to the surface circulation. In addition to direct current measurements, geostrophic calculations have been made using temperature and salinity data collected from oceanographic stations based on ships or on ice-islands and ice-floes.

The circulation that has been deduced is a clockwise gyre centred over the Canadian Basin, with the main surface outflow being the East Greenland Current (Figure 5.20). This current carries southwards not only the pack ice but also icebergs which have calved from glaciers reaching the east coast of Greenland. Off the southern tip of Greenland, the East Greenland Current converges with the warm Irminger Current and most of the ice melts. Some, however, may be carried around to the west coast of Greenland where it is supplemented by large numbers of icebergs from glaciers reaching the west Greenland coast. The ice circulates in Baffin Bay and the Labrador Sea and eventually travels southwards in the Labrador Current. Off the Newfoundland Grand Banks, the Labrador Current converges with the Gulf Stream (Figure 4.26) and even the largest icebergs gradually break up and melt.

5.5.2 THE SOUTHERN OCEAN

The more or less zonal distribution of the ice cover around Antarctica, shown in Figure 5.21(d), reflects the predominantly zonal current flow. As shown in Figure 5.22, there are cyclonic subpolar gyres in the Weddell and Ross Seas but the predominant feature of the circulation of this region is the Antarctic Circumpolar Current—the only current to flow around the globe without encountering any continuous land barrier.

Given that the overlying winds are essentially westerly, in which direction would you expect the sea-surface to slope in the region of the Antarctic Circumpolar Current?

As this is in the Southern Hemisphere, Ekman transport is to the left of the wind, and the sea-surface slopes down towards the Antarctic continent. This sea-surface slope generates a geostrophic slope current to the east, i.e. flow in the same direction as the wind but extending to greater depths than the surface wind-driven layer (*cf.* Figure 3.24 for a similar situation in the subtropical gyres). Below the wind-driven layer,

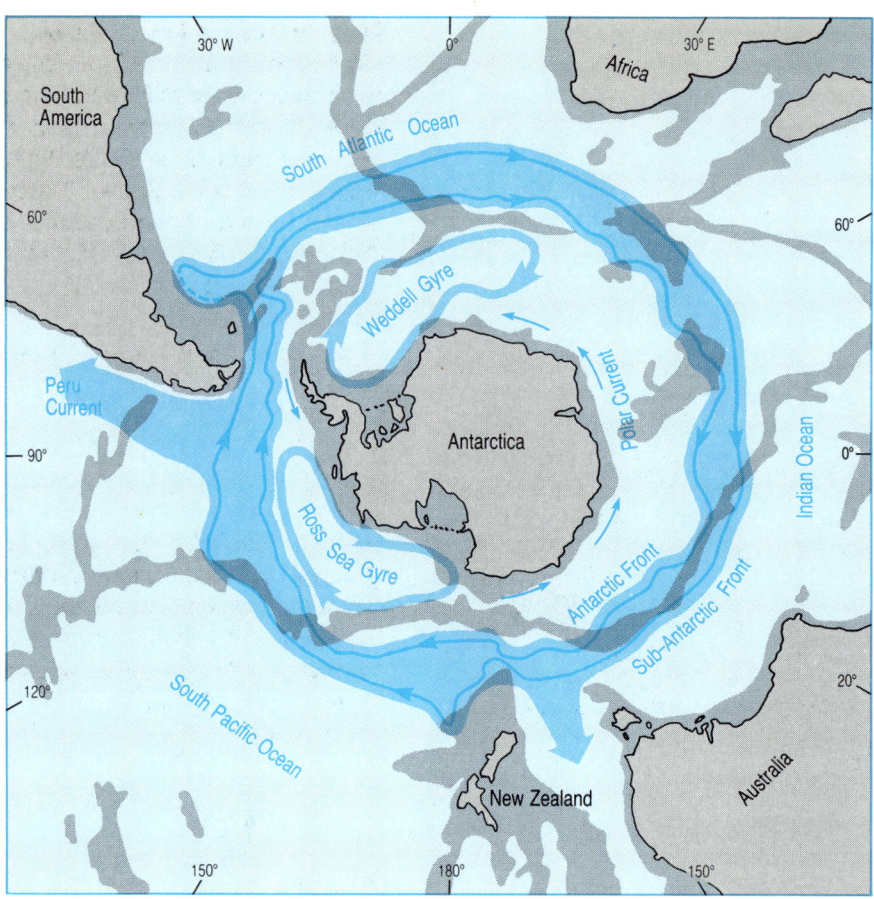

Figure 5.22 Schematic map showing the circulation in the Southern Ocean. The path of the Antarctic Circumpolar Current is shown by the blue tone; the two dark blue lines represent the average positions of the Antarctic Front and the Sub-Antarctic Front which are discussed in the text. Note that a significant part of the current branches northward and flows up the west coast of South America as the Peru Current; there is also a branch of the current flowing northwards below the surface between Australia and New Zealand. The approximate positions of the gyres in the Weddell Sea and Ross Sea are also shown, as is the path of the Polar Current. The Antarctic Divergence is between the Polar Current and the Antarctic Circumpolar Current. Blue-grey shading indicates water depths less than 3 000 m.

the density distribution is such that, in general, the horizontal pressure gradient force and the Coriolis force balance and geostrophic equilibrium is maintained.

In surface layers, the direct effect of the wind stress, combined with the Coriolis force, leads to a northward component of flow, and a region of convergences forms within the strongest part of the Antarctic Circumpolar Current. This region of convergences was until recently thought to be a single convergence, and was named the **Antarctic Convergence**. It is now known to consist of a series of convergences, or fronts, and has been renamed the **Antarctic Polar Frontal Zone** (APFZ). The fronts in the APFZ are associated with strong zonal current jets, with velocities reaching $0.5 - 1.0 \, \text{m s}^{-1}$. Two major jets occur in association with the northern and southern boundaries of the APFZ—the Sub-Antarctic Front and the Antarctic Front (also known as the Polar Front); average positions of these jets are shown in Figure 5.22.

Despite its great length—about 24000 km—the Antarctic Circumpolar Current has remarkably similar characteristics wherever it is observed; furthermore, the Sub-Antarctic Front and the Antarctic Front persist throughout the extent of the current, although the width between them is very variable (Figure 5.22). Figure 5.23 shows temperature and salinity sections across the Antarctic Circumpolar Current in the Drake Passage, between South America and the islands that lie to the north of the Antarctic peninsula (*cf.* Figure 5.22). The positions of frontal zones are indicated by a blue tone. The southernmost front is a boundary between oceanic waters and colder, fresher water which originated in the Weddell Sea. The other two frontal zones are the Antarctic Front and the Sub-Antarctic Front.

Figure 5.23 Sections of (a) temperature (°C) and (b) salinity across the Drake Passage. The isotherms and isohalines sloping up to the south from a depth of about 3 000 m delineate a 'wedge' of warmer, less saline water flowing over colder, more saline water. The sections were made during the southern summer. The blue toned regions are frontal zones. At the surface, the sharpest changes in salinity and, particularly, temperature occur at the Antarctic Front; for this reason, the Antarctic Front was the first to be observed and hence identified as the Antarctic Convergence.

Why do the frontal zones indicated on Figure 5.23 imply locally stronger current flow (the 'jets' referred to above)?

As mentioned above, flow in the Antarctic Circumpolar Current is generally in geostrophic equilibrium. The isopycnic surfaces slope up towards the south, and the steeper their slope the greater the velocity of the eastwards geostrophic current. The frontal zones on Figure 5.23 are characterized by steeper slopes in the isotherms and isohalines, and hence in the isopycnals; they are therefore also characterized by faster geostrophic currents.

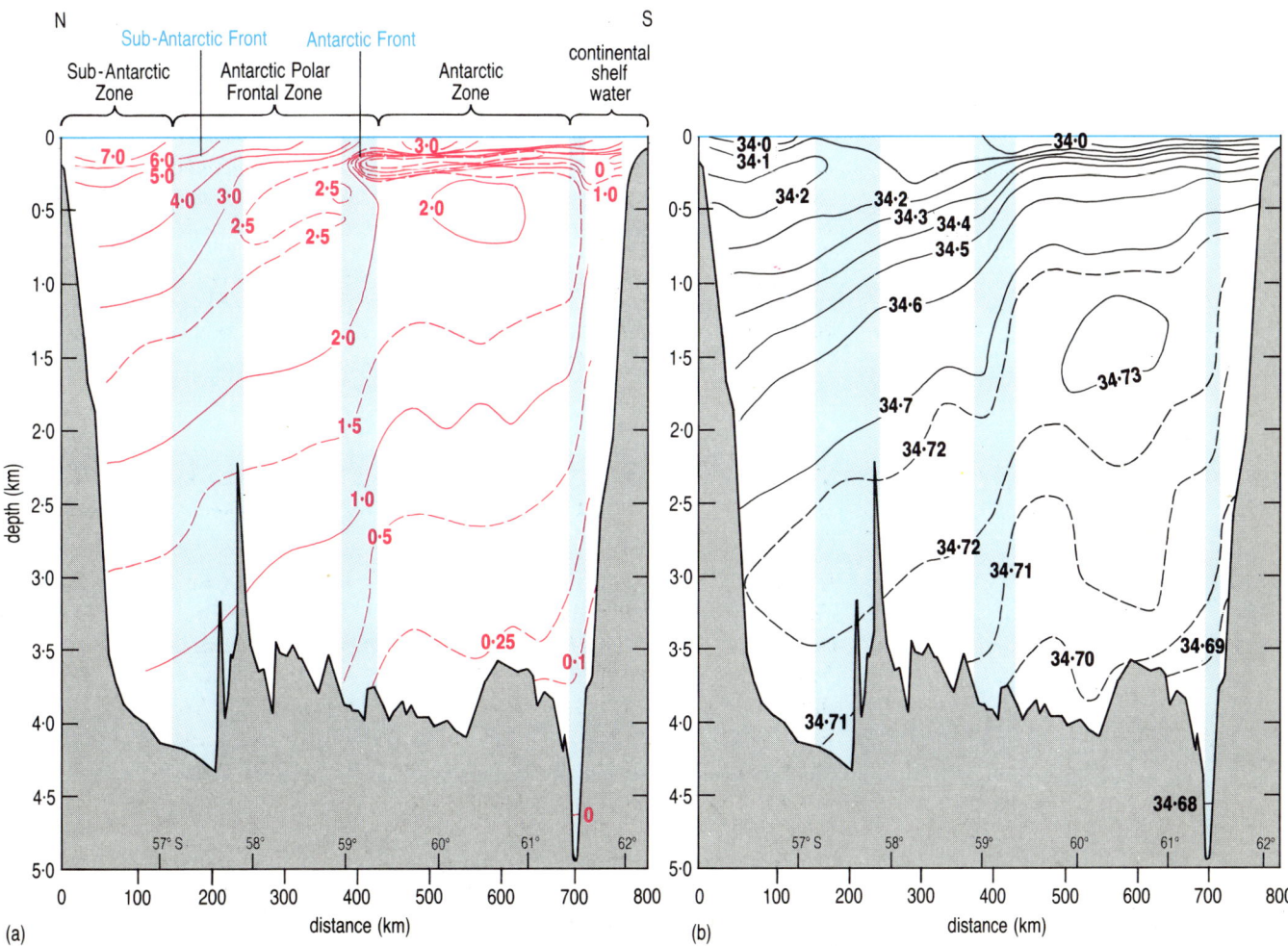

Flow within the APFZ is complex, with many eddies and meanders. The current jet associated with the Antarctic (Polar) Front has been observed to meander southwards and capture colder water, and there are numerous cold-core eddies within the frontal zone. You may remember from Section 3.5.2 that these eddies are a very important heat-transfer mechanism, enabling cold water to move meridionally across the Antarctic Circumpolar Current.

In general, the Antarctic Circumpolar Current is not particularly fast: south of the Antarctic Front, surface speeds are about $0.04\,\text{m s}^{-1}$, while in the faster region, to the north of the Antarctic Front, they may reach $0.15-0.2\,\text{m s}^{-1}$. However, the current is very deep, extending to the sea-floor at about 4000m depth, and its volume transport is therefore enormous. Estimates suggest that the average transport is about $110-140 \times 10^6\,\text{m}^3\text{s}^{-1}$, although during the course of a year it may vary from $28 \times 10^6\,\text{m}^3\text{s}^{-1}$ to $240 \times 10^6\,\text{m}^3\text{s}^{-1}$. In terms of volume transport, it is certainly the mightiest current in the oceans.

One of the puzzles confronting oceanographers is: why is the Antarctic Circumpolar Current not faster than it is, given that it travels uninterrupted around the globe under the cumulative influence of the westerlies? Part of the answer may lie in its interaction with sea-floor topography which, as indicated by Figure 5.22, may also account for some of the variability in the path of the current.

Turbulence adjacent to and within the Antarctic Circumpolar Current itself must provide significant frictional forces which will help to balance the eastward wind stress; another contribution might come from the friction experienced by the current as it flows through the restricted Drake Passage. A further factor that might be important is the accumulation of water that occurs upstream of this restricted passage, leading to a zonal sea-surface slope and a horizontal pressure gradient force acting in opposition to the wind.

So far, we have only been considering horizontal motion. However, as discussed above, the APFZ is a region of *convergence* of surface waters; and between the eastward-flowing Antarctic Circumpolar Current and the westward-flowing Polar Current is a *divergence* of surface water—the **Antarctic Divergence**.

What significance do such divergences and convergences have for the three-dimensional circulation of the ocean?

They must lead to vertical motion: upwelling at the divergences and sinking at the convergences. The Antarctic Divergence is biologically one of the most productive regions of open ocean in the world. The nutrient-rich water upwelled there leads to high primary productivity which supports large populations of zooplankton, and bigger organisms ranging from krill to whales.

The convergences in the Antarctic Polar Frontal Zone are an important source of cold deep sub-thermocline water for the world ocean. Even colder 'bottom water' is formed off the Antarctic continent. These deep and bottom waters will be discussed further in Chapter 6.

5.6 SUMMARY OF CHAPTER 5

1   The major components of equatorial current systems are westward-flowing North and South Equatorial Currents, one or more eastward-flowing Counter-Currents, and an eastward-flowing Equatorial Undercurrent, which is generally centred on the Equator. Flow in the North and South Equatorial Currents is partly directly driven by the Trade Winds (in near-surface layers) and is partly geostrophic flow resulting from horizontal pressure gradient forces.

2   The equatorial current system is best developed in the Pacific Ocean, where the surface waters are under the cumulative influence of the prevailing Trade Winds over the greatest distances. In the Atlantic, the equatorial circulation is affected by the shape of the ocean basin and, indirectly, by the effect of the continental masses on the ITCZ. In the Indian Ocean, the circulation is monsoonal, most resembling that in the other tropical oceans in the northern winter.

3   The Intertropical Convergence Zone is generally displaced north of the Equator so that the South-East Trade Winds blow across it. As a result, divergence of surface waters, and upwelling, occur just south of the Equator. There is a convergence of surface waters at about 4° N.

4   The prevailing easterly winds over the tropical ocean cause the sea-surface to slope up (and the thermocline to slope down) towards the west. As a result, there is an eastward horizontal pressure gradient force and the Equatorial Counter-Current(s) flow down this gradient towards the east in zones of small westward wind stress (the Doldrums).

5   This eastward horizontal pressure gradient force also drives the Equatorial Undercurrent, which flows in the thermocline below the mixed surface layer. The Equatorial Undercurrent is a ribbon of fast-flowing water, many hundred times wider than it is thick. It is generally aligned on the Equator; if it is displaced away from the Equator, the Coriolis force turns it equatorwards again. The Equatorial Undercurrent has a significant volume transport, particularly in the Pacific.

6   In the Pacific and the Atlantic, the most extensive areas of upwelling occur just south of the Equator, in association with the South Equatorial Current. There is also coastal upwelling along the eastern boundaries—either year-round or seasonal—as a result of the Trade Winds blowing along the shore.

7   Surface divergence and upwelling may occur below the ITCZ because it is a region of low pressure and cyclonic winds. When the ITCZ is over certain regions of doming isotherms (apparently associated with flow in counter-currents), the doming intensifies and 'protrudes' into the thermocline. These thermal domes seem to be a feature of the eastern sides of oceans. Indeed, in the Pacific and the Atlantic, *all* types of upwelling occur most readily in the eastern side of the ocean, because there the thermocline is at its shallowest, and the mixed layer at its thinnest.

8   The winds over the Indian Ocean change seasonally as a result of the differential heating of the ocean and the Asian land mass. During the North-East Monsoon (northern winter), the winds are from Asia and are

dry and cool; during the stronger South-West Monsoon, the winds carry moisture from the Arabian Sea to the Indian continent. Because the winds over the equatorial zone change over the course of the year, there is no more or less permanent slope of the sea-surface up to the west as there is in the other oceans. As a result, the Equatorial Undercurrent is only a seasonal feature of the circulation.

9   The most dramatic seasonal change in the surface circulation of the Indian Ocean is the reversal of the Somali Current which flows south-westwards during the North-East Monsoon but is a major western boundary current during the South-West Monsoon. At that time of year, the North Equatorial Current reverses and becomes the South-West Monsoon Current. During the South-West Monsoon, there are regions of intense upwelling along the coast of Somalia and in the Arabian Sea.

10   The ocean can respond to the winds in distant places by means of large-scale disturbances that travel as waves. These waves may propagate along the surface (in which case they may be described as barotropic waves) or along a region of sharp density gradient such as the thermocline (baroclinic waves); surface waves, in particular, travel very fast. Two of the most important types of waves are Kelvin waves and Rossby (or planetary) waves. Rossby waves result from the need for potential vorticity to be conserved. They occur in zonal currents and airstreams and, relative to the flow, only travel westwards. Kelvin waves may travel along coasts (with the coast to the right of the direction of travel in the Northern Hemisphere and to the left in the Southern Hemisphere) or they may travel eastwards along the Equator as a double wave. In these cases, the coast and the Equator, respectively, are acting as wave guides. Wave guides can support a type of wave motion that would not otherwise be possible (as is the case for Kelvin waves) or they may simply 'channel' or 'trap' disturbances, enabling them to propagate more efficiently. Because of the equatorial wave guide, the ocean in low latitudes can respond much more rapidly to changes in the overlying wind than can the ocean at higher latitudes.

11   As a result of the contrasting distributions of land and sea in northern and southern high latitudes, both the type of ice cover and the current pattern of the two regions are very different. A large proportion of Arctic pack ice is several years old, while most Antarctic ice is re-formed yearly. The main circulatory pattern in the Arctic Sea is an anticyclonic gyre with cross-basin flow between the Bering Straits and the Fram Strait, where the outflow becomes the East Greenland Current.

12   The major current feature of the Southern Ocean is the Antarctic Circumpolar Current which, by virtue of its great depth, has an enormous volume transport. Various mechanisms have been proposed to explain why the velocity of the current is not greater, given that it is under the cumulative influence of the westerly winds blowing around Antarctica. The strongest currents flow along fronts in the Antarctic Polar Frontal Zone (formerly known as the Antarctic Convergence); there, current jets often form meanders and eddies. The Antarctic Polar Frontal Zone is a region where surface water converges and sinks; the Antarctic Divergence, between the Antarctic Circumpolar Current and the Antarctic Polar Current, is a region of upwelling

Now try the following questions to consolidate your understanding of this Chapter.

QUESTION 5.9 Figure 5.24 shows the mean zonal current velocity (in m s^{-1}) across 23° 30′ W, as determined by direct measurements during July and August 1974. The blue shaded areas correspond to westward flow. By comparison with Figure 5.1(b), can you identify (i) the South Equatorial Current, (ii) the North Equatorial Current, (iii) one or more Equatorial Counter-Currents, (iv) the Equatorial Undercurrent?

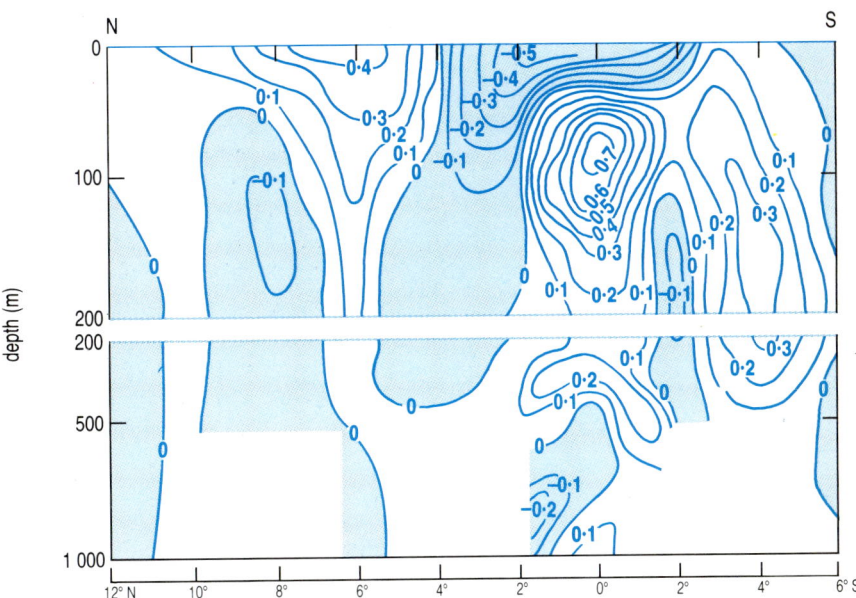

Figure 5.24 Mean zonal current velocity (in m s^{-1}) across 23° 30′ W, as determined by direct measurements during the 'GATE' experiment in July and August 1974. (GATE stands for *G*ARP (Global Atmospheric Research Programme) *A*tlantic *T*ropical *E*xperiment.) The blue shaded areas correspond to westward flow. (Note the change in the vertical scale below 200 m.)

QUESTION 5.10 In Section 5.1.1 we described how the Equatorial Undercurrent flows eastwards in the thermocline in response to the eastward pressure gradient force that results from the basin-wide slope in the sea-surface. Bearing in mind the discussion about the Antarctic Circumpolar Current in Section 5.5.2, can you suggest what might counteract the eastward pressure gradient force, hence ensuring that the Equatorial Undercurrent does not speed up indefinitely?

QUESTION 5.11 Monsoon winds are sometimes described as land breezes and sea breezes, on a very large scale. Would you say that this is a fair description of them?

QUESTION 5.12 At the beginning of Section 5.2.2, we say: 'As you might expect, the surface circulation of the northern Indian Ocean . . . most resembles that of the other two oceans in the northern winter'. However, there is a significant feature of the circulation in the northern *summer* that is shared by the other two oceans. What is it?

QUESTION 5.13 How do the locations of the main regions of upwelling in the Indian Ocean differ from those in the Pacific and Atlantic Oceans?

QUESTION 5.14 What, approximately, is the Rossby radius of deformation for the equatorial Kelvin wave shown in Figure 5.16, given that 1° of latitude is about 110 km?

QUESTION 5.15 As you saw in Section 5.3.2, Rossby waves result from the need for potential vorticity $(f + \zeta)/D$ to be conserved. Bearing that in mind, can you suggest why similar waves (known as 'shelf waves') occur over the region of relatively sharp depth change between the continental shelf and the deep sea-floor? Begin by imagining a current flowing above (say) the continental slope, parallel to a particular depth contour. What happens if, for some reason, the flow is displaced either towards or away from the coast? (You do not need to go into details.)

QUESTION 5.16 The vessel *Fram* took three years to drift in the ice from north of the Bering Straits to Spitsbergen (Figure 5.20).

(a) With which current did the *Fram* travel?

(b) Given that the distance involved is roughly 4000 km, approximately what average speed is implied for this current?

CHAPTER 6
GLOBAL FLUXES AND THE DEEP CIRCULATION

In Chapters 1 and 2, we saw that the ocean–atmosphere system is a huge heat engine, for which solar radiation is the power source. Heat is redistributed over the surface of the globe by winds and surface currents and by deep flow in the thermohaline circulation. We begin this Chapter about the three-dimensional circulation of the oceans by considering conditions at the sea-surface, because it is there that the cold, dense waters of the deep ocean originate.

6.1 THE OCEANIC HEAT BUDGET

In Section 1.1, we considered the heat gained and lost by the Earth–atmosphere system as a whole (Figure 1.4); here we will consider the heat gained and lost by the ocean.

6.1.1 SOLAR RADIATION

Figure 6.1 shows how the amount of solar radiation received annually varies over the Earth's surface. Intuitively, one might expect that the contours would be parallel to lines of latitude. This is clearly not the case.

Figure 6.1 The amount of solar radiation received annually at the surface of the Earth, in $J\,m^{-2}\,yr^{-1}$.

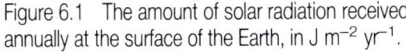

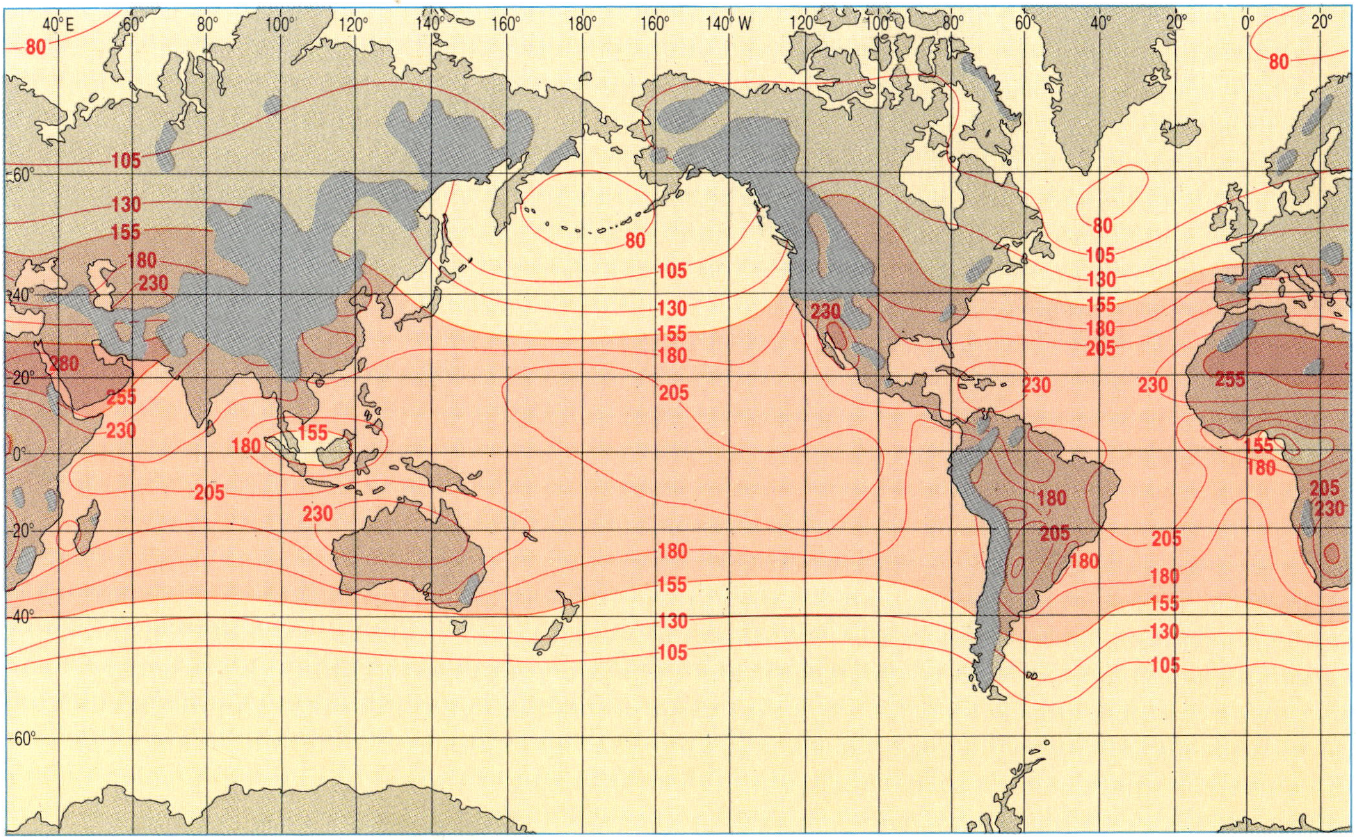

In general, at a given latitude, are contour values greater over the oceans or over the continents? Why might this be?

Insolation—the amount of incoming solar radiation reaching the Earth's surface—is generally greater for continental areas than for the oceans. The atmosphere over the oceans contains a large amount of water as vapour and in clouds, particularly in low latitudes; this water, along with gases such as CO_2 and SO_2, absorbs on average about 30% of incoming solar radiation (Figure 6.2). Over land, the atmosphere tends to be drier, and more cloud-free, particularly in tropical and subtropical areas. Clouds especially have a marked effect as they not only absorb radiation but also reflect it back to space.

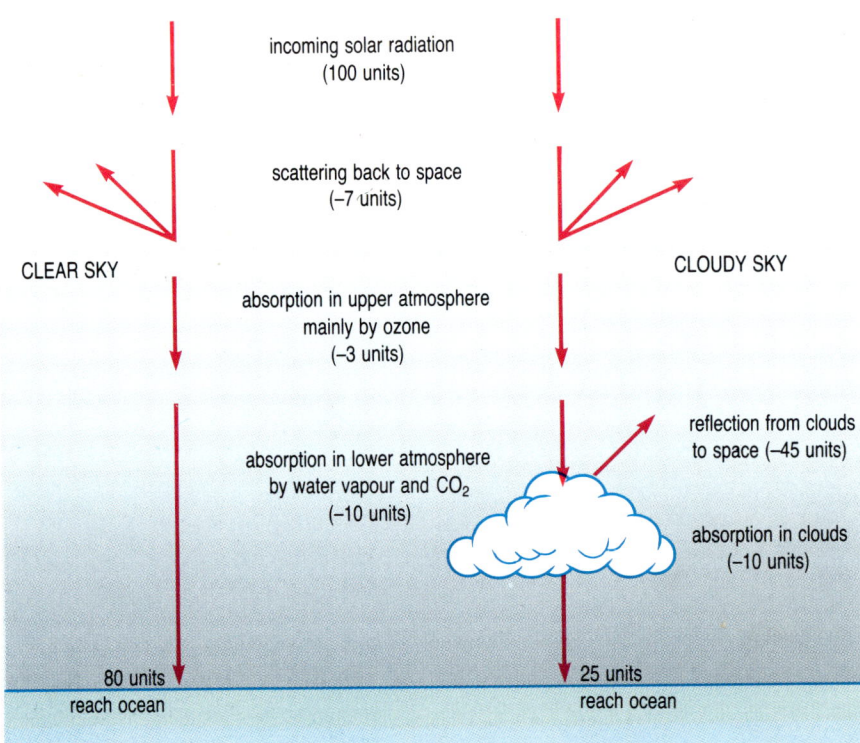

Figure 6.2 Diagram to illustrate the effect of the atmosphere on the radiation budget of the ocean. The values are very approximate and represent average conditions; for example, in low latitudes as much as 40% of incoming solar radiation may be absorbed by the atmosphere over the oceans.

QUESTION 6.1 Would you expect the effect of water in the atmosphere in reducing the amount of solar radiation reaching the ocean to be greater in low or high latitudes? Is this borne out by Figure 6.1?

In studying Figure 6.1 you may have noticed a number of land areas with unusually low values of insolation. These are generally areas where the supply of moisture to the atmosphere is high because the atmospheric pressure is low and there is vigorous convection and cloud formation (as is the case over the Malaysian and Indonesian archipelagos; *cf.* Section 5.4), and/or there are tropical rainforests.

Although by far the most important at the present time, water vapour and clouds are not the only factors that affect incoming solar radiation. The amount of solar radiation received at the Earth's surface may also be reduced by the presence in the atmosphere of ash, smoke, dust and gases from volcanoes and industrial complexes, as well as dust from arid regions.

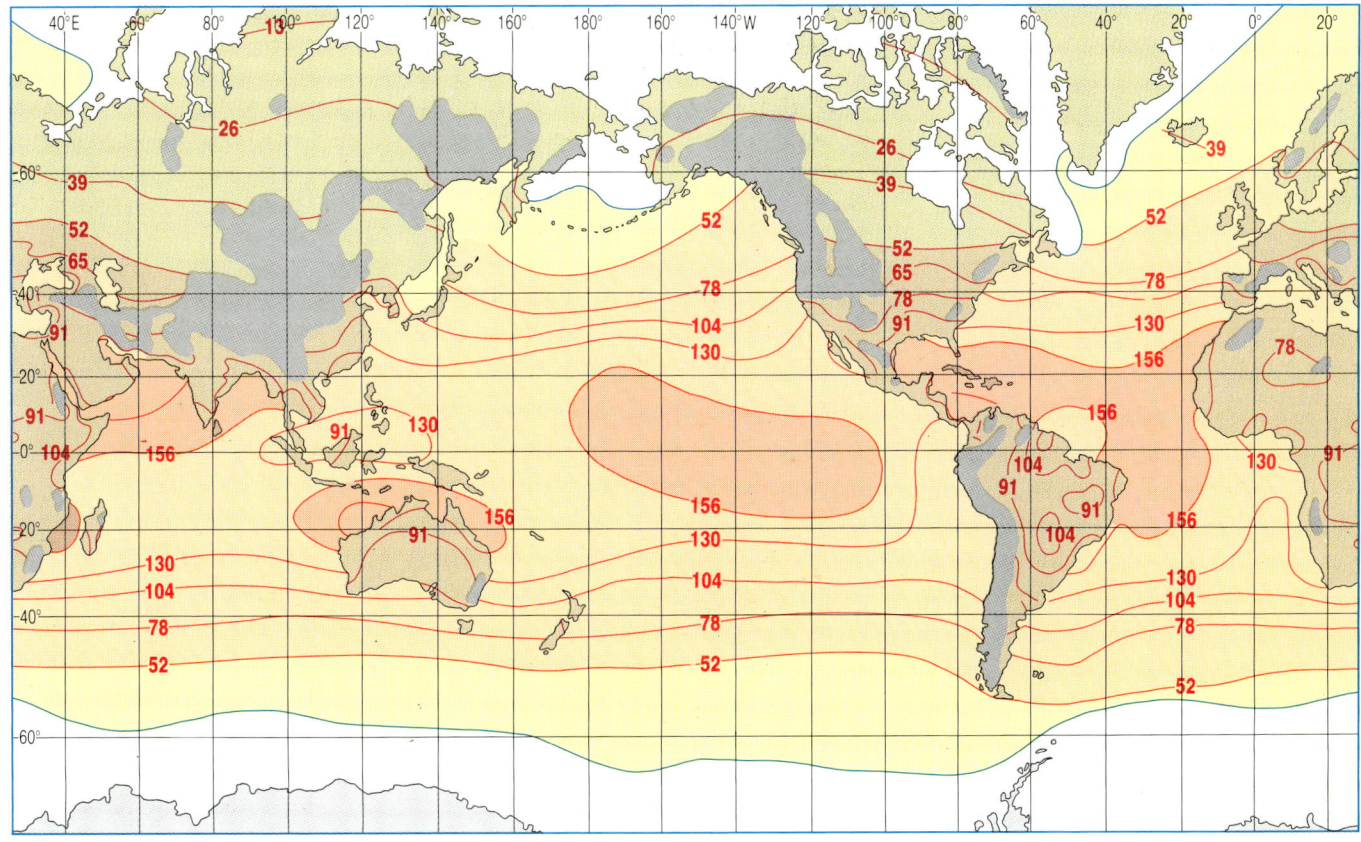

Figure 6.3 The annual radiation balance ($Q_s - Q_b$) at the Earth's surface, in J m^{-2} yr^{-1}. Values have been converted from non-SI units; and contours have been omitted over high ground. The white area shows the approximate winter limit of sea-ice cover.

We mentioned in Section 1.1 that the Earth as a whole not only receives short-wave radiation from the Sun, but also *re*-emits long-wavelength radiation. This is because *all* bodies with a temperature above absolute zero emit radiation: the higher the temperature of the body concerned, the greater the total amount of radiant energy emitted, and the more the radiation spectrum is shifted towards shorter wavelengths*. Thus, the surfaces of the oceans and continents not only absorb and reflect the incoming short-wave solar radiation that has penetrated the atmosphere but also re-emit radiation which is mostly of a much longer wavelength, because of the relatively low temperatures involved. This long-wave radiation is either *reflected* back to the Earth by clouds, or is *absorbed* by clouds, water vapour and other gases—especially carbon dioxide and ozone—all of which re-emit long-wave radiation in all directions (Figure 6.2). In calculating the amount of radiant energy absorbed by the oceans, we therefore have to consider not only the incoming short-wave (<4 μm) radiation (given the symbol Q_s) but also the *net* emission of long-wave radiation (also known as back-radiation, and so given the symbol Q_b). For all latitudes, estimates of $Q_s - Q_b$ are generally positive, i.e. the oceans absorb more radiant energy than they emit (Figure 6.3). Nevertheless, at higher latitudes the value of $Q_s - Q_b$ varies significantly with the time of year.

*In fact, the intensity (I) of the radiation emitted increases in proportion to the *fourth* power of the absolute temperature (T); i.e. $I = \sigma T^4$. This is known as Stefan's law and the constant σ is known as Stefan's constant.

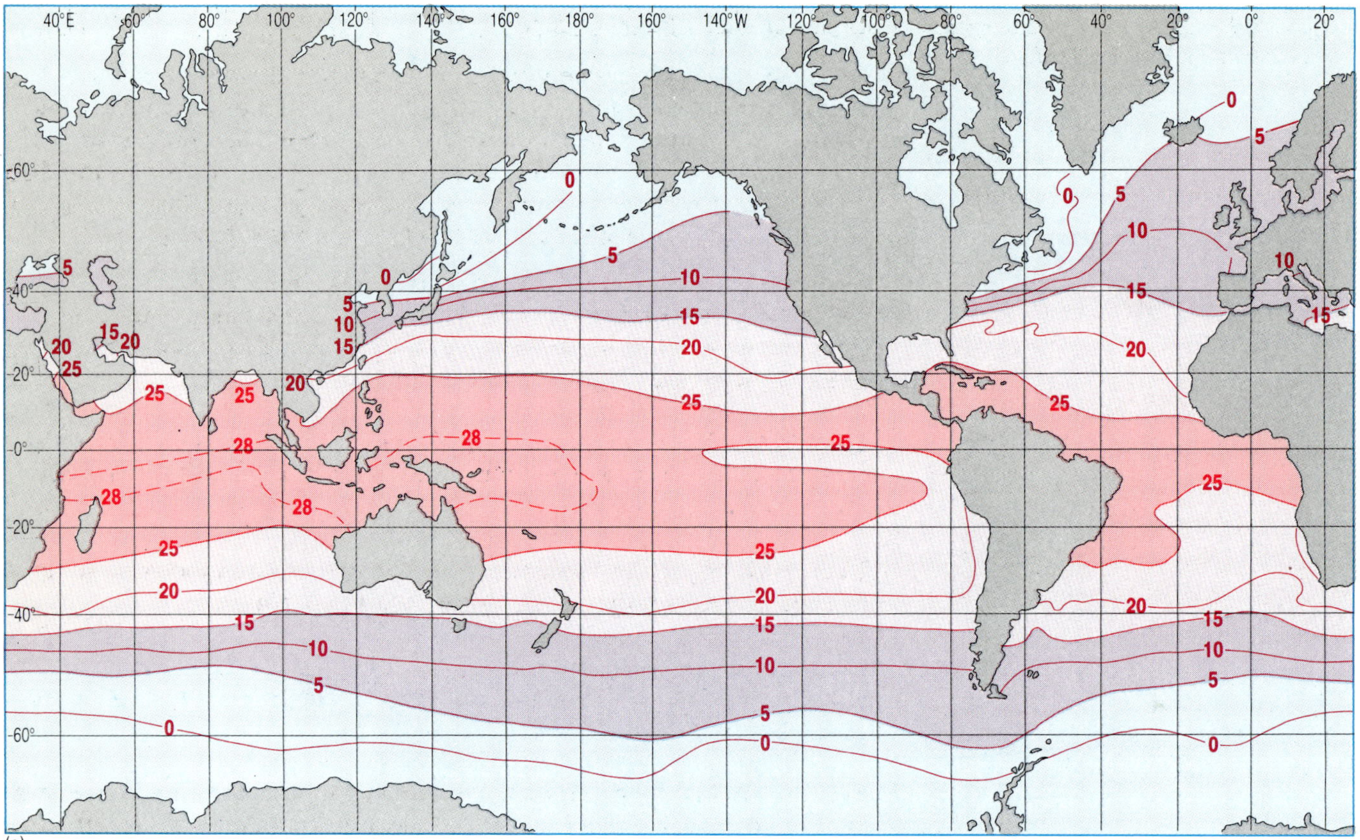
40°E
60°
80°
100°
120°
140°
160°
180°
160°
140°W
120°
100°
80°
60°
40°
20°
0°
20°
60°
40°
20°
0°
20°
40°
60°
0
5
10
15
20
25
28

(a)

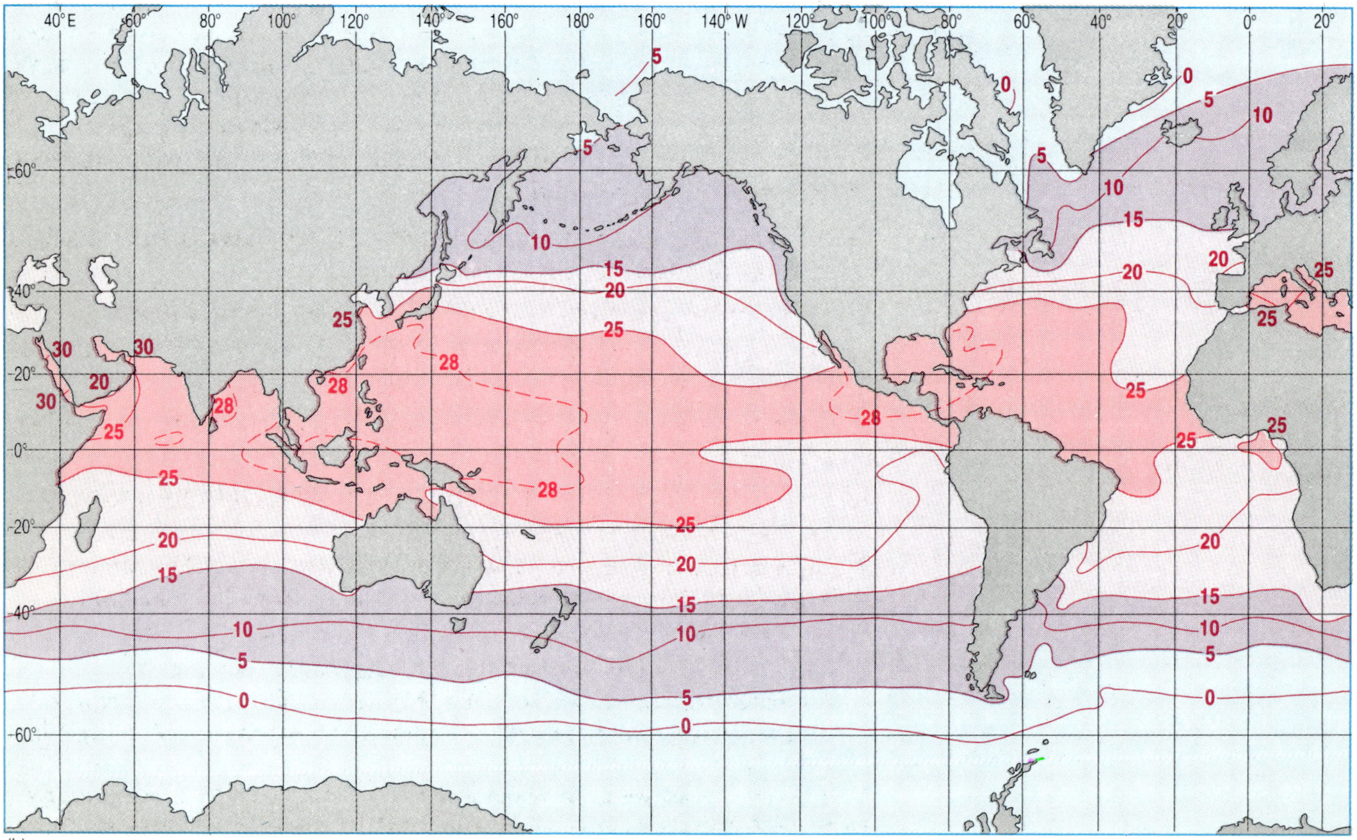
40°E
60°
80°
100°
120°
140°
160°
180°
160°
140°W
120°
100°
80°
60°
40°
20°
0°
20°
60°
40°
20°
0°
20°
40°
60°
0
5
10
15
20
25
28
30

(b)

Figure 6.4 (Opposite) The global distribution of sea-surface temperature (°C) (a) in February, (b) in August.

6.1.2 THE HEAT-BUDGET EQUATION

Of the total amount of energy received from the Sun by the world's oceans, about 41% is lost to the atmosphere and, indirectly, to space, as long-wave radiation, and about 54% is lost as latent heat by evaporation from the sea-surface. A relatively small amount—about 5%—is lost to the overlying atmosphere by conduction. Heat loss by evaporation is generally given the symbol Q_e and heat loss by conduction Q_h.

Temperature is a measure of the thermal energy possessed by the oceans, and if the average temperature of the oceans is to remain constant, the gains and losses of heat must even out over a period. In other words, the **heat budget** must balance*.

QUESTION 6.2 Using the symbols Q_s, Q_b, Q_e and Q_h, write down the heat-budget equation (i.e. an equation of the form: heat gained = heat lost) for the oceans as a whole.

Heat is not only being continuously gained and lost from the oceans, but also redistributed within them, by currents and mixing.

Figure 6.4 shows the global distribution of sea-surface temperature. Would you say that this distribution reflects the influence of ocean currents (Figure 3.1)?

Yes. In particular, the poleward displacement of isotherms on the western sides of oceans may be correlated with the western boundary currents carrying warm water from equatorial regions; the effect of the Gulf Stream in transporting relatively warm water across the Atlantic can also be clearly seen. By contrast, the eastern boundary currents carrying water from higher latitudes cause temperatures on the eastern sides of oceans to be somewhat lower than they would otherwise be. Heat brought into a region of ocean by currents and mixing, i.e. by **advection**, is given the symbol Q_v. (For the ocean as a whole, Q_v is, of course, zero as it refers to the redistribution of heat *within* the ocean.)

In summary, the heat-budget equation for any part of the ocean should include the following terms:

Q_s—solar energy, received by the ocean as short-wave radiation.

Q_b—the *net* loss of energy from the surface of the ocean as long-wave (back-) radiation.

Q_e—the heat lost by evaporation from the surface, less any heat gained by condensation at the surface.

Q_h—the net amount of heat transferred to the atmosphere by conduction across the air–sea interface.

Q_v—the net amount of heat gained from adjacent parts of the ocean by advection (including upwelling or sinking of water) and mixing; when heat is lost by advection, this term will be negative.

Q_t—the amount of surplus heat actually available to increase the temperature of the water; when there is a heat deficit, this term will be negative, and there will be a fall in the temperature of the water.

*The development and decline of ice ages is evidence that the Earth's heat budget does not always balance in the long term, but in the context of this Volume we are not concerned with events occurring over such long time-scales.

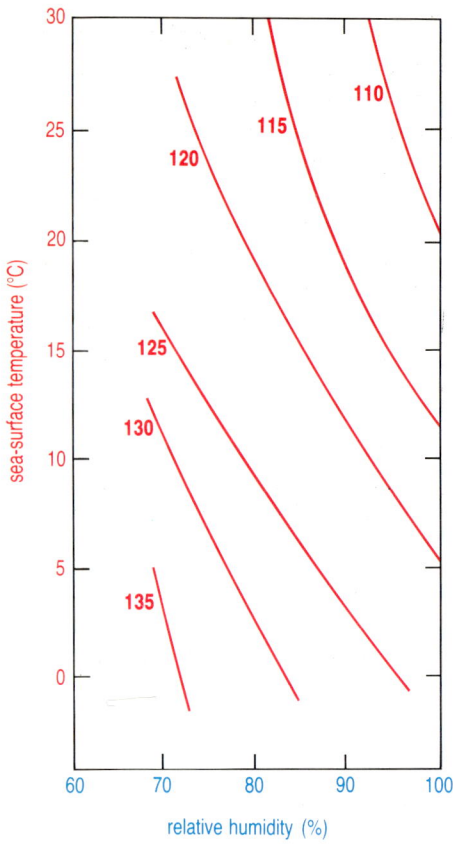

Figure 6.5 Curves showing how the net back-radiation (Q_b) (in (Wm^{-2}) × 10^3), from the sea-surface to a clear sky, varies as a function of the sea-surface temperature and the relative humidity at an altitude of a few metres. (Relative humidity is a measure of the degree of saturation of the air. It is defined as: actual water vapour pressure/saturation water vapour pressure at the ambient temperature, expressed as a percentage.)

The full heat-budget equation for a *part* of the oceans is therefore:

$$Q_s + Q_v = Q_b + Q_h + Q_e + Q_t \qquad (6.1)$$

It is difficult to assess the relative sizes of the quantities in equation 6.1. They are hard to measure directly, and there is a lack of consensus about how to determine them theoretically because so many interacting variables are involved. However, we can review the principles behind the methods for estimating them.

Values for radiation gain (Q_s) can be estimated from knowledge of incoming solar radiation (*cf.* Figures 1.4(a) and 6.1), and values for radiation loss (Q_b) can be estimated using the temperature of the surface skin of the ocean (*cf.* footnote on p. 161). For such estimates to be meaningful, we need accurate information about cloud cover and about atmospheric concentrations of those variable constituents that absorb and re-emit radiation, in particular water vapour and aerosol droplets. Empirical relationships have been derived using those variables which are regularly determined, especially mean cloudiness, relative humidity immediately above the surface of the water, and the surface temperature. Figure 6.5 illustrates the empirically determined relationship between these last two variables and Q_b.

Given that the intensity of radiation (i.e. the amount of energy) emitted by a body increases as its temperature increases, is the variation of Q_b shown in Figure 6.5 what we might expect?

At first sight, no: according to Figure 6.5, for a given relative humidity Q_b *decreases* with temperature. The explanation for this apparent anomaly lies in the fact that the warmer the air over the oceans, the more water it can hold before becoming saturated. Thus, a given relative humidity value at a high temperature corresponds to a greater atmospheric water vapour content than the same relative humidity at a lower temperature. The more water vapour there is in the atmosphere, the more long-wave radiation is absorbed by it, and the more is radiated *back* to the sea-surface, thus decreasing the *net* loss of long-wave energy *from* the sea-surface.

Earlier, we defined Q_h as the amount of heat removed from the sea by *conduction* across the air–sea interface. However, if *only* conduction were involved, Q_h would be very small. As discussed in Section 2.2.2, over most of the oceans, especially in winter and/or in windy conditions, the atmosphere is unstable and subject to turbulent convection. Thus, air warmed by the underlying sea is swiftly removed allowing more cool air to come into contact with the sea-surface. The same argument may be applied to Q_e: air above the sea-surface may become saturated with water vapour as the result of evaporation, but turbulent convection causes it to be quickly replaced by new, drier air.

For these reasons, both Q_h and Q_e depend critically on the degree of turbulent convection taking place in the atmosphere above the sea-surface. In addition, Q_h depends on the vertical temperature gradient immediately above the sea-surface, while Q_e depends on the vertical gradient in atmospheric water vapour content. In most areas of the ocean, the sea-surface is warmer than the overlying air and heat is lost from the sea by conduction and convection. Similarly, over most of the ocean, the water content of the atmosphere decreases with increasing distance from the surface and so evaporation takes place. In fact, as long as the sea-surface is warmer than the overlying air by more than about 0.3°C, heat will be lost from the sea by evaporation and Q_e (defined as a *loss*) will be positive. One of the few regions of the ocean where Q_e is negative is the Grand Banks off Newfoundland. Here the sea temperature is generally less than the air temperature and the relative humidity is such that water vapour condenses onto the sea-surface, leading to a gain of (latent) heat by the sea. The fogs for which the Grand Banks are famous are also a result of condensation occurring as a result of warm, moist air coming into contact with a cold sea-surface; this type of fog is known as **advection fog**.

QUESTION 6.3 Figure 6.6 shows the variation in annual mean values of Q_h and Q_e for the Pacific Ocean; negative values indicate a gain of heat by the sea and positive values indicate a loss of heat from the sea. To what phenomena do you ascribe:

(a) the relatively high positive values of both Q_h and Q_e in the western ocean off Japan?

(b) the negative values of Q_h and the relatively low positive values of Q_e in the eastern equatorial Pacific?

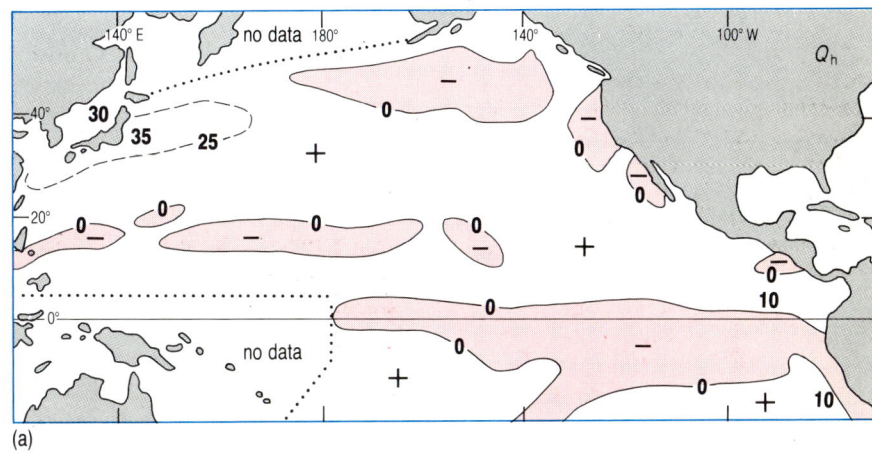

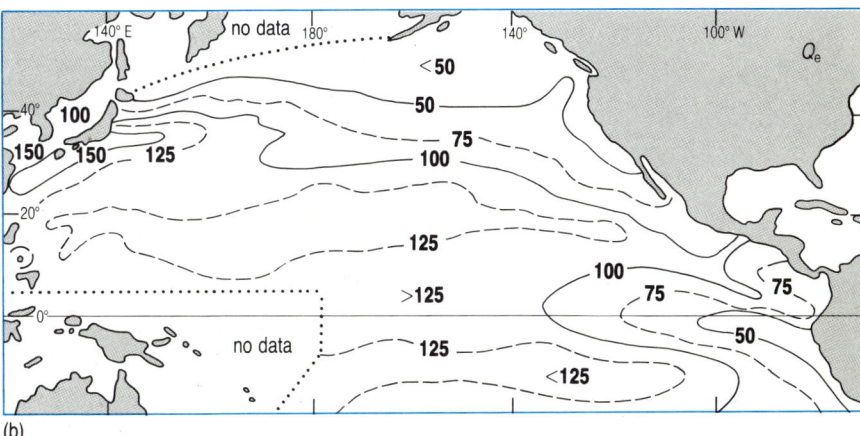

Figure 6.6 The annual mean values (in Wm^{-2}) for (a) Q_h and (b) Q_e for the Pacific Ocean. Negative values (in pink) indicate a gain of heat by the sea; positive values indicate a loss from the sea.

Earlier, we emphasized that the sizes of both Q_h and Q_e depend critically upon the degree of atmospheric turbulence above the sea-surface. In extremely calm conditions, processes occurring at the molecular level become more important; in general, however, conditions at the ocean–atmosphere interface are turbulent, and so Q_h and Q_e vary together—in other words, their *ratio* remains constant. The ratio Q_h/Q_e is known as **Bowen's ratio**, R. It is generally of the order of 0.1 in low-latitude regions and increases to about 0.45 at 70° N.

QUESTION 6.4 'In low latitudes, the amount of heat lost from the oceans by conduction/convection is generally greater than the amount lost by evaporation; in high latitudes, the reverse is the case.' According to the values of Bowen's ratio given above, is this statement true or false?

Bowen's ratio is useful because it gives us an indirect way of estimating Q_h and Q_e whatever the degree of turbulence prevailing in the situation under consideration. Q_h and Q_e *can* be estimated directly by somehow *quantifying* the effect of the turbulence on the transfer of heat and water, but this is difficult. If, however, we have a value for R—or can calculate one using the vertical gradients of heat and atmospheric water content—we can calculate values for Q_h and/or Q_e from the heat-budget equation. This is done by assuming that Q_v and Q_t are zero, in which case, from equation 6.1:

$$Q_s - Q_b = Q_h + Q_e$$

Dividing by Q_e,

$$\frac{Q_s - Q_b}{Q_e} = \frac{Q_h}{Q_e} + \frac{Q_e}{Q_e}$$

Q_h/Q_e is Bowen's ratio, R, so:

$$\frac{Q_s - Q_b}{Q_e} = R + 1$$

and

$$Q_e = \frac{Q_s - Q_b}{1 + R} \qquad (6.2a)$$

In a similar way we can show that:

$$Q_h = \frac{Q_s - Q_b}{1 + (1/R)} \qquad (6.2b)$$

So, as long as the radiation terms in the heat budget are available, we can estimate the terms that depend on the degree of turbulence above the sea-surface.

Figure 6.7 shows the variation with latitude in the Northern Hemisphere of the mean annual values of some of the components of the heat-budget equation. The heat-budget terms represented on Figure 6.7 are the radiation terms (Q_s and Q_b) and those involved in the direct transfer of heat between atmosphere and ocean—either sensible heat, i.e. heat that gives rise to an increase in temperature that can be detected or 'sensed' (Q_h), or latent heat (Q_e). For the oceans as a whole, the heat gained is equal to the heat lost, at least over periods of several years. The same is true at any given location, and so on average, $Q_t = 0$.

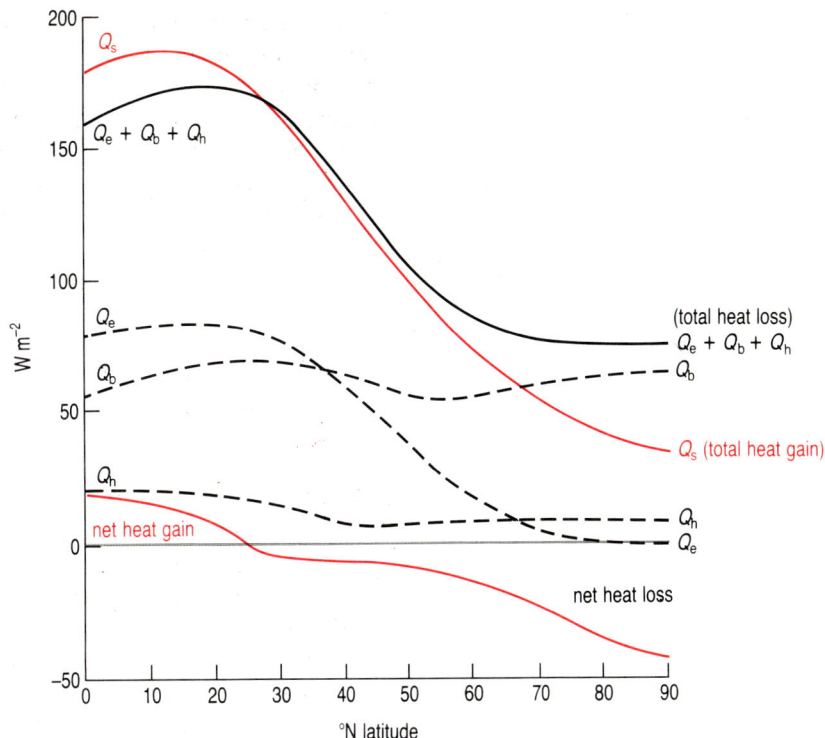

Figure 6.7 The variation with northern latitude of the mean annual values (per unit area) for heat-budget terms relating to heat transfer across the air–sea interface, in the Northern Hemisphere.

How *can* the average value of Q_t for any given location be zero if—as Figure 6.7 shows—there is a net *gain* of heat for the ocean equatorward of about 23° N, and a net *loss* poleward of that latitude?

Figure 6.7 does not include the Q_v term for heat transport *within* the oceans. As discussed in Chapters 1 to 3, ocean currents transport heat from low to high latitudes; the advective loss of heat from low-latitude regions ensures they do not continually heat up, while the advective gain of heat by high-latitude regions ensures they do not continually cool down. Over periods of years, therefore, the mean temperature of the water remains constant and Q_t is zero.

As a result of the poleward transport of heat in warm currents, the surface of the sea is generally above the freezing point of seawater ($\sim -1.9°C$) except at very high latitudes. If sea-ice *does* form, however, the radiation balance is changed dramatically. The **albedo** of the surface— i.e. the percentage of incoming radiation that is reflected—increases, perhaps to as much as 80–90%. Thus Q_s is greatly reduced; however, Q_b for ice is much the same as it is for water, and so $Q_s - Q_b$ is significantly decreased. Once ice has formed, therefore, it tends to be maintained. On the other hand, it has been estimated that the balance in the Arctic Sea is fairly fine, so that if the ice cover were to *melt*, the resulting increase in $Q_s - Q_b$ might *keep* the sea ice-free.

When ice cover increases, heat loss by conduction/convection (Q_h) and by evaporation (Q_e) are reduced, but the temperature of surface waters is still likely to be lowered until a new heat balance is attained. During this period, the input of heat from adjacent parts of the ocean (Q_v) is likely to increase substantially. A small initial decrease in surface temperature in regions that are already close to freezing point can have a considerable effect on the heat budget of both the overlying atmosphere and a very much wider area of ocean. In Section 6.3.2 we will see that the interaction between atmosphere, ocean and ice is even more complex than this might suggest.

6.2 CONSERVATION OF SALT

In considering the heat budget of the ocean, we have assumed that, over periods of several years at least, the Earth's heat supply—solar radiation—remains constant, and that as a result the ocean is neither heating up nor cooling down. We have in effect been applying the principle of conservation of energy. In Section 4.2.3 we introduced the principle of continuity which is another way of expressing the principle of conservation of mass which, because seawater is nearly incompressible, approximates to the conservation of volume. Another conservation principle that is very important in oceanography is the principle of conservation of salt.

The principle that the amount of salt in the oceans remains constant may, at first sight, appear to be seriously flawed.

How is salt added to the ocean?

By rivers, which every year bring about 3.6×10^9 tonnes of salt to the oceans. However, there is a consensus among marine scientists that the rates of input of dissolved substances to the oceans are balanced by their rates of removal to the sediments, so that the oceans are in a **steady state**. The salt content of the oceans, and hence the average salinity of seawater, therefore change little with time. So, for all practical purposes, the principle of conservation of salt is a valid one.

Of course, at a given location in the ocean, the salinity may be changed. Within the oceans, this takes place by mixing of waters with different salinities to produce water with an intermediate salinity. At the surface of the ocean, salinity is increased by evaporation and decreased by the precipitation of rain and snow, and, occasionally, by condensation on the sea-surface. Figure 6.8(a) shows the distribution of the mean surface salinity and Figure 6.8(b) demonstrates the correlation between surface salinity and the balance between evaporation and precipitation ($E-P$).

QUESTION 6.5 (a) In general, are surface salinities higher in the Pacific Ocean or the Atlantic Ocean?

(b) In what way does Figure 6.8(b) demonstrate the effect of the Intertropical Convergence Zone?

Awareness of the global variations of such factors as sea-surface salinity, local evaporation–precipitation balances—and indeed of the various heat-budget terms—is essential if we are to quantify fluxes of water (and heat) across the ocean–atmosphere boundary. Nevertheless, these global variations of themselves can give us only a limited view of what is happening. Figure 6.8 contains no information about how water lost from a particular area of sea-surface is redistributed within the oceans, or how the sea-surface salinity at a particular location is maintained at its average value.

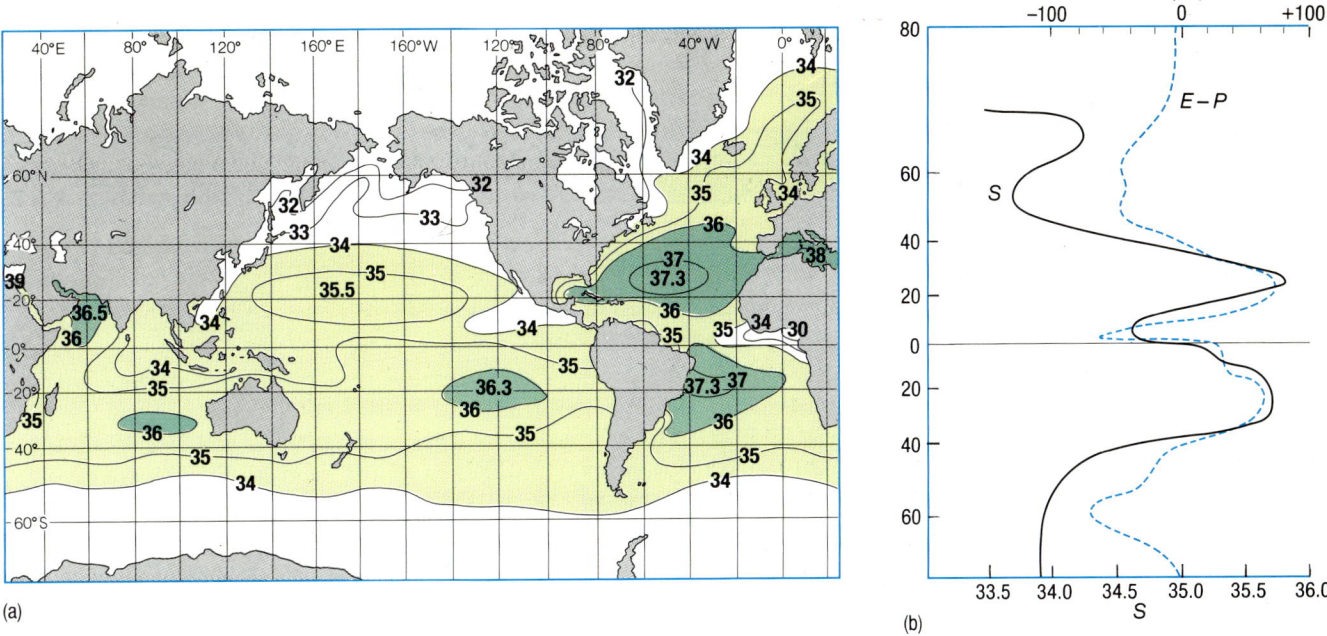

Figure 6.8 (a) The mean annual distribution of surface salinity.

(b) Average values of salinity, S (black line) and the difference between average annual evaporation and precipitation ($E–P$) (blue line), plotted against latitude.

The redistributions of salt and heat within the ocean are often studied together, by monitoring the movement of bodies of water with characteristic combinations of temperature and salinity. These bodies of water, known as water masses, are the subject of Section 6.3.

First, however, let us see how the principle of conservation of salt may be applied on a relatively small scale.

6.2.1 PRACTICAL APPLICATION OF THE PRINCIPLES OF CONSERVATION AND CONTINUITY

In practice, the principle of conservation of salt is most often used, together with the principle of continuity, to study the flow, or the evaporation–precipitation balance, of relatively enclosed bodies of water, with limited connections with the main ocean. These might be fjords, estuaries, or semi-enclosed seas like the Mediterranean or Baltic.

Figure 6.9 represents a channel, or some other semi-enclosed body of water, bounded by two vertical transverse sections with areas A_1 and A_2. Water enters the channel through A_1 at a rate of $V_1 \mathrm{m^3 s^{-1}}$ and leaves it

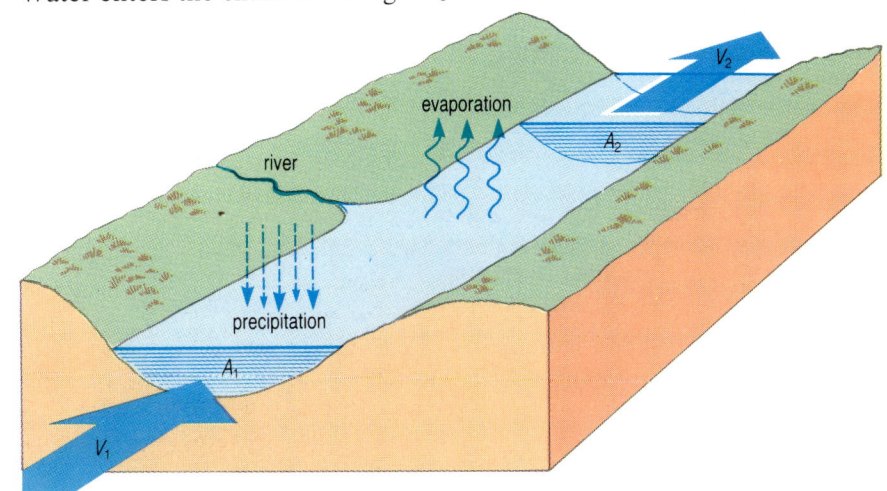

Figure 6.9 The flow of water into and out of a length of channel. Water flows into the channel through the section with area A_1 at a rate of V_1 m³ s⁻¹, and out through the section with area A_2 at a rate of V_2 m³ s⁻¹.

through A_2 at a rate of $V_2 \text{m}^3\text{s}^{-1}$. Between A_1 and A_2, water also enters the channel as a result of precipitation and run-off, and is removed by evaporation. The *net* rate at which water is being added by these processes is represented by F, which is also measured as volume per unit time (m^3s^{-1}). If we assume that the volume of water in the portion of channel under consideration remains constant over a given period of time (and that water is incompressible), we may equate the total volume of water entering the portion of the channel with that leaving, i.e.

$$V_1 + F = V_2 \tag{6.3}$$

If we further assume that the average salinity between the sections remains constant, the amount of salt entering through A_1 must equal that leaving through A_2, because the processes of precipitation, run-off and evaporation do not involve any *net* transfer of salt. Because salinity is measured in parts per thousand by weight, the mass of salt in a kilogram of seawater is the density (in kg m^{-3}) times the salinity. Thus the mass of salt transported across A_1 and A_2 per second must be $V_1\rho_1 S_1$ and $V_2\rho_2 S_2$, and $V_1\rho_1 S_1 = V_2\rho_2 S_2$ where ρ_1, ρ_2 and S_1, S_2 are, respectively, the mean densities and salinities of the water at A_1 and A_2. Proportional changes in density are very small compared with those of salinity—for example, a change in salinity from 30 to 36 (i.e. over almost the complete range of salinities found in the oceans) results in an increase in density of less than 0.5%. We may therefore ignore changes in density and write:

$$V_1 S_1 = V_2 S_2 \tag{6.4}$$

We now have two equations that we may solve for either V_1 or V_2. From equation 6.4:

$$V_1 = \frac{S_2}{S_1} V_2 \tag{6.5}$$

hence, substituting for V_1 in equation 6.3:

$$\frac{S_2}{S_1} V_2 + F = V_2$$

$$F = V_2 \left(1 - \frac{S_2}{S_1}\right)$$

$$V_2 = \frac{F}{\left(1 - \frac{S_2}{S_1}\right)} \tag{6.6a}$$

and, similarly,

$$V_1 = \frac{F}{\left(\frac{S_1}{S_2} - 1\right)} \tag{6.6b}$$

These equations give us a means of calculating the rate of flow across A_1 and A_2, if the mean salinity at each section and the rates of evaporation, precipitation and run-off are known. Also the average volume of water flowing through the sections per unit time is equal in each case to the cross-sectional area × the mean current velocity, i.e. $V_1 = A_1 u_1$ and $V_2 = A_2 u_2$. Thus, if areas A_1 and A_2 are known, the average currents flowing through the sections, u_1 and u_2, may be estimated.

In deriving equations 6.4 to 6.6, we have assumed that salt is carried into and out of the channel only by advection of the mean current; the effect of, say, an eddy of exceptionally high or low salinity could not be taken into account. Also, in estimating average values of the different parameters, variations resulting from tidal flow have to be allowed for. Nevertheless, this use of the principles of continuity and of conservation of salt provides an extremely useful tool in the study of semi-enclosed bodies of water. Here, we will demonstrate how the principles may be applied to flow into and out of the Mediterranean.

At the Straits of Gibraltar, Atlantic water of relatively low salinity flows into the Mediterranean Sea, while high-salinity Mediterranean water flows out at depth.

By reference to Figure 6.10, what are the values of S_1, the mean salinity of water flowing into the Mediterranean, and S_2, the mean salinity of water flowing out?

Figure 6.10 suggests that S_1 is between 36.25 and 36.5, and that S_2 is between 37.0 and 38.0. The average values are about 36.3 for S_1 and 37.8 for S_2.

QUESTION 6.6 (a) Evaporation exceeds freshwater input to the Mediterranean (i.e. input from rivers and precipitation) by $7 \times 10^4 \, m^3 s^{-1}$. Use this information, along with equations 6.6(a) and (b), to calculate values for V_1 and V_2, the rates of inflow and outflow through the Straits of Gibraltar, in $m^3 s^{-1}$.

(b) The Mediterranean contains about $3.8 \times 10^6 \, km^3$ of water. Use the value of V_1 you have calculated to estimate roughly how long it would take for all this water to be replaced once.

Direct current measurements in the upper layers of water in the Straits of Gibraltar indicate that V_1 is of the order of $1.75 \times 10^6 \, m^3 s^{-1}$.

The value you calculated in Question 6.6(b) indicates that it takes about 70 years for all the water in the Mediterranean to be replaced. This is the **residence time** of water in the Mediterranean. The term 'flushing time' is also used, particularly in connection with flow in estuaries; it is useful because it gives a measure of the extent to which pollutant substances are likely to accumulate.

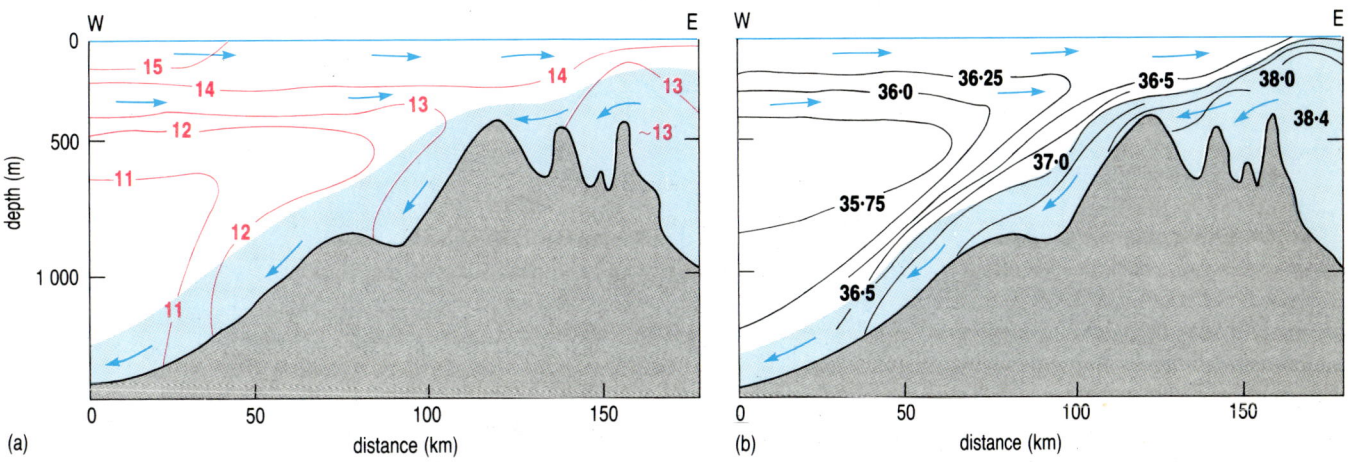

Figure 6.10 West–east sections across the Gibraltar sill of (a) temperature (°C) and (b) salinity, showing the inflow of Atlantic water in the upper layers and the outflow of Mediterranean Water (shown in blue) at depth. (The vertical exaggeration is about × 75.)

The evaporation–precipitation cycle is not the only mechanism whereby salinity may be changed. At high latitudes, the formation of ice and the release of meltwater have a significant effect on the salinity of seawater. In Section 6.3 we will see how the removal of freshwater and/or heat from surface water drives the deep thermohaline circulation.

6.3 OCEANIC WATER MASSES

Most of what is known about the three-dimensional circulation of the ocean has been deduced from the study of bodies of water that are identifiable because they have particular combinations of physical and chemical properties. Such bodies of water are referred to as **water masses**, and the properties that are most often used to identify them are temperature and salinity.

Temperature and salinity can be used to identify water masses because they are **conservative properties**, that is, they are altered *only* by processes occurring at the boundaries of the ocean; *within* the ocean, changes occur only as a result of mixing with water masses with different characteristics. **Non-conservative properties**, on the other hand, are subject to alteration by physical, chemical or biological processes occurring within the oceans.

Water masses that form in semi-enclosed seas provide particularly clear examples of bodies of water with recognizable temperature and salinity characteristics. As discussed in the previous Section, deep water leaving the Mediterranean Sea through the Straits of Gibraltar is of unusually high salinity (Figure 6.10). This **Mediterranean Water** forms in the north-western Mediterranean in winter. Intense cooling and higher than normal evaporation, associated particularly with the cold, dry Mistral wind, increase the density of surface water to such an extent that there is vertical mixing, or convection, right to the sea-floor at more than 2000m depth. The homogeneous water mass so formed has a salinity of more than 38.4 and a temperature of about 12.8°C. The Mediterranean Water leaves the Straits of Gibraltar at depth, below incoming Atlantic water (Figure 6.10), and intense mixing occurs at the interface between them. The least-mixed layer of Mediterranean Water in the adjacent Atlantic has a salinity of 36.5 and a temperature of 11°C. Because of its relatively high density, it sinks down to about 1000m depth where it becomes neutrally buoyant, and it spreads out at this level. Although it is being continually modified by mixing, Mediterranean Water can be recognized throughout much of the Atlantic Ocean by its distinctive signature of high temperature combined with high salinity (Figure 6.11).

There are a large number of water masses, each characterized by temperature and salinity values reflecting a particular set of surface conditions, and generally considered to originate in a particular source region. We have seen in Section 6.1.2 that the temperature of surface waters at any location in the ocean depends on the relative sizes of the components of the heat budget in that region; similarly, the salinity will depend upon the relative importance of the various factors discussed in Section 6.2. However, a water mass with particular temperature and salinity values will only result if water is subject to specific meteorological influences over a significant period, during which it remains in the mixed

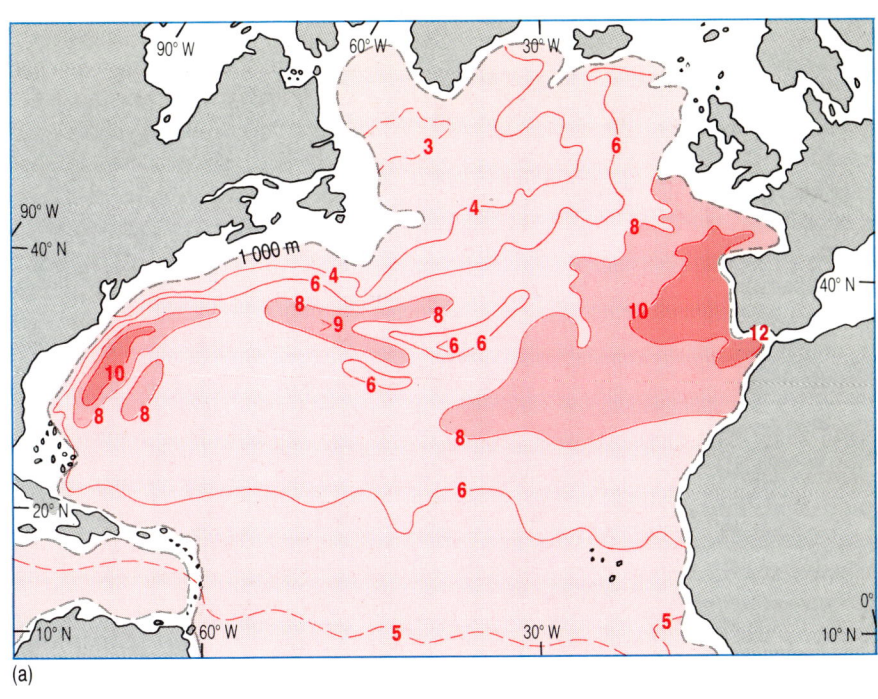

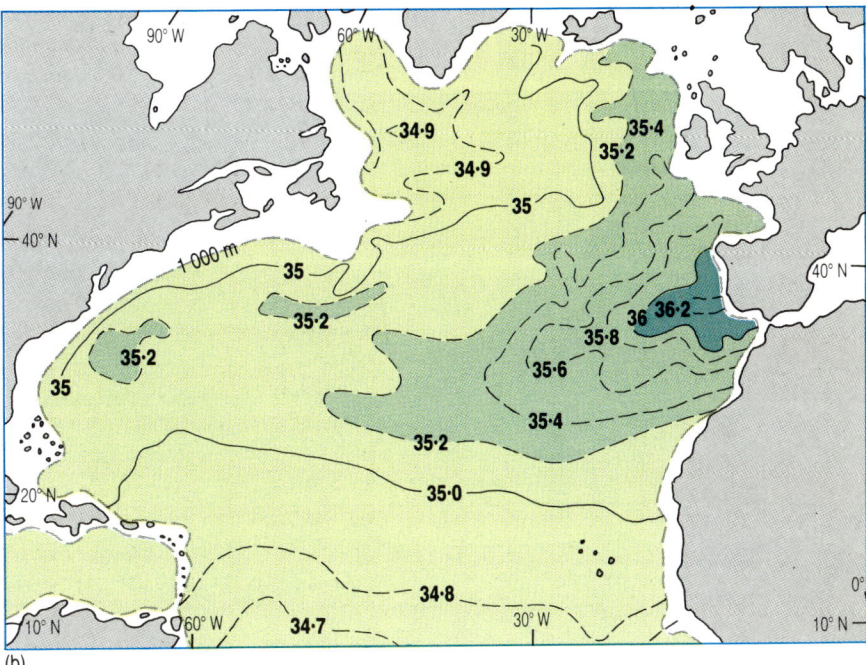

Figure 6.11 Distribution of (a) temperature (°C) and (b) salinity at 1 000 m depth in the North Atlantic, showing the spread of Mediterranean Water. The broken black line is the 1 000 m isobath.

surface layer. Furthermore, if the water is eventually to become isolated from the atmosphere, it must sink down from the sea-surface. These necessary conditions are satisfied in regions where surface water converges.

QUESTION 6.7 Suggest *two* such regions of convergence of surface water: one in mid-latitudes and the other at high latitudes.

6.3.1 UPPER AND INTERMEDIATE WATER MASSES

If you compare Figure 6.12 with Figure 3.1, you will see that the geographical distribution of the world's *upper water masses* is strongly influenced by the pattern of surface currents. Upper water masses are generally considered to include both the mixed surface layer and the upper part of the permanent thermocline. These upper water masses are of varying thickness. If, as is the case in the region of the Equator, salinity is kept low by high precipitation and the temperature is high, the density of surface water will be low; the upper water column will therefore be stable, and only a very shallow water mass can form.

By contrast, the water masses that form in the subtropical gyres—also known as Central Waters (*cf.* Figure 6.12)—are upper water masses of considerable thickness. As discussed in Section 3.4, in regions of convergence like the subtropical gyres (*cf.* Question 6.7) the sea-surface is raised and the thermocline depressed, leading to a thickening of the mixed surface layer (Figure 3.23(c) and (d)). Water sinks from the surface continually, but in winter, cooling of surface water leads to instability and vigorous vertical mixing occurs. As a result, water is alternately brought into contact with the surface and then carried deep down, so that a thick and fairly homogeneous water mass is formed.

The central water mass formed in the Sargasso Sea in winter (labelled Western North Atlantic Central Water on Figure 6.12) has a temperature range of 20.0–7.0°C.

What is the approximate lower depth limit of this water mass according to Figure 4.20 (a temperature section across that part of the Sargasso Sea between the North American coast and Bermuda)?

It is about 1000 – 1100m. Figure 4.20 also demonstrates two other relevant points. The first is that a large volume of water in the North Atlantic subtropical gyre has a temperature close to 18°C. This '18°C water' is an example of a **mode water**, that is, a volume of water associated with large numbers of temperature measurements of the same

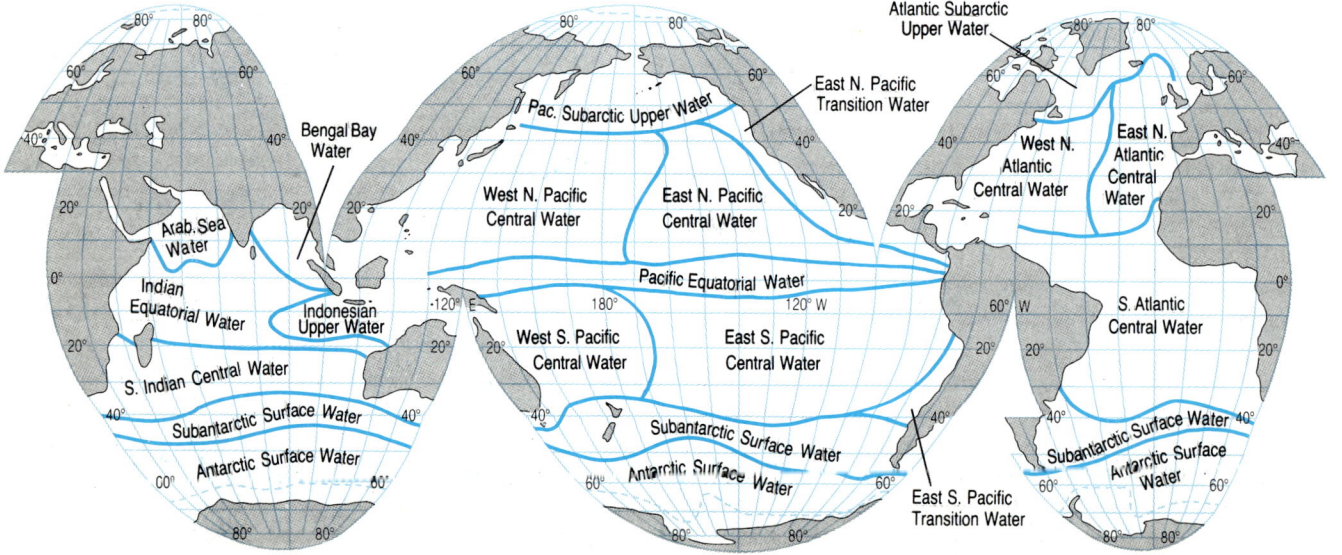

Figure 6.12 The global distribution of upper water masses. (You need not *remember* the details of this map.)

value. The concept of a mode water is intimately related to the second point shown by Figure 4.20, which is that within the main body of North Atlantic Central Waters, the isotherms are widely spaced; in other words, the waters are characterized by a thermostad or pycnostad (Section 5.1.1).

The temperature–salinity characteristics of Western and Eastern North Atlantic Central Waters are very similar. In common with Central Water Masses in the other oceans, North Atlantic Central Waters have moderately high temperatures and above-average salinities.

Bearing in mind that the Central Water Masses form below the anticyclonic subtropical wind systems, can you explain why their salinities are above average?

The subtropical anticyclones are regions where dry air subsides (*cf.* Figure 2.17), and net evaporation ($E–P$) is high, leading to high salinities in the mixed surface layer (*cf.* Figure 6.8) and hence in the water mass as a whole. However, there are differences between the Western and Eastern North Atlantic Central Waters, which reflect the differing environmental conditions on the two sides of the ocean. Of the two water masses, that on the eastern side of the ocean has salinities that are on average about 0.1 to 0.2 higher. One reason for this is the stronger influence of Mediterranean Water on the eastern side of the ocean; another possibility is that because mixing occurs down to deeper levels on the western side of the ocean, the upper water mass is brought into close contact with the low-salinity Western Atlantic Sub-Arctic Water that underlies it.

Western Atlantic Sub-Arctic Water and Mediterranean Water are examples of *intermediate water masses*, which flow between the upper water masses and the deep and bottom water masses. Of the two, Western Atlantic Sub-Arctic Water is the more typical because, like most Intermediate Water, it forms in subpolar regions (*cf.* Figure 6.13) where precipitation exceeds evaporation, and its salinity is therefore low.

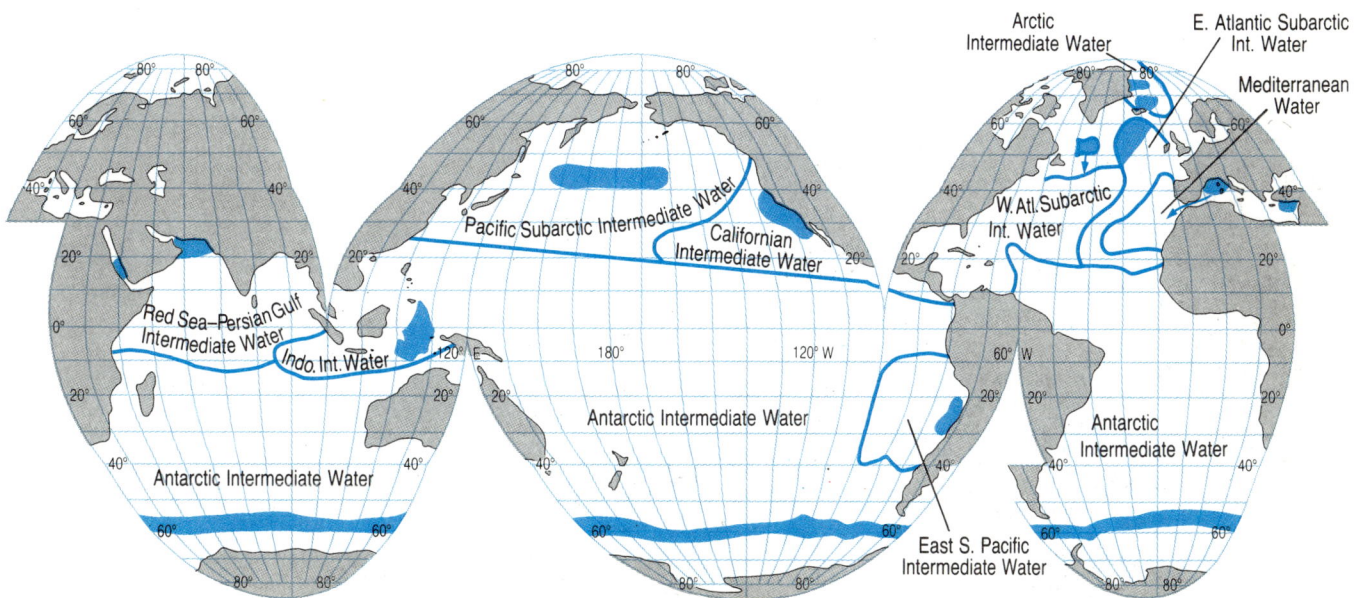

Figure 6.13 The global distribution of intermediate water masses (between about 550 and 1 500 m depth). The source regions of the water masses are indicated by dark blue tone. Note that Antarctic Intermediate Water is by far the most widespread intermediate water mass. (You need not *remember* the details of this map.)

Western Atlantic Sub-Arctic Water consists largely of **Labrador Sea Water** which apparently forms in a cyclonic gyre on the offshore side of the Labrador Current (Figure 5.20). In summer, the density of surface water is lowered by the addition of freshwater from melting sea-ice and icebergs (Section 5.5.1). In winter, however, the surface water in the Labrador Sea is cooled by the pack ice and by cold, dry Arctic air masses that have passed over northern Canada.

Which heat-budget loss terms are increased by the passage of these air masses?

Heat will be lost to the cold air masses by conduction and convection and, because they are also dry, by evaporation. Q_h and Q_e are therefore both increased. The density of surface water in the Labrador Sea is therefore increased through decrease in temperature and increase in salinity. The increased salinity is still fairly low in absolute terms (~34.9) and the decreased temperature is relatively high (~3°C) but their combined effect is sufficient to increase the density of surface water above that of the underlying water. The upper water column is therefore destabilized and vertical convection occurs, with denser surface water sinking and displacing less dense subsurface water, which rises to the surface. Surface water circulating in the gyre is thus repeatedly subjected to vertical mixing, so that eventually a water mass about 1500m thick may be formed. The destabilization process is enhanced by the fact that, because circulation is under the influence of cyclonic winds, there is a divergence of surface water, and the isotherms bow upwards towards the middle of the gyre (*cf.* Figure 3.23(b)) so that relatively dense water is brought close to the surface. (This combination of cooling by cold, dry air and cyclonic circulation also characterizes the conditions associated with the formation of Mediterranean Water.)

After sinking, Labrador Sea Water spreads out at mid-depths both eastward across the northern Atlantic, and southward along the western boundary of the ocean. Labrador Sea Water is very distinctive with temperatures of 3–4°C and low salinities (<34.92). It is therefore considered to be a subpolar mode water.

The subpolar mode water that forms in the Antarctic Polar Frontal Zone (*cf.* Question 6.7) is known as **Antarctic Intermediate Water (AAIW)** and is the most widespread intermediate water mass in the oceans (Figure 6.13). It is characterized by temperatures of 2–4°C and salinities of about 34.2, and is therefore significantly fresher than Western Atlantic Sub-Arctic/Labrador Sea Water.

QUESTION 6.8 Can Antarctic Intermediate Water be identified on Figure 5.23?

After sinking at convergences in the Frontal Zone, Antarctic Intermediate Water spreads northwards throughout the Southern Hemisphere; in the Atlantic Ocean, its low salinity enables it to be traced to at least 20° N (Figure 6.14). The relationship between the flow of Antarctic Intermediate Water and that of the major deep and bottom water masses of the Atlantic Ocean is shown in Figure 6.15.

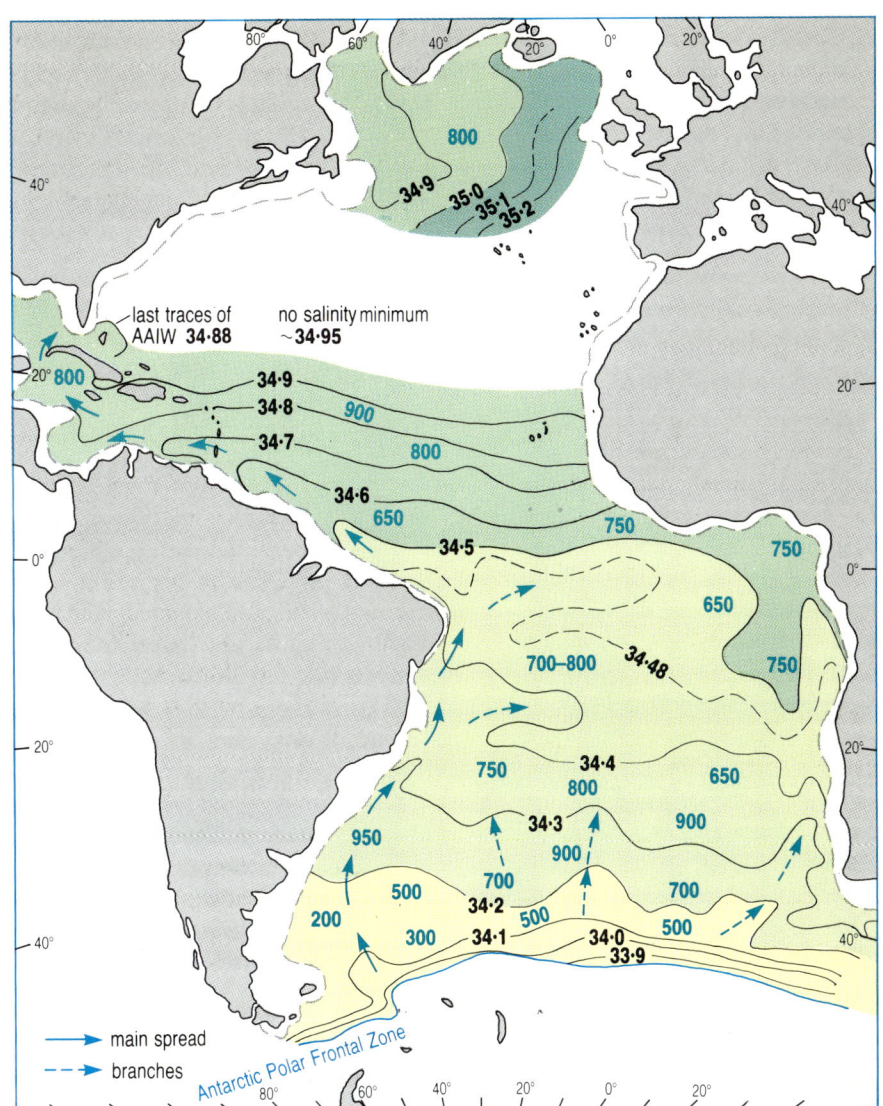

Figure 6.14 The spread of the least-mixed layer of Antarctic Intermediate Water in the Atlantic, as shown by the position of the salinity minimum at about 500–900 m depth. The numbers on the contours are salinity values; the numbers in blue give the approximate depth of the salinity minimum (in m). The thin dashed line is the 600 m isobath. The salinity contours in the north-western Atlantic show the spread of Western Atlantic Sub-Arctic/Labrador Sea Water.

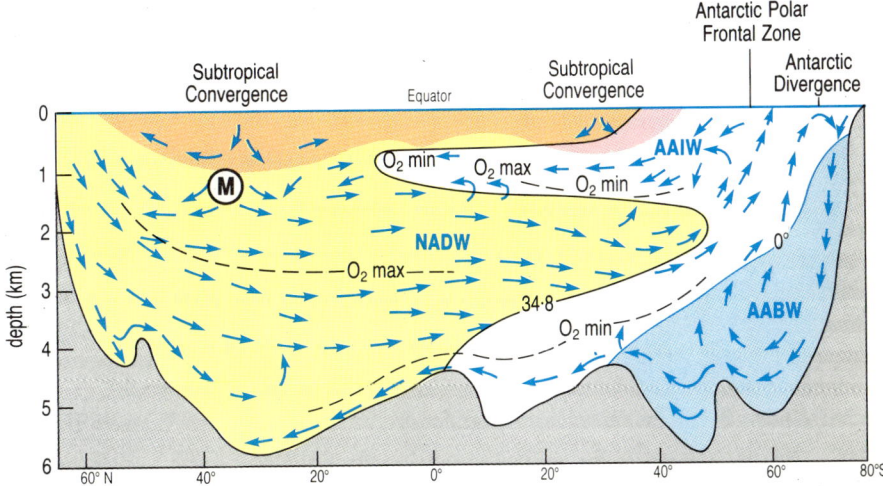

Figure 6.15 Meridional cross-section of the Atlantic Ocean, showing movement of the major water masses; NADW = North Atlantic Deep Water; AAIW = Antarctic Intermediate Water; AABW = Antarctic Bottom Water.
Water with salinity greater than 34.8 is shown yellow; note how the low salinity tongue of AAIW extends northwards from the Antarctic Polar Frontal Zone, to overlie the more saline NADW. The M at about 35°N indicates the inflow of water from the Mediterranean. Water warmer than 10 °C is shown pink/orange, and the cooler than 0 °C (corresponding approximately to the distribution of AABW) is shown blue. The oxygen maxima and minima will be explained in Section 6.5.

6.3.2 DEEP AND BOTTOM WATER MASSES

Deep water masses flow between the surface and intermediate water masses and the sea-bed; if the deepest water in contact with the sea-bed is distinguishable from overlying water, it is referred to as *bottom water*.

North Atlantic Deep Water

As shown by Figure 6.15, the major deep water mass of the Atlantic is **North Atlantic Deep Water (NADW)**. The main source of North Atlantic Deep Water is believed to be the cyclonic subpolar gyres in the Norwegian and Greenland Seas. As shown in Figure 5.20, there is a fairly free passage of surface water between these Norwegian and Greenland Seas and the North Atlantic, with water flowing into the Seas mainly between Scotland and Iceland, and out mainly between Iceland and Greenland. However, the irregular plateau extending from Scotland to Greenland (and passing through the Faeroe Islands and Iceland) presents a major obstacle to flow at depths greater than about 400m and a complete barrier at depths greater than about 850m (Figure 6.16). Furthermore, the bottom topography isolates the Norwegian and Greenland Seas from the deepest water in the Canadian and Eurasian Basins (Figure 5.20).

The Atlantic water that flows into the Norwegian and Greenland Seas is exceptionally warm and saline for these latitudes, with temperature and salinity values generally in excess of 8°C and 35.25 respectively.

Can you identify the origin of this water?

This is water from the North Atlantic Drift, the downstream continuation of the Gulf Stream. While circulating in the Norwegian and Greenland Seas, this water is cooled considerably. It is presumed that deep convection occurs and that the water sinks, filling up the basin—which in places is 3km deep—to the level of the submarine ridge. The deep water masses that accumulate in the Norwegian and Greenland Seas have a salinity of ~34.9 and temperatures below 0°C.

What has caused the salinity to decrease from the 35.25 characteristic of North Atlantic Drift water?

Like the Labrador Sea, the region is generally one of excess precipitation, sufficient to drastically reduce the salinity of the water while it is in the mixed surface layer. However, the salinity of the *deep* waters in the Norwegian and Greenland Seas is *increased* as a result of a deep high-salinity outflow from the Arctic Sea. This high-salinity water is believed to result from the seasonal formation of shelf ice around the margins of the Arctic, which abstracts freshwater and leaves behind dense 'brine' (this process will be discussed in more detail in connection with ice-formation in the Southern Ocean).

The Deep Water accumulated in the Norwegian and Greenland basins intermittently overflows the submarine ridge. It does so at various specific locations—notably through the Denmark Strait (between Iceland and Greenland), between Iceland and the Faeroe Islands, and through a narrow channel south of the Faeroe Islands (Figure 6.16). As the deep water overflows the plateau and flows down into the depths of the Atlantic, there is extensive turbulent mixing with overlying water, some of which is entrained. The characteristics of the resulting water mass therefore depend not only upon the characteristics of Norwegian Sea and Greenland Sea deep waters, but also on the characteristics of the water above the various outflows and the degree of entrainment that occurs.

Two types of North Atlantic Deep Water can be distinguished; these are called North-East Atlantic Deep Water and North-West Atlantic Bottom Water. As shown schematically in Figure 6.16, both have their main southward flow on the western side of the Mid-Atlantic Ridge. *North-East Atlantic Deep Water* enters the Atlantic between Iceland and Scotland. It consists essentially of a mixture of the original deep water and the overlying North Atlantic Central Water, and so has temperature–salinity characteristics of 2.5°C and 35.03. Much of this North-East Atlantic Deep Water flows through the Gibbs Fracture Zone into the western Atlantic to overlie *North-West Atlantic Bottom Water*. North-West Atlantic Bottom Water consists of the deep water that overflowed the ridge between Iceland and Greenland and then entrained North-East Atlantic Deep

Figure 6.16 Schematic map showing the main paths followed by the two types of North Atlantic Deep Water—North-East Atlantic Deep Water and North-West Atlantic Bottom Water—in relation to the topography of the sea-floor and the flow of Labrador Sea Water.

Water, North Atlantic Central Water and Labrador Sea Water during its passage round southern Greenland into the Labrador Sea.

The influence of Labrador Sea Water on North Atlantic Deep Water may be observed in northern and western parts of the North Atlantic. What other intermediate water mass has a significant influence on the characteristics of North Atlantic Deep Water, particularly on the eastern *side of the ocean?*

Mediterranean Water. As described earlier, the least-mixed layer of Mediterranean Water is generally found at a depth of about 1000 m (Figure 6.11), but its influence extends down to more than 2 000 m.

Antarctic Bottom Water

Antarctic Bottom Water (AABW) is the most widespread water mass in the world and is found in all three ocean basins, particularly in their southern parts (Figure 6.17). It forms in winter around the continent of Antarctica, particularly in the Weddell Sea and Ross Sea (Figure 5.22).

Before we discuss how Antarctic Bottom Water forms, we should emphasize an important aspect of the stratification of polar waters, in the Arctic as well as the Antarctic. Except where ice is actually forming, surface water is relatively fresh, because of the combined effects of excess precipitation and the production of meltwater; this surface water is also very cold. Below the surface layer there is generally a layer of relatively warm water which is nevertheless denser because of its relatively high salinity. The stability of this layering may be destroyed through turbulent mixing caused by wind, combined with an increase in density of the surface waters.

In northern polar latitudes, the main mechanism whereby the density of surface waters is increased is winter cooling by cold winds. In the Southern Ocean, where the seasonal production of ice is more extensive (Figure 5.21), the interaction between ice and surface waters plays a major role in the formation of dense water.

Figure 6.17 The global distribution of deep and bottom water masses (between a depth of about 1 500 m and the sea-floor). As in Figure 6.13, the source regions are shown by dark blue tone. The fine dashed line is the 4 000 m isobath. (The unlabelled regions are to a large extent occupied by Pacific and Indian Ocean Common Water—see text.)

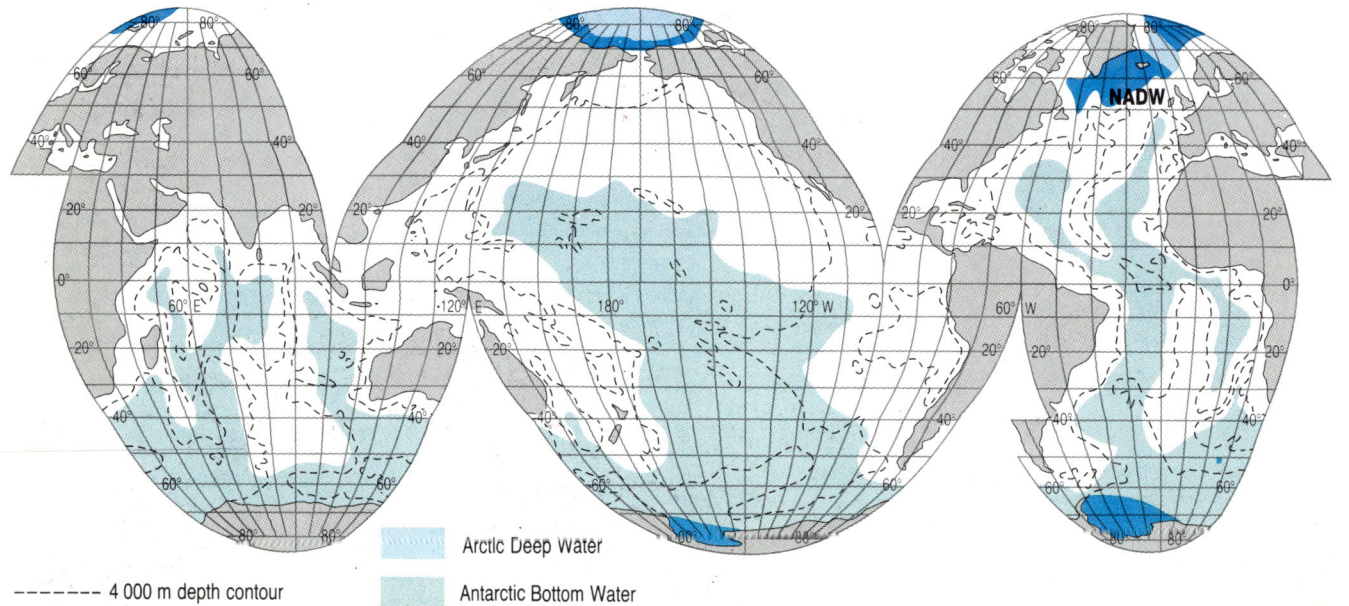

Ice–water interaction is perhaps best illustrated through discussion of **polynyas**, extensive areas of ice-free water within the winter ice cover. There are two types of polynyas: coastal polynyas and open ocean polynyas. *Coastal polynyas* develop when strong winds blowing off the Antarctic continent drive newly formed ice away from the shoreline, exposing a zone of open sea that might be 50–100 km wide. *Open-ocean polynyas* develop far from the coast within the pack ice (both over the continental shelf and in deeper water), and include the largest and most long-lived polynyas. The Weddell polynya was an enormous area of open ocean that reappeared in the Weddell Sea during three consecutive winters (1974–1976); at its largest it measured about 1000 by 350 km. The Weddell polynya reappeared in approximately the same position each year (Figure 6.18), above a sea-bed topographic high, known as the Maud Rise. This implies that its formation may be related to the upward deflection of relatively warm subsurface water, and in fact this is a

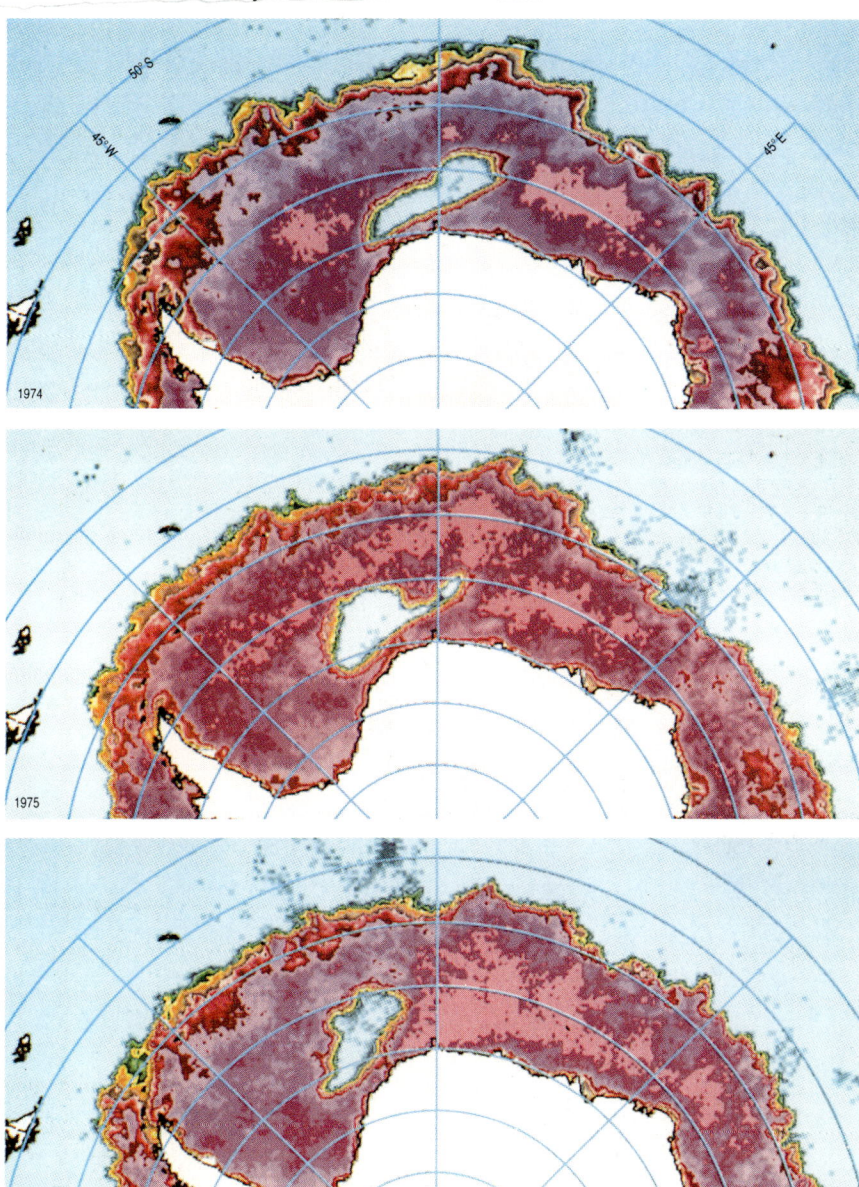

Figure 6.18 *Nimbus* satellite images showing the position of the Weddell polynya during the southern winters of 1974, 1975 and 1976. The westward movement of the polynya has been attributed to the generally westward current flow in the region. As in Figure 5.21, a light blue tone represents regions that are ice-free, while a pinky-red tone represents regions that are more or less completely covered with ice. The Antarctic continent has been superimposed in white. Similar satellite images have shown that polynyas also occur in the Arctic ice cover, particularly in the Greenland Sea.

possible explanation for a number of other open-ocean polynyas.

Both coastal and open-ocean polynyas are sites where surface water may be cooled and destabilized so that deep convection occurs.

QUESTION 6.9 We stated earlier that water in contact with ice is cooled by conduction, until a new heat balance is attained. Why, then, are polynyas—regions with *no* ice cover—characterized by significant heat loss?

In fact, the main mechanisms whereby heat is lost to the atmosphere are different for the two types of polynya. Coastal polynyas have been referred to as 'sea-ice factories': the wind drives sea-ice away from the continent as soon as it freezes, re-exposing the sea-surface to the atmosphere so that more ice can form (Figure 6.19). The continued production of ice removes large amounts of heat from surface water, mainly in the form of *latent heat of freezing*. This is the heat lost to the atmosphere by water while it remains at freezing point ($-1.9°C$ for seawater) as ice crystals are in the process of forming, and is analogous to the latent heat of condensation released when water vapour condenses to form liquid water. Coastal polynyas 'manufacture' ice on an enormous scale, perhaps producing much of the ice in the adjacent ocean. It has been calculated that the heat flux to the atmosphere from a coastal polynya is more than 300 Wm^{-2}, enough to supply a ten-centimetre-thick layer of ice to the adjacent sea each day.

What effect will the continued production of sea-ice have on the salinity of surface water?

As discussed in connection with the formation of North Atlantic Deep Water, the formation of ice results in an increase in the salinity of surface water. When ice forms, some salt is trapped amongst the ice crystals, but sea-ice is generally less saline than the water from which it forms and so the remaining water is correspondingly more saline. This effect is very

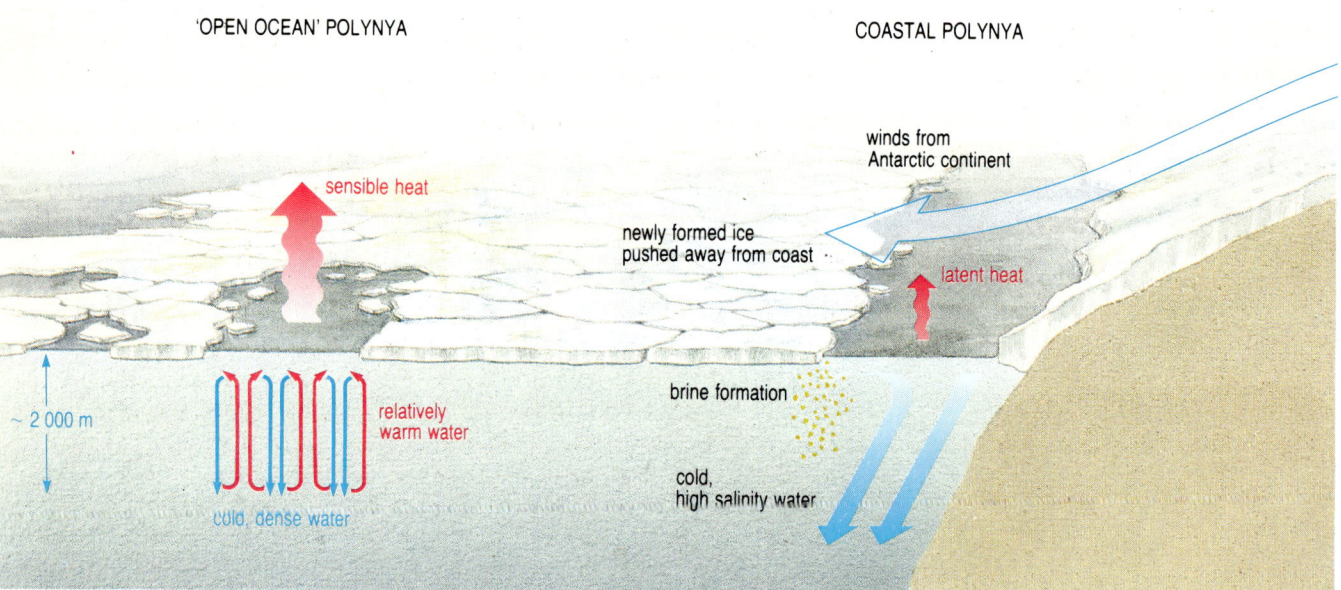

Figure 6.19 The different roles played by coastal and 'open ocean' polynyas in the production of Antarctic Bottom Water.

marked in coastal polynyas where ice is continually forming and being removed from the area. Surface water in coastal polynyas is both saline and cold, and may be dense enough to sink.

In 'open-ocean' polynyas, heat is lost from the sea-surface mainly by conduction/convection. Water cooled at the surface sinks and is replaced by warmer subsurface water which in turn is cooled and sinks, forming deep convection cells (Figure 6.19). After the Weddell polynya had formed, the temperature of deep water changed dramatically, decreasing by 0.8°C all the way down to a depth of 2500m. It is reasonable to assume that the 'missing heat' had been carried to the surface by convection, and it has been estimated that the rate of overturning of water necessary to transport this much heat may have been as much as $6 \times 10^6 m^3 s^{-1}$ during the winter months, when the polynya was active.

Antarctic Bottom Water is believed to form both over the Antarctic continental shelf and in deeper water. Poleward of about 60°S, the difference in density between the surface layer of cold, fresh water and the underlying warmer, saline layer—known as **Circumpolar Deep Water**—is slight, and so the stratification is easily destabilized. Strong winds mix the two layers together, aided by the turbulence generated around ice-floes extending down into the water. This mixing combines with the increase in density of surface waters—particularly as a result of processes occurring in polynyas—to produce a dense water mass which sinks and mixes with underlying waters to produce Antarctic Bottom Water.

The densest Antarctic Bottom Water is formed over the continental shelf, where the shelf waters have salinities of 34.4–34.8 and a temperature of about −2°C (this low temperature can be attained because the freezing point of seawater decreases with increasing pressure). Water sinking from the surface circulates for some time over the shelf before flowing down the continental slope to the deep sea. As it does so, it mixes to some extent with Circumpolar Deep Water and its density is slightly reduced. The resulting water mass is not as dense as the newly formed Deep Waters of the Norwegian and Greenland Seas (or that of the Mediterranean), but it does not have to flow over submarine ridges or pass through narrow straits, where large volumes of less dense water may be entrained through vigorous turbulent mixing. Antarctic Bottom Water is therefore the densest water mass in the open ocean.

After formation, Antarctic Bottom Water circulates eastwards around the Antarctic continent, perhaps a number of times, and then flows northwards, at a rate of about 20×10^6 cubic metres per second. This transport figure includes entrained waters, and the actual production of Antarctic Bottom Water with temperature–salinity characteristics of −0.4°C and 34.66 is probably about $10 \times 10^6 m^3 s^{-1}$. In the western trough of the Atlantic Ocean, the Antarctic Bottom Water flows northwards below southward-flowing North Atlantic Deep Water (readily distinguishable by its higher salinity), but on the eastern side of the Mid-Atlantic Ridge the northward passage of Bottom Water is restricted by the Walvis Ridge, which extends south-westwards from south-west Africa to the Mid-Atlantic Ridge at a depth of less than 3500m. However, the tendency for Antarctic Bottom Water to flow along the western side of the ocean appears to be a general feature of the deep circulation; we will discuss this again briefly in Section 6.5.

Before moving on, we should briefly consider Antarctic Circumpolar Water. This relatively warm (~0.25–2.0°C), relatively saline (>34.6) water may be clearly seen on Figure 5.23.

What water masses are represented by the high-salinity, high-temperature signature?

This combination of relatively high salinity and relatively high temperature is due mainly to the presence of North Atlantic Deep Water. In flowing southward, this has entrained Central Waters formed in the subtropical gyres (themselves relatively warm and saline), and been influenced by the warm saline outflow from the Mediterranean. Figure 6.20 shows how the modified North Atlantic Deep Water/Circumpolar Water flows up under the Antarctic Intermediate Water and rises towards the Antarctic Divergence.

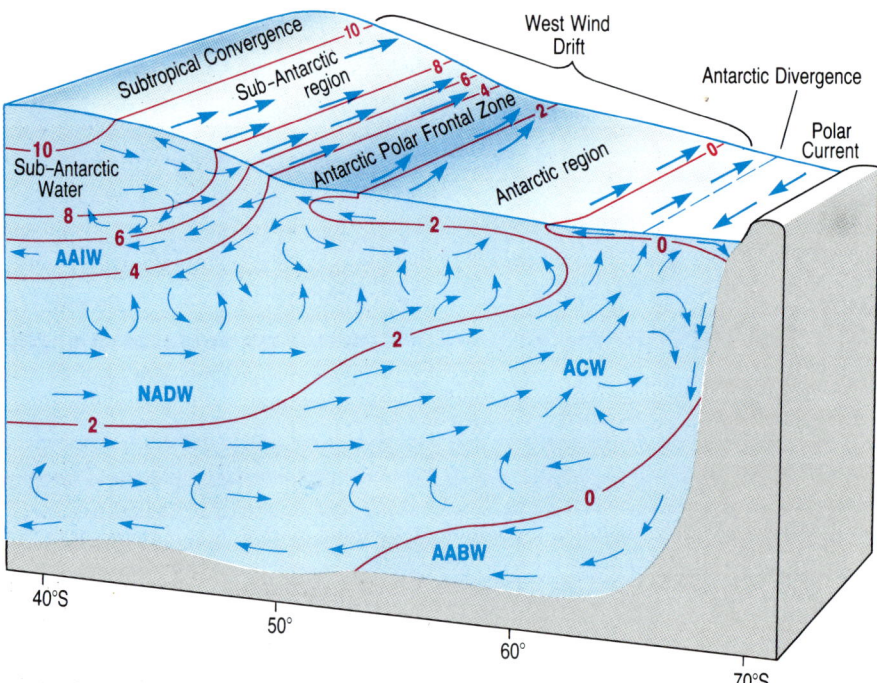

Figure 6.20 Schematic block diagram showing the surface currents and the vertical motion of water masses in the Atlantic Ocean poleward of about 40° S. North Atlantic Deep Water (NADW) becomes Antarctic Circumpolar Water (ACW) and rises to the surface at the Antarctic Divergence. Surface water flowing northwards from the Antarctic Divergence sinks at the Antarctic Polar Frontal Zone (as AAIW), while that flowing southwards may become AABW. The contours show isotherms in °C; this schematic diagram should be compared with Figure 5.23 which shows the actual temperature and salinity distribution measured in the Drake Passage between 56° and 62° S.

Thus, warm water carried downwards in the subtropical gyres (and the Mediterranean) is transported polewards and upwards until eventually it reaches the surface of the ocean around Antarctica, where a large amount of heat is given up to the atmosphere. Cooled water flowing northwards from the Antarctic Divergence will sink at convergences in the Antarctic Polar Frontal Zone in the form of Antarctic Intermediate Water, while that flowing polewards may eventually be converted to Antarctic Bottom Water (Figure 6.20). The cycle of water mass formation is therefore an intrinsic part of the thermohaline circulation which, along with the surface current system, acts to transport heat from low to high latitudes.

Pacific and Indian Ocean Common Water
Pacific and Indian Ocean Common Water is the largest water mass, and accounts for about 40% of the total volume of water in the oceans. There are no source regions of Deep Water in the North Pacific, and Red Sea/Persian Gulf Water—a relatively warm, saline water mass—is the only dense water mass formed in the vicinity of the Indian Ocean (*cf.* Figure 6.13).

How, then, is Pacific and Indian Ocean Common Water formed?

It must be formed by mixing of other water masses; the most likely candidates are Antarctic Intermediate Water, North Atlantic Deep Water and Antarctic Bottom Water. Of these, Antarctic Bottom Water is by far the most important contributor to Pacific and Indian Ocean Common Water. It contributes about half the water, and North Atlantic Deep Water and Antarctic Intermediate Water each contribute about a quarter. As discussed earlier, North Atlantic Deep Water flows into the Southern Ocean in the western basin of the Atlantic; here it mixes with Antarctic Intermediate Water above and Antarctic Bottom Water below. It continues its passage mainly eastwards around the tip of South Africa (still mixing with the other water masses), and then flows northwards into the Indian basin and thence into the Pacific.

In Section 6.4 you will see how the observed temperature–salinity characteristics of Antarctic Intermediate Water, North Atlantic Deep Water and Antarctic Bottom Water may be used to quantify the contributions of these water masses to Pacific and Indian Ocean Common Water.

6.4 OCEANIC MIXING AND TEMPERATURE–SALINITY DIAGRAMS

By identifying the source regions of water masses, and following their subsequent passage through the oceans, we can build up a qualitative picture of the three-dimensional circulation of the oceans. As water masses move away from their source regions, their temperature–salinity characteristics are changed by mixing with adjacent water masses, and we can *use* these changes to assess *how much mixing* has occurred. Before describing how temperature–salinity characteristics may be used in this way, we should first say a little about how mixing occurs in the oceans.

6.4.1 MIXING IN THE OCEANS

Flow in the oceans is turbulent (Section 3.1.1), and mixing is predominantly the result of 'stirring' by turbulent eddies. Turbulent mixing is most pronounced along isopycnic (density) surfaces, where it may occur with the least expenditure of energy, but mixing of water of different densities also occurs.

In Section 6.3, we tended to discuss the spread of water mass characteristics as if water masses always spread away from their source region in an even manner. However, it is becoming clear that mesoscale eddies play a significant part in the redistribution of heat and salt.

Although their motion may act to even out inhomogeneities within water masses, eddies also transport bodies of water with particular temperature and salinity characteristics from one part of the ocean to another (Sections 3.5.2, 4.3.5 and 5.5.2). Eddies containing anomalous bodies of water have now been found in many parts of the world ocean; for example, eddies containing Mediterranean Water—jocularly referred to as 'Meddies'—have been observed as close to the Straits of Gibraltar as the Canary Basin and as far away as the Bahamas. Furthermore, the 'tongue' of Mediterranean Water shown on Figure 6.11 may consist partly or wholly of 'decayed' Meddies.

If the discovery of mesoscale eddies shed new light on the processes contributing to oceanic mixing on a large scale, laboratory experiments demonstrated the existence of some very small-scale mixing processes. Until relatively recently, it was thought that bulk mixing by turbulence was the only significant mechanism whereby water mass characteristics could be changed. According to this view, the behaviour of seawater is entirely determined by its *joint* temperature and salinity characteristics. However, over the past decade or so, it has become clear that variations in temperature and salinity *of themselves* may cause mixing. Even in the absence of turbulent mixing, heat and salt diffuse through seawater as a result of processes occurring on the molecular level. Heat diffuses faster than salt, and the relatively fast transfer of heat from a layer of warmer, more saline water to one of colder, less saline water may be sufficient to cause small-scale instability. The best-known phenomenon associated with the double-diffusion of heat and salt is **salt fingering** (Figure 6.21), which was first observed below the outflow of Mediterranean Water into the Atlantic. It is now recognized that salt fingering and related processes can make a significant contribution to vertical mixing within the oceans generally, and that these very small-scale convective overturns affect the large-scale characteristics of water masses.

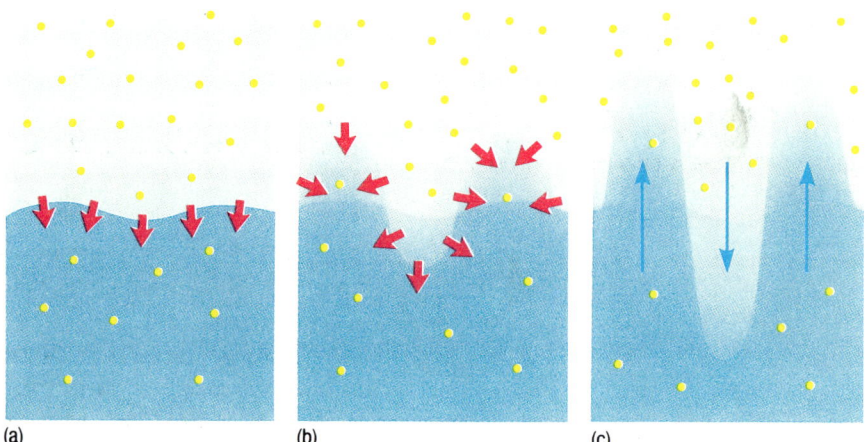

Figure 6.21 (a) When warm saline water overlies cooler and less saline water, the more rapid diffusion of heat (red arrows) than salt (dots), leads to instability and the development of 'salt fingers', as shown in (b) and (c).

However, if we assume that mixing occurs through turbulent processes only, and ignore mixing through processes occurring on a molecular level, we may use joint temperature–salinity characteristics to study both the distribution of water masses and how they mix together.

6.4.2 TEMPERATURE–SALINITY DIAGRAMS

In order to identify the water mass, or water masses, present at a given location in the ocean, a set of observations of temperature and salinity for successive depths at that location are plotted on a graph with temperature on the vertical axis and salinity on the horizontal axis, and the points are joined up in order of increasing depth. The result is known as a **temperature–salinity (T–S) diagram**.

If a water mass is completely homogeneous, it will be represented on a T–S diagram by a single point, and will be described as a **water type**. Observations clustering around such a point indicate the presence of this water type. For example, a number of observations of newly formed Deep Water in the north-western Mediterranean would all plot at or around a point corresponding to $T = 12.8°C$ and $S \approx 38.4$.

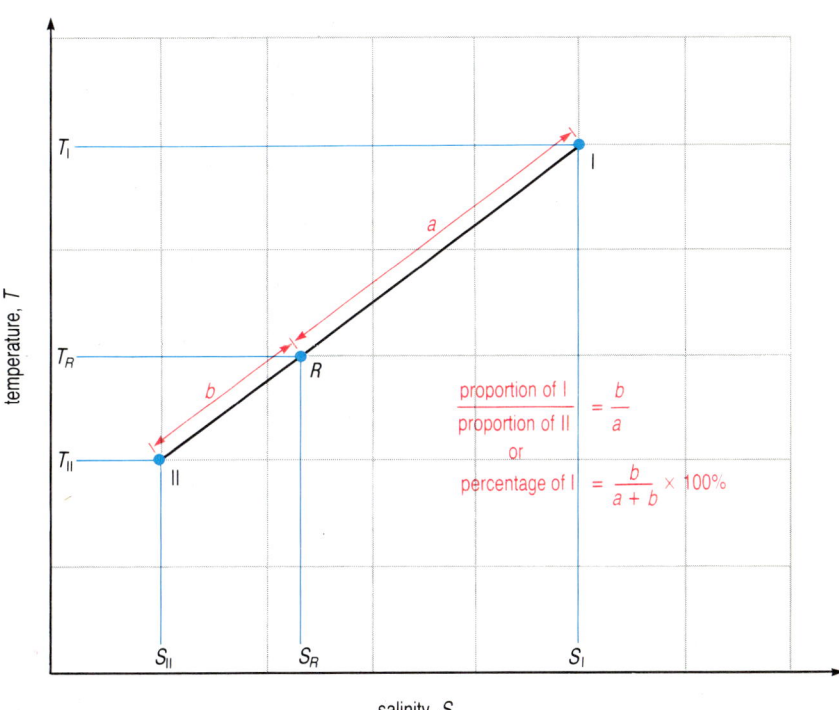

Figure 6.22 A temperature–salinity diagram showing the effect of mixing a water type I having T_I and S_I with water type II having T_{II} and S_{II}. The resulting mixture R (having T_R and S_R) will be represented by a point on the line between I and II, the position of which will be determined by the relative proportions of the two water types in the mixture.

Figure 6.22 shows what happens when two water types of differing temperature and salinity are mixed, and illustrates the following principles.

1 Whatever the relative proportions of the two water types, the point representing the temperature and salinity of any mixture *must* lie on the straight light joining the two water types on the T–S diagram.

2 Following on from 1, the actual position of the point representing the mixture will be determined by the relative proportions of the two water types. In the example shown in Figure 6.22, the point for the mixture, R, lies closer to type II than type I, so the mixture must contain a bigger proportion of type II. The actual proportions of the two water types that are present in the mixture can be determined by measuring the lengths of the segments a and b, as shown in Figure 6.22. (Note that we have been using the term 'water type' rather than 'water mass', because we are assuming that mixing is occurring between two water masses each

represented by a single temperature and salinity value, rather than by a range of values.)

Although *T–S* diagrams can be used to *predict* temperatures and salinities that result from mixing of water masses, the usual application of the method is to determine the relative proportions of different (known) water masses contributing to the water we are interested in, and for which we know the temperature and salinity (from measurements). The following question illlustrates how this can be done.

QUESTION 6.10   If water type I, with a temperature of 5°C and a salinity of 35.5 mixes with water type II with a temperature of 2°C and a salinity of 34.5, to give a mixture with *T–S* characteristics of 3°C and 34.85, what are the proportions of water types I and II in the mixture? (You can use Figure 6.22 by lightly adding scales of your own choosing and then plotting the data directly onto it.)

Before moving on to consider how *T–S* diagrams may be used to interpret more complex situations, we should note that not all straight segments of *T–S* curves reflect mixing *between* water masses—they may indicate variations *within one* water mass. Such variations might result from waters of slightly different *T–S* characteristics forming at different times of year and, according to their densities, sinking to different depths. Alternatively, surface conditions may vary within the source region during the period (usually winter) when the water mass forms. Water mixing down along sloping isopycnic (equal-density) surfaces will eventually become *vertically* stratified (Figure 6.23(a)) and the water mass will be represented by a more or less straight line on a *T–S* diagram (Figure 6.23(b)); the water mass used as an example in the Figure is North Atlantic Central Water.

The procedure described above can be extended to the more complex situation where three water types, I, II and III, are mixing together. In this case, the mixture *R* must lie inside the triangle formed by joining points I, II and III together on the *T–S* diagram (as in Figure 6.24). If we know from measurement the temperature and salinity of R (T_R, S_R), the relative proportions of the intermixing water types contributing to that mixture can be determined by a simple (if slightly tedious) graphical procedure. This is shown in Figure 6.24, which we can use to determine

Figure 6.23   (a) Schematic diagram showing the formation of a water mass by the sinking of surface water along isopycnic surfaces, in a region where environmental conditions at the surface vary. Water with a range of *T–S* values is formed, and distributions of *T* and *S* along a line drawn on the surface is reflected by their distribution with depth (e.g. at station A). The resulting *T–S* plot is a straight line, as shown in (b). The example used here is the formation of North Atlantic Central Water in the North Atlantic subtropical gyre, discussed in Section 6.3.1. Temperature is given in °C

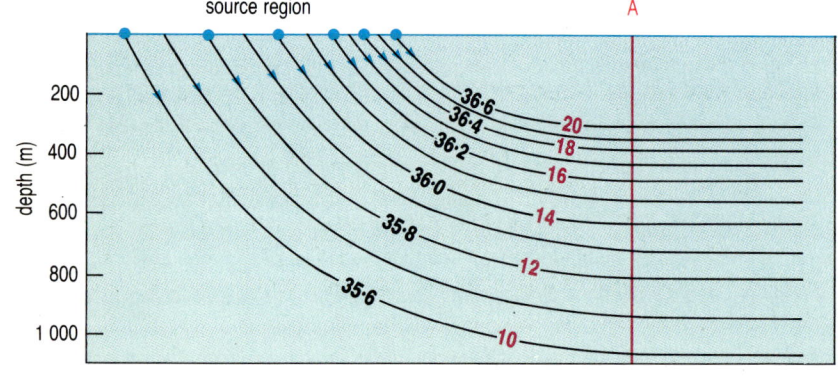

(a)

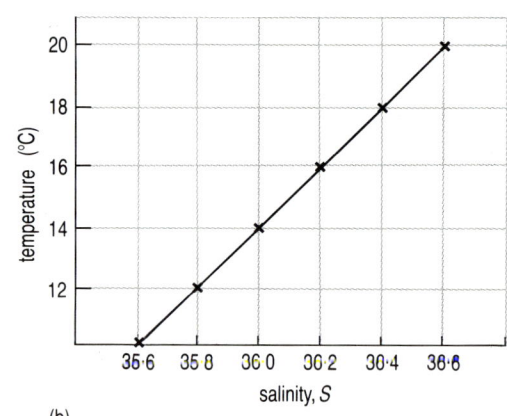

(b)

the relative proportions of water types I, II and III. Using a ruler to measure the segments a–f, and inserting values (here rounded off to the nearest 0.05) in the formula shown on the diagram, we get

I : II : III = 0.40 : 0.45 : 0.15 or 40%, 45%, 15%.

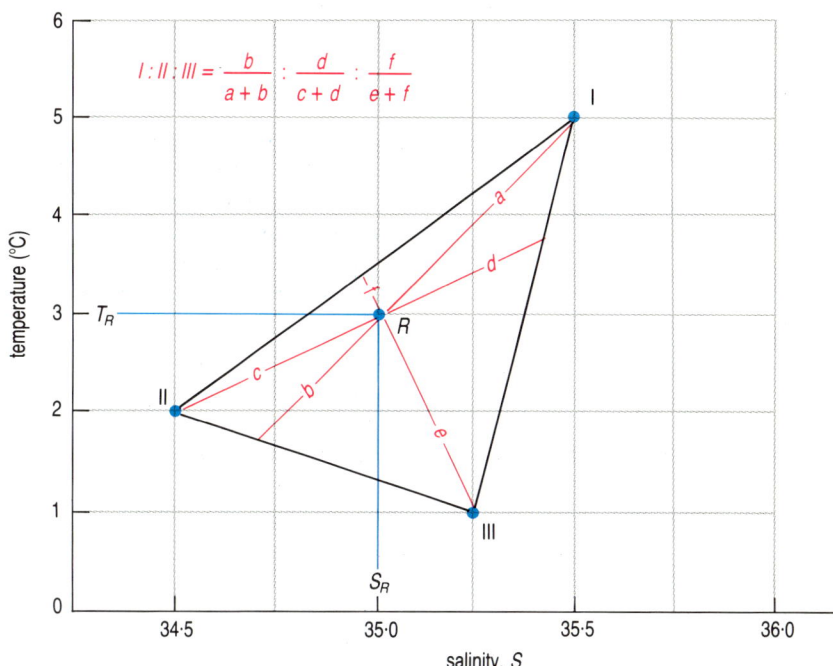

Figure 6.24 A temperature–salinity diagram showing the effect of mixing three water types (I, II and III) to give a mixture R with T = 3 °C and S = 35. The method used to determine the relative proportions of I, II and III in the mixture is shown here graphically and described in the text.

In other words, water type R on Figure 6.24 is the result of the mixing together of approximately 40% of water type I, 45% of water type II and 15% of water type III.

Having seen how the relative proportions of water types contributing to a mixture may be worked out using a temperature–salinity diagram, let us now consider how the distributions of temperature and salinity change as three water masses mix together in the oceans, and how the T–S curve changes as a result.

Imagine three homogeneous water masses overlying one another; one is at fairly shallow depth (200–600m), one at intermediate depth (600–1000m) and one fairly deep (1000–1400m). We will assume that the intermediate and deep water masses have the same temperature but different salinities—this approximates to what is often found in the oceans. Temperature and salinity profiles are shown in Figure 6.25(a) and (b), while the T–S relationships are shown in part (c). The diagrams in 1, at the top, illustrate the situation before any mixing has occurred, while 2 and 3 show subsequent stages as mixing progresses. As mixing occurs between *moving* water masses, the diagrams in 1, 2 and 3 should be seen as representing the situation at three different locations, although from the point of view of the moving water masses they correspond to three different stages in time.

Initially (stage 1), the three water masses are homogeneous and may be represented on the T–S diagram by three points (i.e. they are water types). As mixing progresses (stage 2) the sharp interfaces between the water masses become transition zones, so that the 'corners' of the

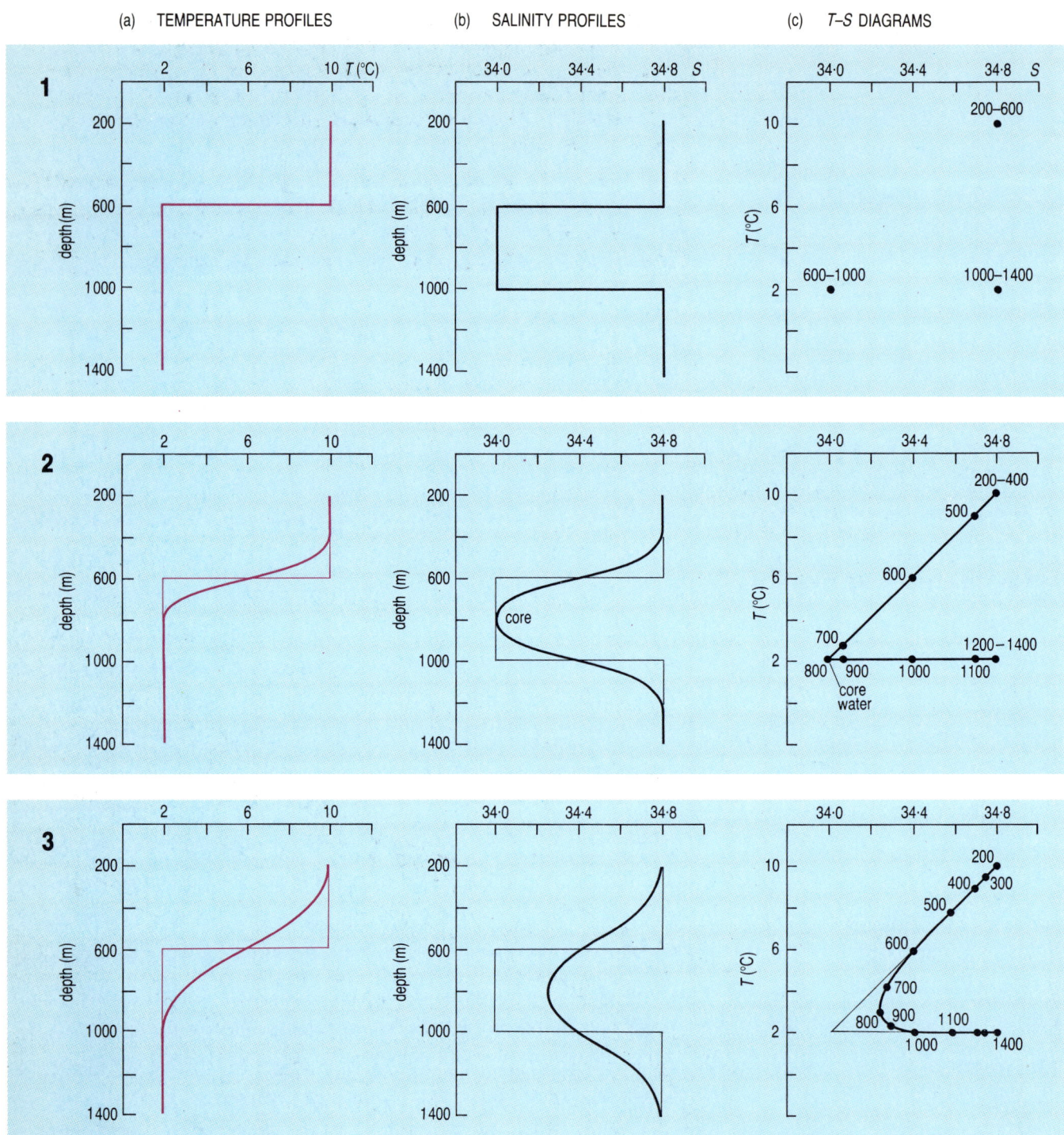

Figure 6.25 Profiles of (a) temperature and (b) salinity, along with (c) the corresponding T–S diagrams to illustrate the mixing of three homogeneous water masses (water types). Stage 1 (top) represents the situation before any mixing has taken place; stage 2 (middle) shows an early stage of mixing when the core of intermediate water is very prominent; by stage 3 (bottom), the core has been eroded.

temperature and salinity profiles become rounded off and water with characteristics between those at 400 m and 800 m, and between those at 800 m and 1200 m, appears on the T–S diagram. A layer of the intermediate water with its original temperature and salinity is still discernible. This is known as **core water** and it shows up on the T–S diagram as a sharp point. As the core water continues to be affected by mixing both above and below, the sharp angle on the T–S diagram is eroded away, and the final *T–S* plot has the curved shape shown in stage 3. The changing shape of the *T–S* plot, from a sharp point to a curve, thus corresponds to the modification of the 'least-mixed layer' of the water mass as it spreads away from its site of formation (as shown by Figures 6.11 and 6.14 for Mediterranean Water and Antarctic Intermediate Water, respectively).

Stage 3 of the hypothetical scenario we have been considering is similar in many ways to what was actually observed at *Meteor* Station 200, at 9°S in the Atlantic Ocean. We may use the *T–S* curve for this station, shown in Figure 6.26, to apply the principles illustrated in Figures 6.24 and 6.25.

QUESTION 6.11 First locate the position of *Meteor* Station 200 on Figure 6.15. Now assume that at this station, the water between 400m and 1800m depth is the result of the mixing of Antarctic Intermediate Water with North Atlantic Deep Water (below it) and with water that has the *T–S* characteristics represented by the 400m point on the *T–S* curve (above it).

(a) Is it possible to identify an eroded core of Antarctic Intermediate Water in Figure 6.26?

(b) What is the percentage of Antarctic Intermediate Water present in the water at 800m depth (to the nearest 10%)? (Assume that the characteristics of Antarctic Intermediate Water and North Atlantic Deep Water correspond to the *centres* of the blue rectangles.)

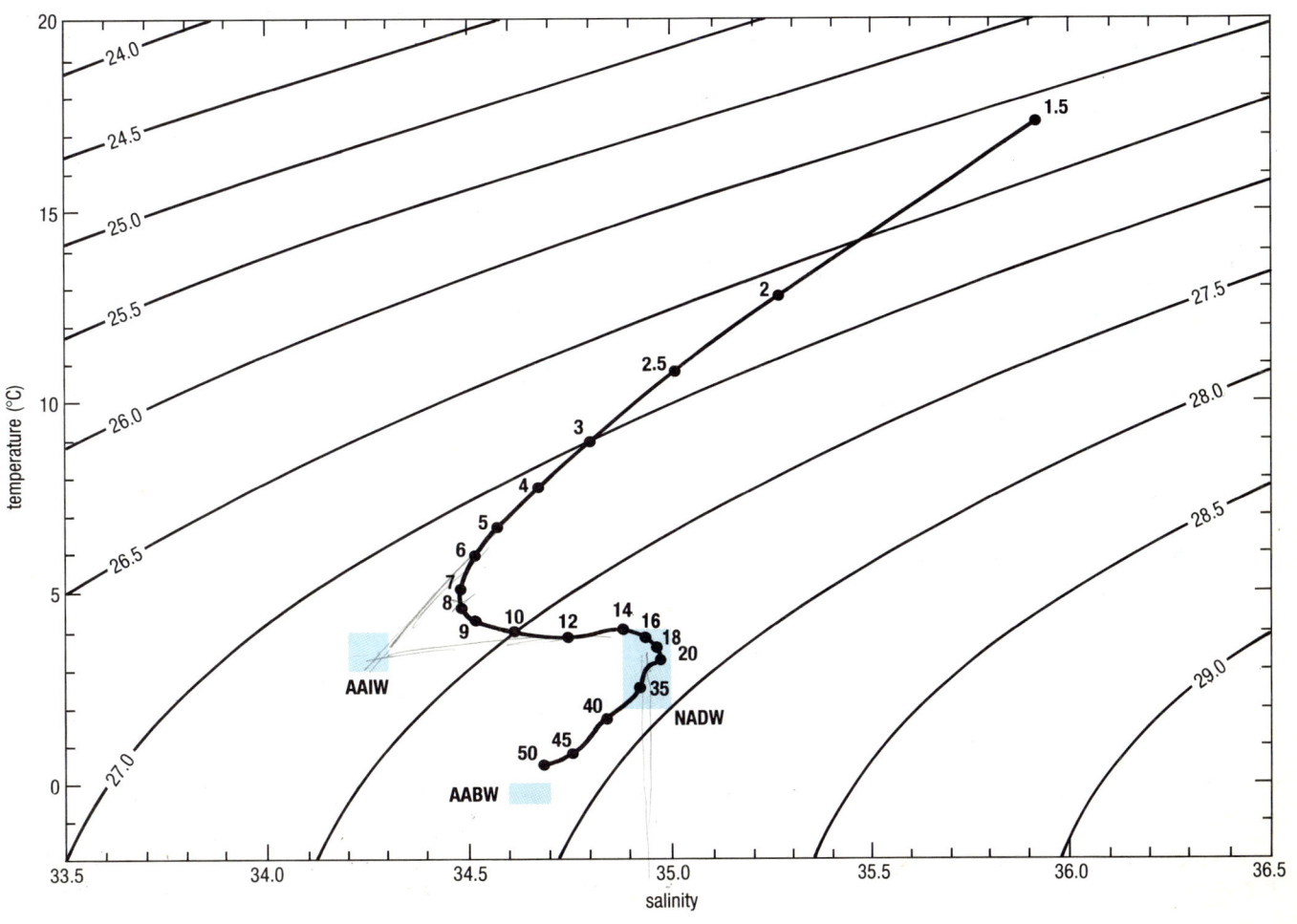

Figure 6.26 The *T–S* curve for *Meteor* Station 200, at 9° S in the Atlantic Ocean. Numbers on the curve give depths in hundreds of metres. Observations were made at intervals from 150 m down to 5 000 m. The *T–S* characteristics of Antarctic Bottom Water (AABW), North Atlantic Deep Water (NADW) and Antarctic Intermediate Water (AAIW) are also shown. (For use in Question 6.11.)

Temperature–salinity diagrams and stability

The density (ρ) of seawater is a function of temperature, salinity and pressure. By assuming constant (atmospheric) pressure, it is therefore possible to draw lines of equal density onto *T–S* diagrams. Variations in the density of seawater are very small—in the oceans as a whole, density at atmospheric pressure ranges between about 1025 and 1028 kg m^{-3}. For convenience, oceanographers generally write density in terms of σ (sigma), where

$$\sigma = \rho - 1000$$

Using this system, densities of 1026.51 and 1026.94 kg m^{-3} are written as 26.51 and 26.94, and may more easily be compared with one another. By convention, these σ values are written without units.

Until relatively recently, the temperature value used to calculate ρ and σ for a sample of water was generally the temperature at depth, i.e. its *in situ* **temperature**. The σ value calculated on the basis of *in situ* temperature, *in situ* salinity and atmospheric pressure is known as σ_t (**sigma-*t***). In Figure 6.26, the equal-density lines used are lines of equal σ_t.

However, the situation is complicated by the fact that seawater is compressible, albeit only slightly.

What implication does this have for the temperature of seawater at depth in the ocean?

The temperature of deep water will be raised through adiabatic compression (*cf.* Section 2.2.2). As a result, the *in situ* temperature of a sample of water is *higher* than the temperature that would be recorded for the same sample of seawater at the surface, under atmospheric pressure.

Adiabatic heating of seawater has two important consequences. The first is that because *in situ* temperature (T) is changed by pressure, it is not strictly a conservative property. On the other hand, **potential temperature** (θ), i.e. *in situ* temperature corrected for compression and heating, *is* a conservative property.

The second consequence of adiabatic heating is that a *T–S* curve may give a misleading impression of the degree of stability of the water column. For a water column to be stable, its density must increase downwards, so that the *T–S* curve representing it would cross σ contours corresponding to successively increasing density values, in the direction of increasing depth. If you look at Figure 6.26, you will see that the part of the *T–S* curve corresponding to the very deepest water, between 4500 and 5000 m, curves upwards, apparently indicating that at these depths density *decreases* with depth. This effect is spurious, and is partly the result of the *in situ* temperatures at depth being increased through abiabatic heating. If we were to plot θ against S, and compare it with contours of **sigma-theta** (σ_θ)—the σ values calculated from the potential temperature, θ, salinity, and atmospheric pressure—we might well find that this water column was stable all the way to the sea-floor. Any apparent instability still remaining would disappear when the direct effect of compression on the density of seawater was accounted for; because σ_t and σ_θ correspond to density *at atmospheric pressure*, the shape of a temperature–salinity curve in relation to contours of σ_t and σ_θ does not give a completely accurate impression of the stability of very deep water.

To overcome some of these problems, water temperatures are increasingly recorded in terms of θ, and water masses are generally defined in terms of their θ–S characteristics, rather than their T–S characteristics. Furthermore, more accurate methods of temperature measurement, combined with continuous profiling techniques, have meant that correcting for the effect of pressure is more likely to reveal interesting aspects of the temperature–salinity curve. Nevertheless, for many purposes, measurements of *in situ* temperature are adequate and so both T and θ are widely used.

To consolidate your understanding of this Section, study Figure 6.27—a θ–S curve for a station to the east of the Azores—and attempt Question 6.12.

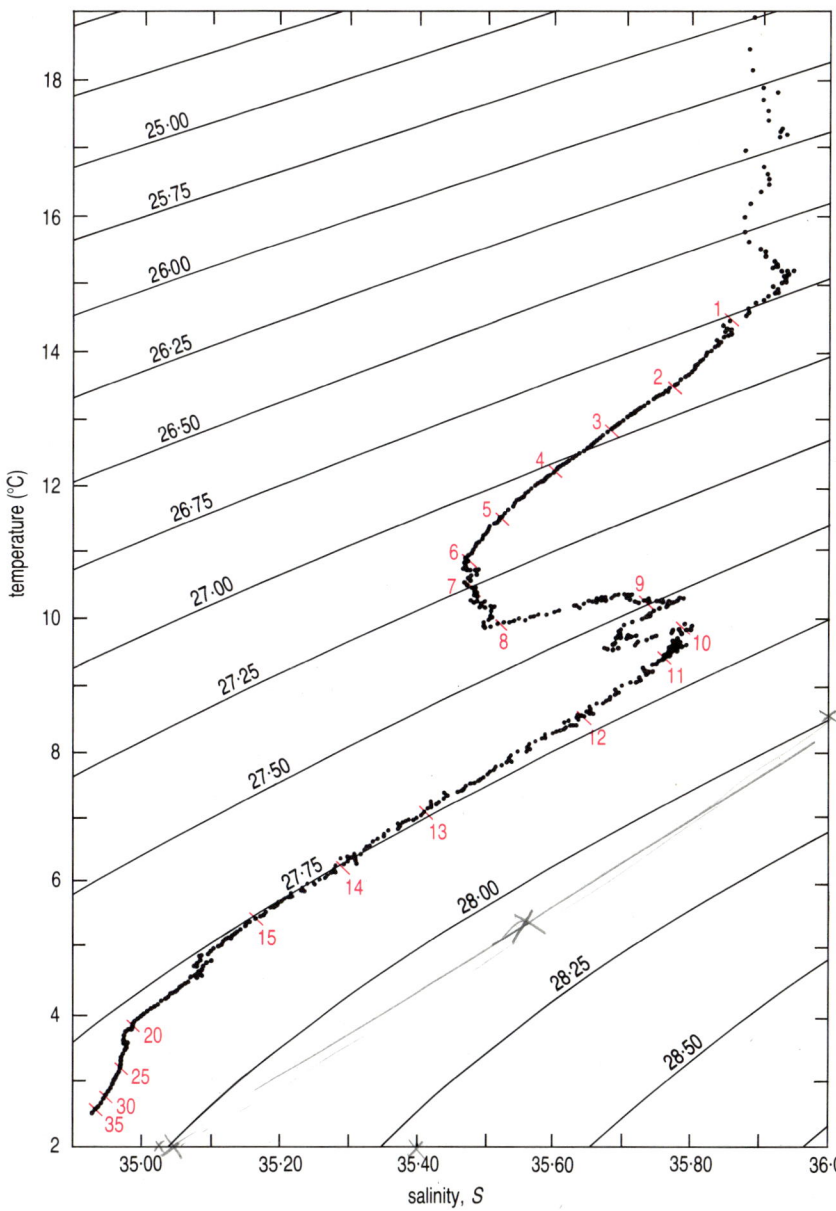

Figure 6.27 θ–S curve for station *Suroit* 1070, to the east of the Azores. The numbers on the curve are in hundreds of decibars, and so approximate to hundreds of metres. The topmost part of the curve, above 100 dbar, corresponds to the thermocline. Equal-density lines are lines of equal σ_θ. (Note the detail visible on this curve, which was computer-generated using θ and S data obtained by continuous profiling techniques. By contrast, the curve in Figure 6.26 was obtained from measurements made mostly at intervals of about 100 m or more.)

QUESTION 6.12 (a) The water mass represented by the top 600m or so of the θ–S curve is North (East) Atlantic Central Water.

(i) What is the water mass below that, between about 800m and 1200m depth, and what is unusual about it at this location?

(ii) What is the deepest water mass represented by the θ–S curve?

(b) Two water masses with θ = 2°C, S = 35.04 and θ = 8.5°C, S = 36.00 mix together in approximately equal proportions. What is particularly interesting about the resulting mixture?

The phenomenon identified in (b), whereby two water masses mix together to form water with a higher density than either of the original contributions, is known as 'cabelling'. It may be important in the production of deep and bottom water masses.

Water mass analysis has been invaluable to oceanographers attempting to build up a three-dimensional picture of large-scale flow within the oceans. Currents at depth are often too slow and/or too variable for their average motions to be observed directly. However, although temperature–salinity diagrams enable us to identify both the depth of the least-mixed (core) layer of a water mass and the direction in which it is spreading, they tell us nothing about the *rate* at which the water is moving. For this, we need to track water masses using a 'time-coded' tracer, or a *non-conservative* property of the water mass; how this may be done is discussed in Section 6.5.

There is, however, another *conservative* property of ocean water that we have not so far mentioned in connection with the tracking of water masses: its potential vorticity. As discussed in Section 4.2.1, away from regions of strong current shear, planetary vorticity f is very much greater than relative vorticity ζ and so, to a first approximation, potential vorticity is given by f/D where D is the thickness of the layer under consideration. In regions where water masses of significant vertical extent are forming, the water column is well mixed, i.e. there is a pycnostad in which isopycnic surfaces are widely spaced (Section 6.3.1). If we take D to be the distance between two selected isopycnic surfaces, then within the water mass f/D is relatively small (Figure 6.28); furthermore it will remain so as the water mass spreads away from the source region,

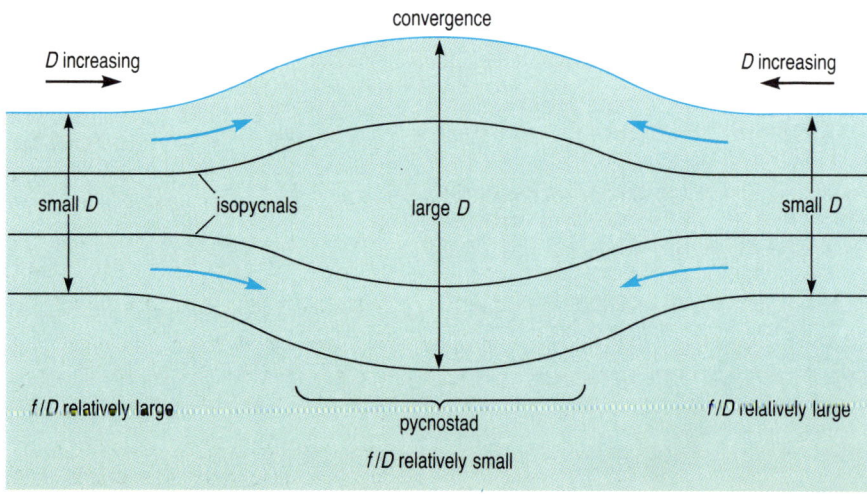

Figure 6.28 Schematic diagram to illustrate why potential vorticity is low in a homogeneous water mass; the example shown here is a Central Water Mass. In the open ocean away from regions of strong shear, $f \gg \zeta$ and so potential vorticity = f/D; if D is large, then f/D is small.

enabling it to be tracked. In Figure 6.29, the depth at which the potential vorticity within a particular water mass is at its lowest has been identified, and the salinity of the water at that depth has been contoured.

The spread of which water mass is shown by Figure 6.29?

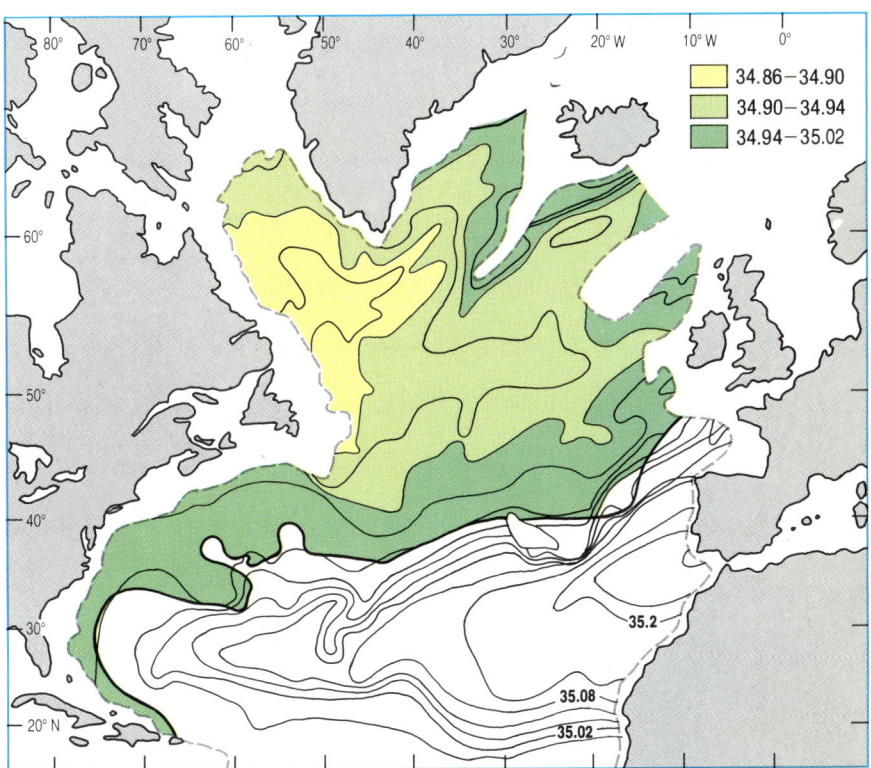

Figure 6.29 Distribution of salinity at the depth of the potential vorticity minimum (largest separation of isopycnic surfaces) associated with a particular water mass (see text). The depth of the potential vorticity minimum varies, but it is about 1 500 m. The heavy line indicates the furthest limit at which the potential vorticity minimum can be identified.

The region of origin of the water mass is the Labrador Sea, and this, combined with the low salinities associated with the water mass, identify it as Labrador Sea Water (cf. Figure 6.14).

6.5 TRACERS

The non-conservative property of ocean water most used traditionally as a tracer is the content of dissolved oxygen. Surface waters are supersaturated with dissolved oxygen because it goes into solution when air is churned down into the water by breaking waves and, in addition, is produced by phytoplankton photosynthesis; however, throughout the water column oxygen is used up in the bacterial oxidation (decomposition) of decaying material and in the respiration of organisms. Thus, the longer a water mass has been isolated from the atmosphere, the lower, in general, will be its content of dissolved oxygen. The oxygen content of surface waters is particularly high in polar regions, because cold water can contain more dissolved gas than can warm water. So, the oxygen maxima on Figure 6.15 correspond to the 'core' regions of cold water masses, while the oxygen minima correspond roughly to boundaries *between* water masses.

Silica is also used successfully as a tracer of certain water masses. The bottom waters of the Southern Ocean contain high concentrations of silica because diatoms in the surface waters use silica to build their skeletons, and much of this silica dissolves when they sink to the sea-floor on death. Figure 6.30 shows the distribution of dissolved oxygen and of silica along about 30° S, between South America and the Mid-Atlantic Ridge.

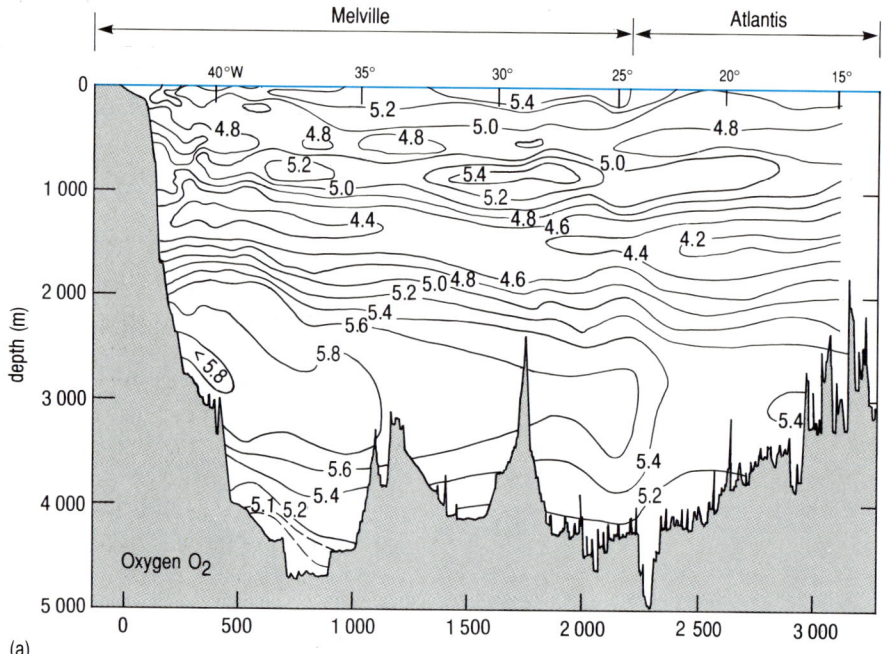

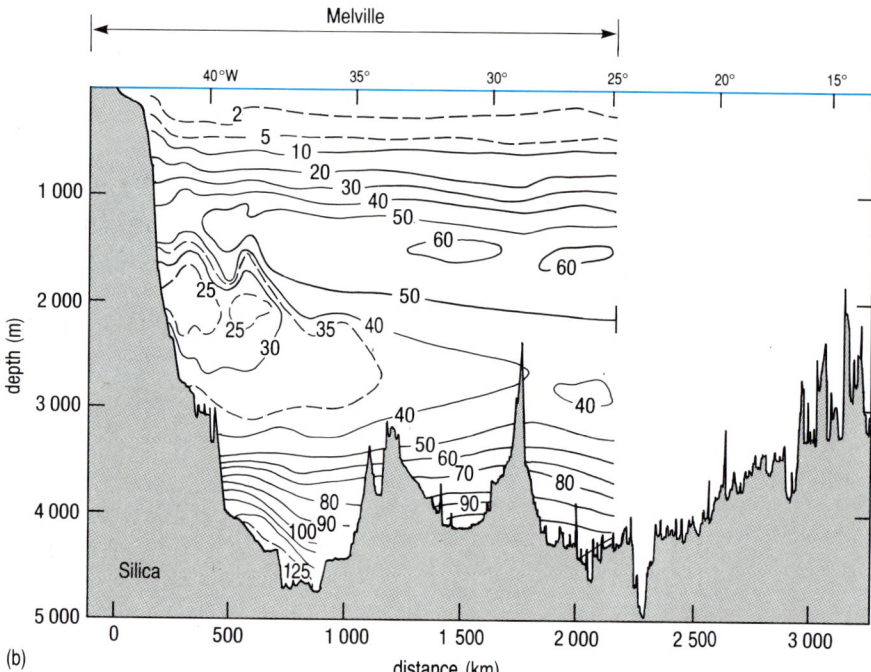

Figure 6.30 Sections of (a) dissolved oxygen concentration (ml l^{-1}) and (b) silica concentration (μmol l^{-1}) between South America and the Mid-Atlantic Ridge. The sections were made along 30° S (approximately) by the research vessels *Melville* and *Atlantis* in November 1976 and May 1959 respectively.

QUESTION 6.13 Given that North Atlantic Deep Water is characterized by low silica concentrations and Antarctic Bottom Water by high silica concentrations, can you identify these two water masses in Figure 6.30? What striking aspect of their flow patterns may be inferred from Figure 6.30?

Silica concentration is particularly useful for studying the mixing of North Atlantic Deep Water with Antarctic waters. This is because the silica concentration of North Atlantic Deep Water is very uniform, while its temperature–salinity characteristics—which reflect the fact that it is the sum of various contributions—are fairly broad.

Dissolved oxygen and silica are of limited use as tracers, because without further information about their production and/or consumption in the water column we cannot use them to get an indication of when the water mass concerned was formed at the surface—that is, of its 'age'. However, during the latter part of this century, oceanographers have been supplied with a number of useful tracers for which they *know* the exact time of entry (or at least the earliest possible time).

One group of substances that began to enter the oceans relatively recently are the chlorofluorocarbons (CFCs), which are used in refrigeration systems and as propellants in aerosols. Chlorofluorocarbons, generally referred to as Freons (their trade name), are particularly useful to the oceanographer because they are relatively easy and cheap to measure at sea, in relatively small amounts of seawater; some tracers—radioactive ^{39}Ar and ^{85}Kr, for example—may only be accurately measured by sampling hundreds of litres of water.

There is no natural source of chlorofluorocarbons, and since they were first manufactured in the 1930s, their total concentration in the atmosphere has increased almost exponentially. Nevertheless, different chlorofluorocarbons have increased in concentration at different rates: the ratio of CCl_3F (known as Freon 11) to CCl_2F_2 (Freon 12) in the atmosphere rose from near-zero in the mid-1940s to approximately 0.6 in the mid-1970s (Figure 6.31(a) and (b)). The Freon 11 : Freon ratio of water now at depth in the ocean is a measure of the relative concentrations of the two Freons in the atmosphere when the water was last at the surface (after allowance is made for their different solubilities). It is thus possible to 'age' the water; that is, to work out when it was last in contact with the atmosphere.

The Freon 11:Freon 12 ratio stopped increasing in about 1973, but significant production of a third Freon—Freon 113—started in 1975 (Figure 6.31(a) and (b)). Since then its atmospheric concentration has increased so rapidly that it can be used to pin-point the actual *year* that a water mass was last at the surface (providing it was post-1975). Freon 113

Figure 6.31 (a) The increases in the atmospheric concentrations of Freon 11 (CCl_3F) and Freon 12 (CCl_2F_2) between 1940 and 1975. Since 1975, concentrations have remained more or less constant and so Freon 11 : Freon 12 ratios are not very useful for 'aging' recently formed water masses. By the end of the period covered by this diagram, Freon 113 was also being produced.

(b) Changes in the Freon 11 : Freon 12 ratio and the Freon 113 : Freon 11 ratio, up until 1987.

(c) Concentration–depth profiles of Freon 11, Freon 12 and oxygen, in the South Atlantic at about 55° S. (1 pmol (= picomol) is 10^{-12} mol.) Note that despite its lower concentrations in the atmosphere, Freon 11 has higher concentrations in seawater than does Freon 12 because it is more soluble.

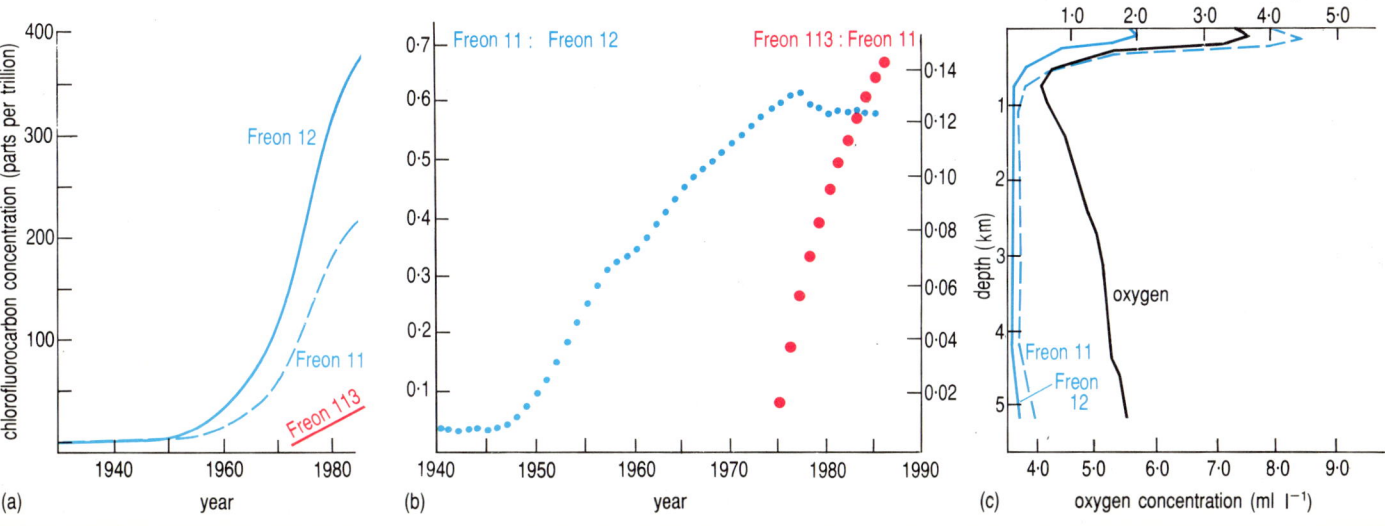

has proved particularly useful for the study of water masses that sink and spread very rapidly—North Atlantic Deep Water, for example.

The concentrations of chlorofluorocarbons may also be used simply as a 'fingerprint'. Figure 6.31(c) shows concentration–depth profiles for Freon 11 and Freon 12 (and, for comparison, dissolved oxygen) for the South Georgia Basin in the South Atlantic. The increased concentrations below 4000m show the presence of Antarctic Bottom Water which has only recently sunk down away from the atmosphere. By tracking these elevated Freon (and oxygen) concentrations, it is possible to observe the northward spread of Antarctic Bottom Water.

Two other tracers that have been used extensively are ^{14}C (carbon-14 or radiocarbon) and ^{3}H, tritium. Both are radioactive and are introduced into the atmosphere by nuclear explosions, whence they are carried into the ocean *via* precipitation and run-off. Both are also produced naturally in the atmosphere by cosmic-ray bombardment. ^{3}H has a relatively short half-life of about 12 years; ^{14}C, on the other hand, has a half-life of 5700 years, which makes it very suitable for studying the time-scales of oceanic mixing. Indeed, naturally occurring ^{14}C is perhaps the tracer most used to calculate ages and mixing times in the oceans.

Chemically, ^{14}C behaves exactly like the dominant isotope of carbon, ^{12}C. It therefore occurs in atmospheric CO_2, and is found in the oceans in dissolved and particulate form in both the hard and soft tissues of marine organisms. The highest ratio of ^{14}C to ^{12}C atoms is found in the atmosphere, and in the surface waters in equilibrium with the atmosphere. When the carbon is removed from the air–sea interface—either carried down in solution by sinking water or through the sedimentation of organic particles—the $^{14}C:^{12}C$ ratio begins to decrease as the radiocarbon decays back to ^{14}N. Therefore, the older the seawater—i.e. the longer it has been away from the surface—the lower its $^{14}C:^{12}C$ ratio will be. The $^{14}C:^{12}C$ ratio of a sample of seawater can be measured using a mass spectrometer and the result used to calculate its age.

Carbon isotope data have been used to estimate a residence time for water in the deep ocean of about 1500 years. However, it is now known that carbon in particulate form sinks from surface to deep waters very much faster than was originally thought, and as a result this estimate has been considerably revised. For example, in the North Atlantic, the mean residence time of water colder than 4°C is probably closer to 200 years. This average conceals large geographical variations: for example, the residence time for water in the European Basin is about 13 years, while that for the Guinea Basin is more than 938 years. Recent calculations based on the ^{14}C distribution in the abyssal waters of the oceans indicate that average residence (or replacement) times for water below 1500m depth are approximately 510, 250 and 275 years in the Pacific, Indian and Atlantic Oceans, respectively. By taking into account the volumes of the water masses involved, these residence or replacement times may be used to estimate the rates at which the various water masses form.

The implication of these finite residence times for the deep water masses is that, ultimately, the water must recirculate to the surface. In summary, the deep and bottom waters spread from the polar regions along the western sides of the ocean basins (Section 6.3.2). They spread eastward from these deep western boundary currents and well up through the main thermocline. Throughout most of the ocean this upwelling is a more or

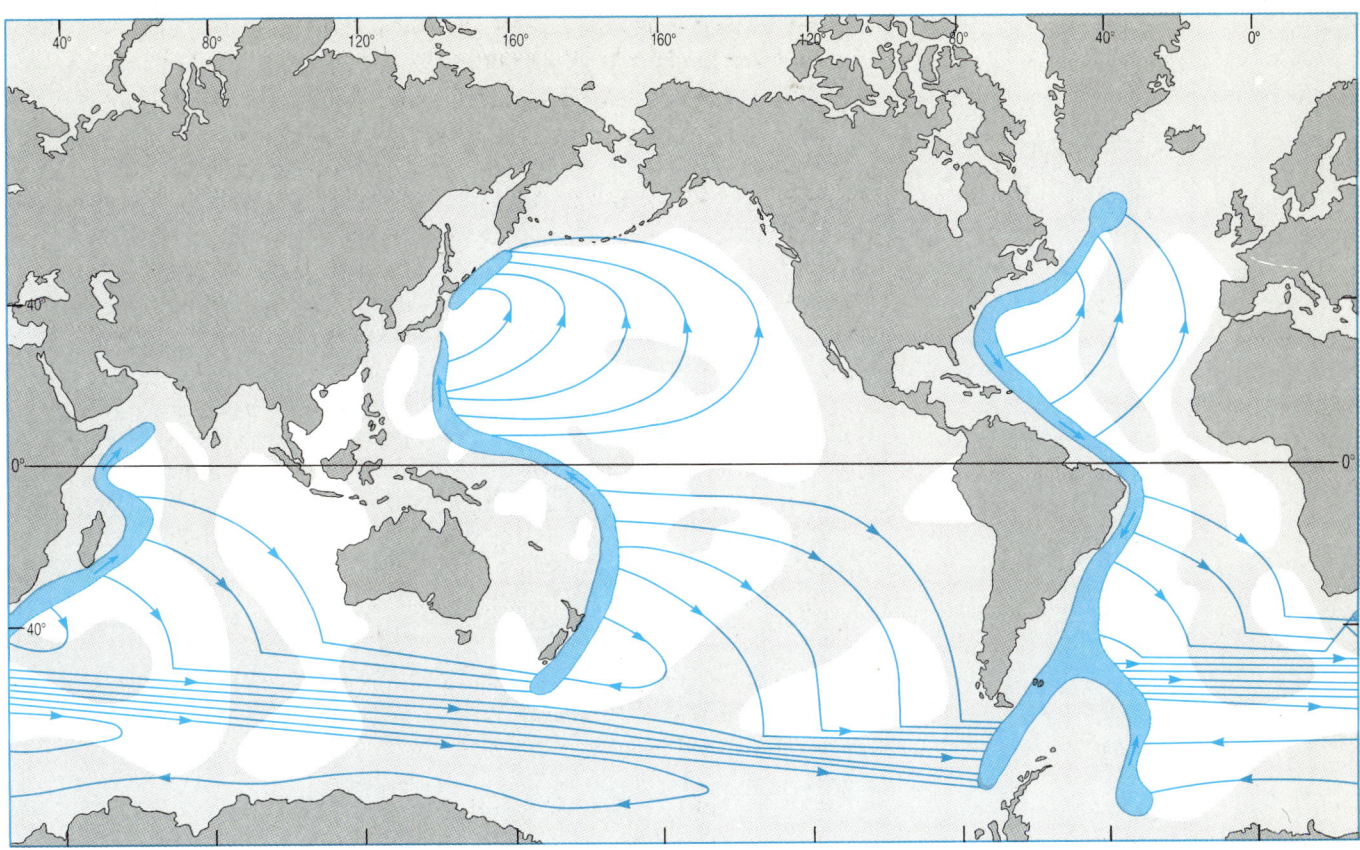

Figure 6.32 Stommel's simplified model of the deep circulation of the world ocean, with source regions of Deep and Bottom Water in the North Atlantic and the Weddell Sea.

less uniform upward diffusion; in some regions, such as the Equatorial and Antarctic Divergences, or where there is coastal upwelling, rates of upward movement of water are enhanced. On returning to the upper layers of the oceans, the water rejoins the wind-driven circulation; eventually, it returns to polar regions where it enters the cycle again.

Figure 6.32 shows the simplified model of the deep circulation, which was derived theoretically by Stommel before any deep boundary currents had actually been observed. You may notice that according to this model the predominant flow of deep water in the Atlantic is from the northern source region, so that the deep western boundary current flows southwards in both the North and South Atlantic. In the North Atlantic, the southward-flowing deep western boundary current has been observed directly using Swallow floats (Figure 3.29), and indirectly by means of geostrophic calculations (Figure 4.19). In the South Atlantic, however, the flow *along the bottom* is actually the northward flow of Antarctic Bottom Water, which is concentrated on the western side of the ocean and over-ridden by the southward-flowing western boundary current of North Atlantic Deep Water. Figure 6.30 showed both these deep western boundary currents most clearly (*cf.* Question 6.13).

6.6 GLOBAL FLUXES OF HEAT AND FRESHWATER

In this Volume we have seen how heat is redistributed over the surface of the globe by winds in the atmosphere and by wind-driven surface currents and density-driven water masses in the ocean. Because water masses carrying heat are characterized by particular salinities, fluxes of heat and salt are often studied together. The global redistribution of heat and salt may also be considered in terms of the global redistribution of heat *and freshwater*. These fluxes are, moreover, not independent: the evaporation–precipitation cycle which results in the transport of freshwater from one part of the globe to another also transfers latent heat.

Figure 6.33 shows estimates of the global transport of heat and freshwater *within* the oceans. The values give, for different latitudes, the *net* transport, taking into account both surface and deep currents. Perhaps the most remarkable aspect of Figure 6.33(a) is the net transport of heat from the South Atlantic to the North Atlantic. This reflects the fact that the Atlantic Equatorial Current system is not symmetrical about the Equator (Figure 3.1) so that the South Equatorial Current carries a large amount of heat from one hemisphere to the other. Furthermore, it is believed that a significant amount of warm water that has come from the Pacific, via the Indonesian seas, flows westward across the Indian Ocean into the Agulhas current system; it then finds its way into the South Atlantic, perhaps by means of Agulhas current 'rings' (*cf.* Section 4.3.5), and thence into the North Atlantic.

We now turn our attention to Figure 6.33(b), showing the global transport of freshwater within the oceans.

QUESTION 6.14 (a) What do the three arrows along the bottom of Figure 633(b) (with values of 65, 90 and 118 $\times$ 10^3 tonnes per second) represent?

(b) By reference to Figure 6.8, can you explain why there is convergence of freshwater fluxes just north of the Equator in the Atlantic?

(c) Bearing in mind the information given in Figure 6.8(a) and discussed in Question 6.5, what might be the link between this convergence (along with the decrease in the flux of freshwater towards the equatorial Atlantic) and the *increase* in the southward flux of freshwater in the Pacific, from 25 $\times$ 10^3 to 787 $\times$ 10^3 tonnes per second?

The *net* transfer of freshwater from the Pacific to the Atlantic, indicated by Figure 6.33(b), is *not* a consequence of the different rates of input of freshwater to the two basins. The input of freshwater to the North Atlantic, through precipitation, rivers and melting ice, is about 104 cm yr^{-1}, while that to the North Pacific is about 91 cm yr^{-1}. The important difference is in the *evaporation* rates—about 103 cm yr^{-1} for the Atlantic and 55 cm yr^{-1} for the Pacific. The resulting *E–P* values— -36 cm yr^{-1} for the North Pacific and -1 cm yr^{-1} for the North Atlantic— not only reflect the net transfer of freshwater from the Atlantic to the Pacific, they also go a long way towards explaining the markedly different salinities in the two oceans. We have already noted the relatively high values for the *surface* waters of the Atlantic and the relatively low values of the *surface* waters of the Pacific (Figure 6.8(a)). This difference in surface salinities reflects the marked difference between the average salinities of the two oceans: about 34.9 for the Atlantic and about 34.6 for the Pacific.

The values given in Figure 6.33 are based on air–sea fluxes of heat and water estimated from ship-based meteorological observations; in other words, they are based on fluxes calculated using empirical formulae similar to those discussed in Section 6.1.2. Another way to obtain

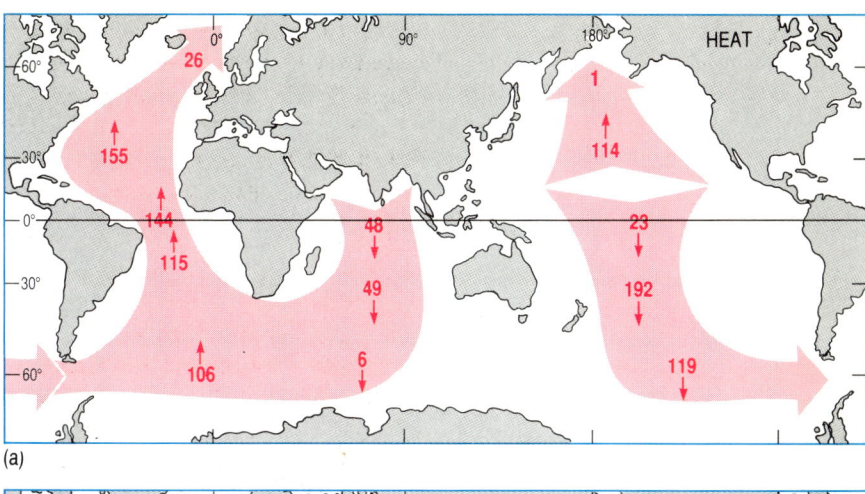

Figure 6.33 Global circulations *within* the ocean of (a) heat (in 10^{13}W) and (b) freshwater (in 10^3 tonnes s^{-1}), based on surface fluxes estimated from shipboard observations.

information similar to that in Figure 6.33 would be to calculate the heat and freshwater contained in water masses and surface currents flowing across ocean-wide sections. This has been done, and the results are shown in Figure 6.34.

How close is the agreement between the values of heat flow shown in Figure 6.34 and those shown in Figure 6.33(a)?

The agreement is not impressive. In the Atlantic some of the values at comparable latitudes are of roughly the same order of magnitude; in the

Figure 6.34 Direct measurements of heat flow within the ocean (in 10^{13}W), based on oceanographic sections and current measurements.

Pacific the only values that can be compared differ by more than an order of magnitude.

There are a number of reasons for such a disparity between the two sets of estimates. Like the mean current maps, global maps of heat and freshwater fluxes are mainly compiled using data collected from commercial or Admiralty ships, rather than research vessels, and—despite the observation code (Section 4.1.1)—observation practice and positioning of the instruments vary from ship to ship. In addition, estimating losses of heat and water from the sea-surface to a turbulent atmosphere is fraught with difficulty (Section 6.1.2). The quality of the data on heat and freshwater fluxes is therefore not as high as might be wished. There are also problems regarding the quantity and, in particular, the distribution of data. Of necessity, observations are concentrated along shipping lanes, and some regions—especially in the Pacific—are completely devoid of reliable data.

The use of temperature–salinity and current velocity data to calculate the flow of heat and water across deep sections is even more restricted by lack of data. As indicated by Figure 6.34, there have been a limited number of 'basin scale' sections; the traverse of the Atlantic at 30° S by the *Melville* and *Atlantis* (Figure 6.30) is one of only a few such sections in the South Atlantic. In the Pacific, where the distances involved are enormous, the situation is even worse.

Where temperature and salinity data are available, it is possible to estimate the velocity across a section using the geostrophic method (Section 3.3). However, if no direct current measurements are available, such estimates are likely to be too low.

By reference to Figure 3.17, can you remember the reason for this?

Geostrophic calculations only provide information about *relative* current velocities and so, without direct current measurements at depth, cannot reveal any 'barotropic' or depth-independent current flow.

During the 1980s and 1990s, worries about climatic change have focused increasing interest upon the movement of heat and water through the atmosphere–ocean system. There is particular concern about global warming, which may be occurring in response to increased concentrations of atmospheric CO_2 produced by the burning of fossil fuels and as a result of deforestation—the so-called **greenhouse effect**.

Predicting the effects of increased atmospheric CO_2 is very difficult. This is partly because the variables involved are interdependent (for instance, the temperature of ocean water affects the amount of CO_2 that can be dissolved in it) but also because the global data base is totally inadequate. In an attempt to remedy this situation, the oceanographic community is launching the **World Ocean Circulation Experiment (WOCE)**, which will take place during the first half of the 1990s. This intensive international oceanographic programme is scheduled to coincide with the launching of the next generation of oceanographic satellites which will provide global coverage of sea-surface conditions.

In summary, the aim of WOCE is to develop the means to predict global climatic variations occurring over periods of the order of ten years. To this end, the following aspects of the world oceanic circulation are being addressed:

1 The large-scale fluxes of heat and freshwater, their redistribution within the ocean, and their annual and inter-annual variability.

2 The dynamic balance of the global circulation, and its response to changes in surface fluxes of heat and freshwater.

3 The different components of oceanic variability, on time-scales from months to years, and on spatial scales of 10^3 km to global scale; also, statistical variability on smaller scales.

4 The modes and rates of formation of water masses, and their subsequent behaviour at depth.

Can you see how the rates of formation and sinking of water masses affect the rate at which temperatures at the surface of the Earth will rise in response to the greenhouse effect?

If there were no sinking of surface water, any increase in the heat content of the atmosphere and ocean would be shared between the atmosphere and the thin mixed surface layer, and the rise in sea-surface temperature would be relatively fast. In fact, surface waters are continually sinking, particularly in polar regions, and so any increase in heat content will be distributed through the atmosphere *and the body of the ocean*, and the rise in temperature at the surface of the sea (and the land) will be relatively slow. However, raised surface temperatures may *inhibit* the formation of deep water masses.

The data collected during the WOCE research cruises can, of themselves, tell us little about future rates of climatic change. An intrinsic part of the WOCE programme is, therefore, the development of powerful computer models that can simulate not only the interaction of the ocean and atmosphere, with its complex feedback loops, but also changes in the radiation balance. It is hoped that, by determining which atmospheric and oceanographic parameters are essential to climate prediction, a cost-effective 'climate observation and prediction' system can be developed.

In the final Section, we will contemplate perhaps one of the most interesting questions to be asked about the deep circulation of the oceans.

6.6.1 POSTSCRIPT: WHY IS NO DEEP WATER FORMED IN THE PACIFIC?

It is sometimes said that there is no deep Pacific water mass comparable to North Atlantic Deep Water because the topography of the Pacific basin at northern high latitudes is so different from that of the Atlantic. It is true that there are no semi-enclosed seas like the Norwegian and Greenland Seas, where water that has acquired characteristic temperature and salinity values can accumulate at depth behind a sill. However, there *is* a well-developed subpolar gyral circulation (Figure 3.1). Might we not therefore expect some cold deep water masses to form and then spread out at depth, rather in the manner of Labrador Sea Water?

The formation of deep water masses depends on the production of relatively dense surface water, through cooling and/or increase in salinity: North Atlantic Deep Water is both cold and relatively saline ($S \sim 35.0$). We have already noted that surface salinities in the Pacific are significantly lower than those in the Atlantic, particularly in the northern part of the basin. We have also noted that this is a result of low evaporation, rather than high precipitation.

In comparison with the North Atlantic (which is supplied with heat from the South Atlantic) the North Pacific has relatively low sea-surface temperatures (Figure 6.4). As discussed in Section 6.1.2, a cool sea-surface cools the overlying air, thereby reducing its ability to hold moisture and so reducing the evaporation rate; furthermore, the effect of a low evaporation rate is to limit the extent to which the density of surface water may be increased through increase in salinity. Paradoxically, therefore, in the North Pacific, a cool sea-surface prevents the surface layers from becoming sufficiently dense to sink, and there can be no North Pacific Deep Water!

6.7 SUMMARY OF CHAPTER 6

1 The temperature and salinity of water in the ocean are determined while that water is at the surface. The temperature of surface water is determined by the relative sizes of the different terms in the oceanic heat budget equation; the salinity is determined by the balance between evaporation and precipitation ($E-P$) and, at high latitudes, by the freezing and melting of ice.

2 The heat-budget equation for a part of the ocean is:

$$Q_s + Q_v = Q_b + Q_h + Q_e + Q_t$$

where Q_s is the amount of heat reaching the sea-surface as incoming short-wave radiation, Q_v is heat advected into the region in currents, Q_b is the heat lost from the sea-surface by long-wave (back-) radiation, Q_h is the heat lost from the sea-surface by conduction and convection, Q_e is the net amount of heat lost from the sea-surface by evaporation, and Q_t is the net amount of heat available to raise the temperature of the water.

3 The net radiation balance ($Q_s - Q_b$) is largely controlled by variations in Q_s; these depend partly on the latitudinal variation in incoming solar radiation but also on the amount of cloud and water vapour in the atmosphere. The amount of long-wave radiation emitted from the sea-surface depends on its temperature. However, Q_b is the *net* loss of heat from the sea-surface, and so is also affected by cloud cover, and the water vapour content, etc., of the overlying air. $Q_s - Q_b$ is generally positive.

4 Q_h and Q_e depend upon the gradients of temperature and water content, respectively, of the air above the sea-surface. Both Q_h and Q_e generally represent a loss of heat from the sea. Both are greatly enhanced by increases in the amount of atmospheric turbulence above the sea-surface, and assuming that the turbulent transfer of heat and water vapour above the sea-surface takes place by identical mechanisms, the ratio of Q_h/Q_e is constant (for a given location and time of year). Q_h/Q_e is known as Bowen's ratio (R) and allows Q_h and Q_e to be calculated if Q_s and Q_b are known. As Q_e is generally several times Q_h, values of Bowen's ratio are less than one.

5 The formation of ice at the sea-surface greatly influences the local heat budget; in particular, it leads to an increase in the albedo and a substantial decrease in Q_s, while Q_b is not much affected. Thus, once formed, ice tends to be maintained.

6   The principle of conservation of salt, combined with the principle of continuity, may be used to make deductions about the volume transports into and out of semi-enclosed bodies of water or, alternatively (if these are known), about the evaporation–precipitation balance in the region concerned.

7   Water masses are bodies of water that are identifiable because they have certain combinations of physical and chemical characteristics. The properties most used to identify water masses are temperature (strictly potential temperature) and salinity, because away from the sea-surface they may only be changed through mixing, i.e. they are conservative properties. Deep and bottom water masses, and 'thick' upper water masses, are formed in regions of convergence, and where deep convection results from the destabilization of surface waters through cooling and/or increase in salinity. Whether a water mass can form in this way depends not on the absolute density of surface waters but on their density relative to that of underlying water.

8   Central water masses are 'thick' upper water masses that form in winter in the subtropical gyres. They are characterized by relatively high temperatures and relatively high salinities. The water mass that forms in the Sargasso Sea has a remarkably uniform temperature of about 18°C. This '18°C water' is an example of a mode water.

9   The most extensive intermediate water mass is Antarctic Intermediate Water which forms in the Antarctic Polar Frontal Zone. Like other intermediate water masses formed in subpolar regions (e.g. Labrador Sea Water), Antarctic Intermediate Water is characterized by low temperature and low salinity. By contrast, Mediterranean Water is characterized by high temperature and high salinity.

10   North Atlantic Deep Water is formed in winter, mainly through cooling of surface waters in the Norwegian and Greenland Seas. It accumulates behind the Greenland–Scotland ridge, whence it intermittently overflows, cascading down into the deep Atlantic and mixing with the overlying water masses.

11   Antarctic Bottom Water is the most widespread water mass in the oceans. Abstraction of freshwater into ice plays an important part in its formation, particularly in coastal polynyas; cooling of surface waters is also important, especially in 'open ocean' polynyas (e.g. the Weddell polynya). Antarctic Bottom Water formed over the continental shelf in the Weddell and Ross Seas is the densest water in the open oceans. After flowing down the continental slope, it circulates eastwards around the Antarctic continent a number of times and then flows northwards, mainly along the western sides of the three ocean basins.

12   The deep water mass with the largest volume is Pacific and Indian Ocean Common Water; it is a mixture consisting of about half Antarctic Bottom Water, and a quarter each of Antarctic Intermediate Water and North Atlantic Deep Water.

13   Dense water tends to sink along isopycnic surfaces, and mixing along isopycnic surfaces occurs with a minimum expenditure of energy. However, mixing does occur between water of different densities. Furthermore, mixing also occurs as a result of double-diffusive processes occurring on a molecular scale (e.g. salt fingering).

14 Temperature–salinity diagrams may be constructed for different locations in the ocean. Assuming that mixing occurs through turbulent processes only, temperature–salinity diagrams may be used to determine the proportions of different water masses contributing to the water at a given depth. Sequences of temperature–salinity diagrams may be used to trace the least-mixed (or core) layer of water mass, as it spreads through the ocean. Comparison of the trend of a θ–S curve with contours of σ_θ may be used to evaluate the degree of stability of a water column.

15 Because it is a conservative property, potential vorticity may be used to track water masses. Oxygen and silica concentrations, which are non-conservative properties, may be used to identify as well as track water masses. Certain substances that enter the ocean at the sea-surface are particularly useful for estimating the 'ages' of water masses. These include radioactive isotopes such as ^{14}C (radiocarbon) and ^{3}H (tritium)—which are formed both naturally and in nuclear tests—and chlorofluorocarbons, i.e. Freons (which have been manufactured since the 1940s). Using radiocarbon data, residence times for water in the deep oceans have been estimated to be of the order of 200–500 years.

16 If we are to predict climatic change, such as might result from the greenhouse effect, it is vital that global fluxes of heat and freshwater are more accurately quantified. This is one of the aims of the World Ocean Circulation Experiment. The other, related, aim is the development of sophisticated computer models of the ocean–atmosphere system.

Now try the following questions to consolidate your understanding of this and earlier Chapters.

QUESTION 6.15 Which of these statements about various heat-budget terms are true, and which are false?

(a) At night, $Q_s - Q_b$ is negative, and there is a net loss of radiative heat from the sea.

(b) The fogs over the Grand Banks are caused by advection of heat (Q_v) by the Gulf Stream.

(c) Mediterranean Water forms partly as a result of large heat losses, as both Q_h and Q_e, from the sea-surface in the north-western Mediterranean.

(d) It is to be expected that along the coast of Somalia during the South-West Monsoon both Q_h and Q_e would represent a gain of heat by the sea.

QUESTION 6.16 It is thought that one of the factors which determine whether an open-ocean polynya will persist or freeze over is the production of meltwater from the surrounding ice. Why might the production of meltwater cause an open-ocean polynya to 'die', i.e. stop convecting?

QUESTION 6.17 As mentioned in the text, the Antarctic Convergence was renamed the Antarctic Polar Frontal Zone. What features of the zone mean that the use of the term 'frontal' is appropriate?

QUESTION 6.18 Figure 6.35, **1** is a computer-generated three-dimensional θ–S diagram for all ocean water with θ less than 4°C, and Figure 6.35 **2–4** show similar information for the three individual ocean basins. In each diagram, the elevation of the surface corresponds to the volume of water with given θ–S characteristics.

Figure 6.35 Computer-generated θ–S diagrams for water colder than 4 °C (for use with Question 6.18). The elevation of the surface corresponds to the volume of water with given combinations of θ and S, the volumes having been calculated for θ–S classes of 0.1 °C × 0.01. Diagram **1** is for the world ocean, **2** is for the Indian Ocean, and **3** and **4** are for the Pacific and Atlantic Oceans (not necessarily in that order). In **1** and **3**, the elevation of the highest peak corresponds to 26×10^6 km^3 of water, in **2** it corresponds to 6.0×10^6 km^3 of water, and in **4** to 4.7×10^6 km^3.

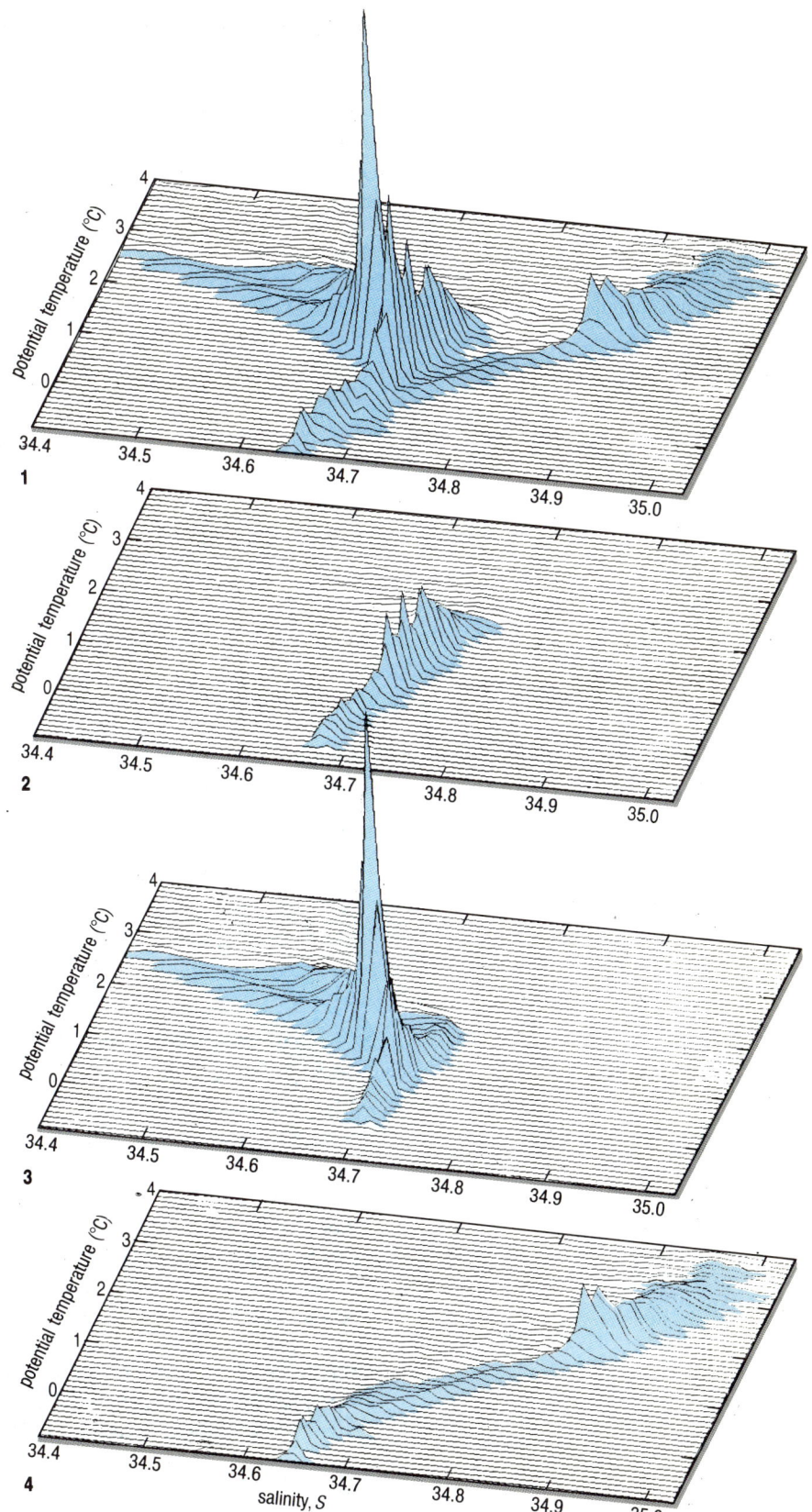

(a) Diagram **2** corresponds to the Indian Ocean. Given the characteristics of North Atlantic Deep Water, can you say which of the remaining diagrams (**3** and **4**) corresponds to the Atlantic and which to the Pacific?

(b) What is the water mass, distinguishable in all the diagrams, which has potential temperatures less than 0°C?

(c) To what is the peak (visible in diagrams **1**, **3** and, to some extent, **2**) attributable?

QUESTION 6.19 In Section 6.3.2, we stated that Pacific and Indian Ocean Common Water consists of approximately 50% Antarctic Bottom Water and about 25% each of North Atlantic Deep Water and Antarctic Intermediate Water. Use Figure 6.36, along with the principles outlined in Section 6.4.2, to check these estimates.

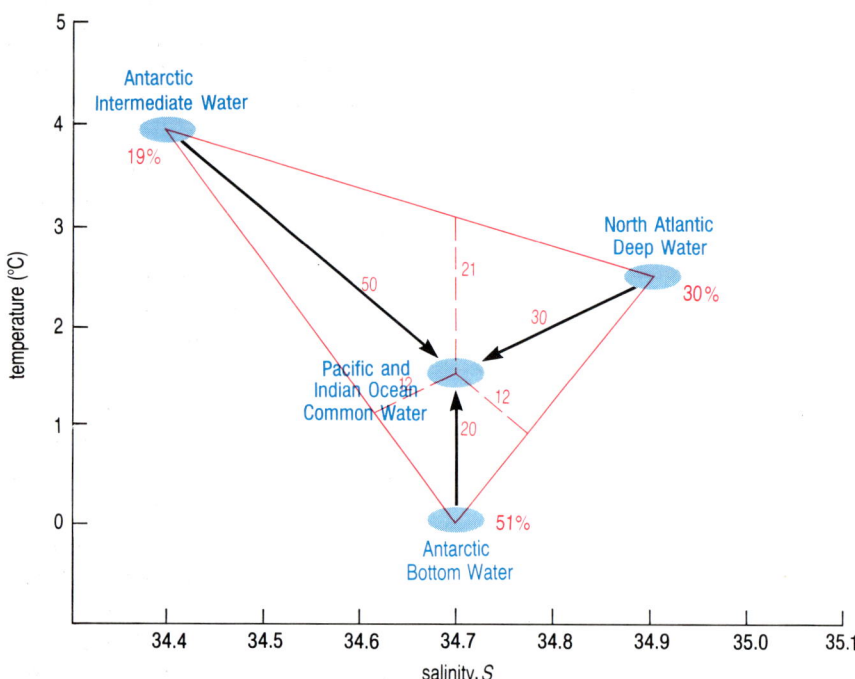

Figure 6.36 *T–S* diagram showing the characteristics of the water masses that mix together to form Pacific and Indian Ocean Common Water. (For use with Question 6.19.)

QUESTION 6.20 What feature of the circulation of the Southern Ocean is particularly responsible for the high concentrations of dissolved silica in the bottom waters there?

QUESTION 6.21 *Briefly* explain the greenhouse effect by reference to *one* of the terms in the heat-budget equation (equation 6.1).

QUESTION 6.22 To be effective, the atmosphere–ocean computer models being developed for the WOCE programme will have to have a sufficiently fine spatial resolution to be able to reproduce mesoscale eddies. Why are mesoscale eddies so important in this context?

SUGGESTED FURTHER READING

GORDON, A. L. AND COMISO, J. C. (1988) Polynyas in the Southern Ocean, in *Scientific American*, **257**, pp. 70–77. An informative article discussing not only how polynyas form and are maintained, but also their importance for the global climate.

HARVEY, J. G. (1976) *Atmosphere and Ocean: Our Fluid Environments*, The Artemis Press. An overview of the physical properties and circulation of the atmosphere and the ocean; has a small amount of mathematics.

KNAUSS, J. A. (1978) *Introduction to Physical Oceanography*, Prentice-Hall. This book begins with a theoretical background about the balances of forces involved in ocean circulation, and then provides an overview of ocean currents, tides and surface waves.

NIEUWOLT, S. (1977) *Tropical Climatology: An Introduction to the Climates of the Low Latitudes*, John Wiley and Sons. Although written from a meteorological viewpoint, this book illustrates the importance of the ocean for the tropical climate.

PERRY, A. H. AND WALKER, J. M. (1977) *The Ocean–Atmosphere System*, Longman. A mainly descriptive book concerning interactions between ocean and atmosphere.

PICKARD, G. L. AND EMERY, W. J. (1982) *Descriptive Physical Oceanography: An Introduction*, 4th (SI) enlarged edition, Pergamon Press. An introduction to the physical properties of the ocean and to oceanographic techniques (current measurement, etc.) along with descriptions of hydrography and circulation, region by region.

POND, S. AND PICKARD, G. L. (1978) *Introductory Dynamic Oceanography*, Pergamon Press. A discussion of the mathematical relationships used in physical oceanography, with explanations of their physical bases and descriptions of how they may be manipulated.

STOMMEL, H. (1965) *The Gulf Stream: A Physical and Dynamical Description*, 2nd edition, University of California Press/Cambridge University Press. A classic text, which describes and explains ideas and theories about the Gulf Stream and hence about ocean circulation in general; contains a fair amount of mathematics.

WELLS, N. (1986) *The Atmosphere and Ocean: A Physical Introduction*, Taylor and Francis. This text covers a wide range of atmospheric and oceanic phenomena at a fairly advanced level.

WHITWORTH III, T. (1988) The Antarctic Circumpolar Current, in *Oceanus* **31** (Summer) pp. 53–58. A lively discussion of the Antarctic Circumpolar Current in the context of the Antarctic Polar Frontal Zone and water masses. (This particular issue of *Oceanus* concentrates on legal, scientific and political problems concerning Antarctica.)

WIEBE, P. H. (1982) Rings of the Gulf Stream, in *Scientific American*, **246** (March), pp. 50–60. A well-illustrated article, describing the formation and subsequent history of a number of warm-core and cold-core rings, along with general information about the impact of such rings on the oceanic environment.

ANSWERS AND COMMENTS TO QUESTIONS

CHAPTER 1

Question 1.1 On the real Earth, the eastwards velocity of the surface is greatest at the Equator and zero at the poles. For the hypothetical cylindrical Earth, the eastwards velocity of the surface would be the same whatever the distance from the poles, and so there would be no apparent deflection and no Coriolis force.

Question 1.2 (a) According to Figure 1.4, the radiation balance changes from a net surplus to a net loss at about 38° of latitude.

(b) In Figure 1.4(b), the areas labelled 'net surplus' and 'net loss' are approximately equal, so the total amount of radiation energy gained by the Earth–atmosphere system is equal to that lost. Note that this only works because the horizontal axis is increasingly compressed towards higher latitudes, thus compensating for the decreasing area of the globe within given latitude bands.

Question 1.3 (a) The area under the dashed blue curve in Figure 1.5 is about half as big again as that under the solid blue curve, indicating that the atmosphere transports polewards about one-and-a-half times as much heat as the ocean.

(b) Negative values indicate southwards (i.e. equatorward) transport of heat. So, south of about 5°N and poleward of about 56°N the oceans are responsible for a southwards heat transport.

Question 1.4 (a) In addition to its southwards firing velocity, the missile has the eastwards velocity of the surface of the Earth at the Equator. As the missile travels southwards, the eastwards velocity of the Earth beneath it becomes less and less, so that *in relation to the Earth* the missile is moving not only southwards but eastwards. Thus a missile fired southwards from the Equator is deflected towards the east.

(b) (i) In the Northern Hemisphere, the Coriolis force deflects currents to the right, and so would tend to deflect southwards any current flowing initially eastwards.

(ii) On the Equator the Coriolis force is zero and there is no deflection.

Question 1.5 False. At the Equator, day and night are of comparable length, being equal at the equinoxes. The high *average* insolation is the result of relatively large amounts of radiation being received all year.

Question 1.6 Like that in Figure 1.4(a) and (b), the horizontal axis of (c) is scaled according to surface area, but in this case it is not the area of the Earth's surface that is relevant but the area of *ocean* surface. The proportion of the Earth's surface that is ocean is much smaller in the Northern Hemisphere than in the Southern Hemisphere, which between 30 and 65 degrees of latitude is largely ocean. The horizontal axis is therefore even more compressed at northern high latitudes than southern high latitudes.

CHAPTER 2

Question 2.1 (a) A parcel of dry air displaced upwards will cool at the adiabatic lapse rate of 9.8°C km^{-1}. The rate of decrease of temperature with height in the column of air is *less* than that i.e. 20°C in 2.5 km, or 8°C km^{-1}. The air that has been displaced upwards will become cooler than its surroundings and sink back to its original level, and the situation will be stable.

(b) However, if the rising air becomes saturated, its decrease in temperature as a result of adiabatic cooling is offset by the latent heat released when water vapour condenses, and so it will cool according to the *blue* curve. Thus, its rate of cooling is *less* than that of the surrounding air, it will be warmer than its surroundings and will continue to rise. The situation will therefore be *unstable*.

Question 2.2 (a) Easterly waves and cyclones are generated in the region of the equatorial low pressure zone/Intertropical Convergence Zone (zone of highest sea-surface temperature). Figure 2.3 shows that the ITCZ is generally displaced northwards with respect to the geographical Equator; in the Atlantic and eastern Pacific it does not lie more than 5° south of the Equator, even in the southern summer.

(b) The development of cyclones depends on air flowing towards an atmospheric low while under the influence of the Coriolis force, so that a cyclonic spiral forms. The Coriolis force is zero at the Equator and very small within about 5° to either side of it.

Question 2.3 (a) False. In areas of high surface pressure, air is sinking; air is rising in areas of low surface pressure.

(b) False. *Rising* air cools adiabatically and cannot contain as much water vapour as at ground/sea-level, so that clouds and precipitation result. Thus, the ITCZ, for example, is characterized by high precipitation.

(c) True; see Figure 2.3.

(d) False. As discussed in the text in connection with monsoonal circulations, the flow of air over Eurasia is cyclonic (inwards and upwards) in the northern summer (Figure 2.3(a)); and anticyclonic (downwards and outwards) in the northern winter (Figure 2.3(b)).

(e) False. The Trade Winds of the two hemispheres meet along the ITCZ, which is a zone of *low* pressure.

(f) False. Figure 2.3(b) shows that in polar regions the tropopause (the top of the troposphere) is about 9–10 km above the surface of the Earth. It *is* about 17 km above the surface of the Earth in the tropics.

Question 2.4 (a) Your sketch should look something like Figure A1 (overleaf). You probably identified area 1, shaded in blue, as the most dangerous. Other regions you may have considered dangerous are area 2, where ships are in danger of being blown into the cyclone's path, and area 3, where the sea has become confused in the wake of the cyclone.

(b) The 'fuel' that drives a tropical cyclone could be said to be water vapour, because it is the change in state from gas (water vapour) to liquid (cloud droplets) that provides the latent heat which enables the upward cyclonic movement of air around the central region of the cyclone to continue.

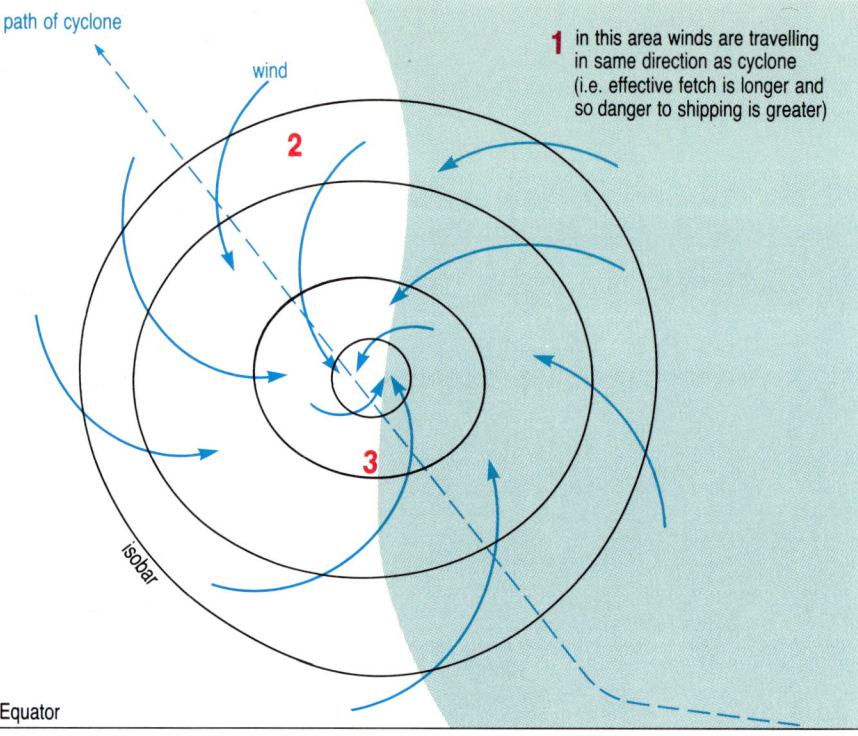

Figure A1   Answer to Question 2.4(a).

Question 2.5   See Figure A2 (opposite). Note that the subtropical high pressure regions are dark in Figure 2.17 because they are regions where *dry* air sinks, in contrast to the ITCZ which is a region where moist air rises.

Question 2.6   The two examples given in this Chapter are the large-scale undulations of the jet stream in the upper westerlies (Section 2.2.1) and the easterly waves in the Trade Wind belts (Section 2.3.1).

CHAPTER 3

Question 3.1   (a) (i)   The main cool surface currents in the Pacific are the California Current and the Peru or Humboldt Current, while (ii) in the Atlantic they are the Canaries Current and the Benguela Current.

(b)   All these currents flow equatorwards along the *eastern* boundaries of the oceans. (The characteristics of these eastern boundary currents will be discussed in Chapter 4.)

Question 3.2   Using equation 3.1 ($\tau = cW^2$), and substituting for the values of τ and W given, we have:

$$0.2 = c(10)^2$$

$$\therefore c = \frac{0.2}{100}$$

$$= 2 \times 10^{-3}$$

The units of c are $\dfrac{\mathrm{N\,m^{-2}}}{(\mathrm{m\,s^{-1}})^2} = \dfrac{(\mathrm{kg\,m\,s^{-2}})\mathrm{m^{-2}}}{\mathrm{m^2\,s^{-2}}} = \mathrm{kg\,m^{-3}}$,

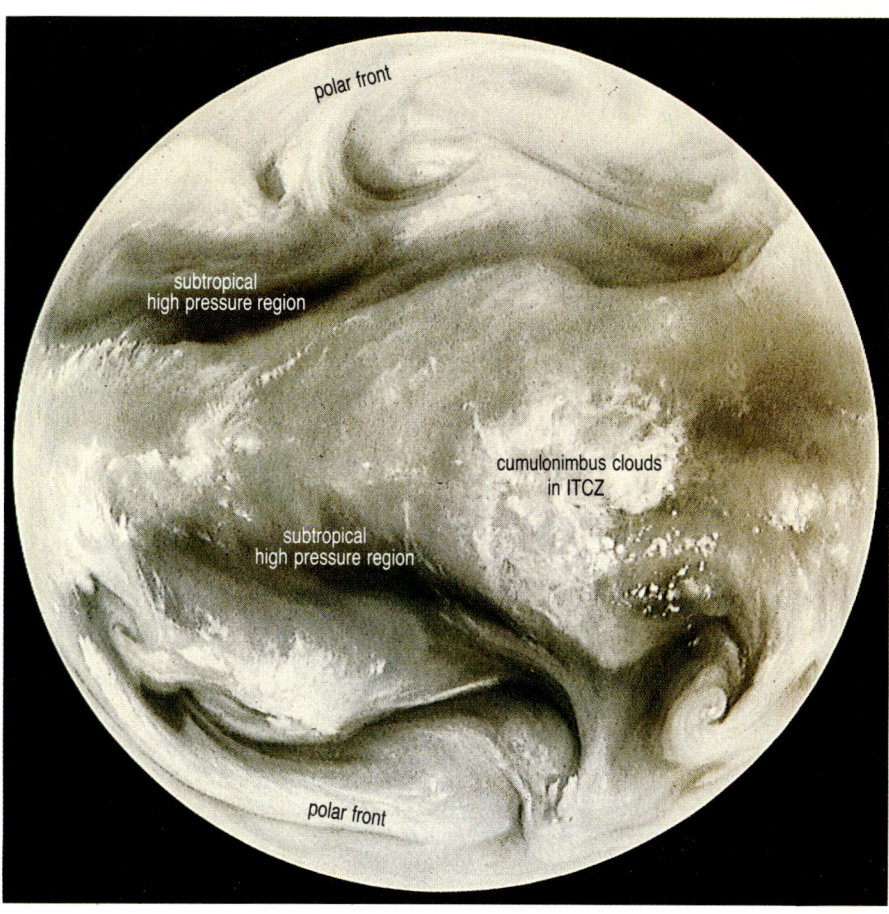

Figure A2 Answer to Question 2.5.

i.e. the units of density. Indeed, c is usually regarded as being composed of ρ_a, the density of the overlying air (about 1.3kg m^{-3}) multiplied by C_D, a dimensionless drag coefficient (of the order of 1.5×10^{-3}).

Question 3.3 In the thermocline, temperature decreases markedly with depth and, as a result, density *increases* markedly with depth. The water is well stratified and very stable and so mixing is inhibited. This is in contrast to the situation near the sea-surface where water is well mixed by wind and waves. The degree of turbulence, and hence the eddy viscosity, is therefore much greater in the mixed surface layer than in the thermocline.

Question 3.4 (a) The Coriolis parameter $f = 2\Omega \sin \phi$.

At 40° of latitude this is equal to: $2 \times (7.29 \times 10^{-5}) \times \sin 40° \text{s}^{-1}$

$$= 2 \times 7.29 \times 10^{-5} \times 0.643$$
$$\approx 9.4 \times 10^{-5} \text{s}^{-1}$$

(b) (i) According to Ekman's theory, the surface current speed is given by:

$$u_0 = \frac{\tau}{\sqrt{A_z \rho f}} \qquad (3.3)$$

Using the values given, and putting f equal to $10^{-4}\,\text{s}^{-1}$, to make the sums easier:

$$u_0 = \frac{0.1}{\sqrt{10^2 \times 10^3 \times 10^{-4}}}$$

$$= \frac{0.1}{\sqrt{10}} = \frac{0.1}{3.16} \approx 0.03\,\text{m s}^{-1}$$

(ii) According to Ekman's theory, the surface current should be 45° *cum sole* of the wind direction. The situation under consideration is in the Southern Hemisphere, and so 45° *cum sole* is 45° to the left. The wind is westerly (towards the east) and so the surface current should be towards the north-east (i.e. north-easterly).

(c) According to the 'rule of thumb', the surface current is typically about 3% of the wind speed. In this case, the wind speed is $5\,\text{m s}^{-1}$ and so the surface current might be expected to be about $(3/100) \times 5 = 0.15\,\text{m s}^{-1}$. This is about 5 times the value calculated in (b); however, in that calculation we used a value for A_z at the upper end of the likely range of values (10^{-2}–$10^2\,\text{kg m}^{-1}\text{s}^{-1}$) and a smaller value for A_z would have resulted in a larger value for u_0.

Question 3.5 The period of an inertia current is given by equation 3.7:

$$T = \frac{2\pi}{f} \quad \text{where } f = 2\Omega \sin\phi$$

Now, $\Omega = 2\pi/24\,\text{hr}^{-1}$, and at the poles $\phi = 90°$ and $\sin\phi = 1$.

Therefore, $T = \dfrac{2\pi}{2 \times (2\pi/24) \times 1} = \dfrac{24}{2} = 12$ hours.

At a latitude of 30°, $\sin\phi = \frac{1}{2}$ and so

$$T = \frac{2\pi}{2(2\pi/24) \times \frac{1}{2}} = 24 \text{ hours.}$$

Question 3.6 (a) Pressure gradients act from areas of high pressure to areas of low pressure, so the horizontal pressure gradient force must be represented by the arrow pointing to the left, and the Coriolis force by the arrow pointing towards the right.

(b) We are told that the current is flowing into the page, and we know that the Coriolis force acts *cum sole* of the direction of motion, i.e. to the right in the Northern Hemisphere and to the left in the Southern Hemisphere. In this case, the Coriolis force is acting to the right of the direction of motion, and so the situation illustrated must be in the Northern Hemisphere.

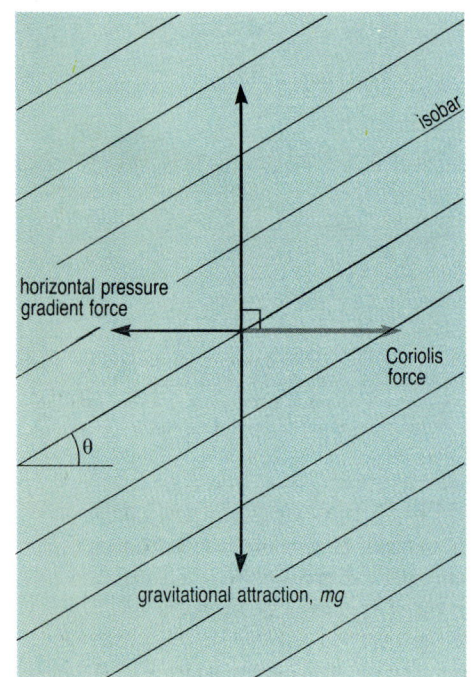

Figure A3 Answer to Question 3.6(a).

Question 3.7 (a) The first step is to calculate f. At 30° of latitude, $f = 2\Omega \sin 30° = 2 \times 7.29 \times 10^{-5} \times 0.5 = 7.29 \times 10^{-5}\,\text{s}^{-1}$. In this case, $z - z_0$ is $2000 - 1000 = 1000\,\text{m}$, and we are assuming that h_B is the same.

$\rho_B/\rho_A = 1.0262/1.0265 = 0.9997$.

$L = 100 \times 10^3\,\text{m}$ and $g = 9.8\,\text{m s}^{-2}$.

Therefore, substituting into equation 3.13:

$$u = \frac{gh_B}{fL}\left(1 - \frac{\rho_B}{\rho_A}\right)$$

$$u = \frac{9.8 \times 1000\,(1 - 0.9997)}{7.29 \times 10^{-5} \times 10^5}$$

$$= \frac{9800 \times 0.0003}{7.29}$$

$$= \frac{2.29}{7.29}$$

$$\approx 0.40\,\text{m s}^{-1}$$

(b) According to equation 3.12:

$$u = \frac{g}{f}\left(\frac{h_B - h_A}{L}\right)$$

So $\quad \dfrac{fu}{g} = \dfrac{h_B - h_A}{L}$

and $h_B - h_A = \dfrac{Lfu}{g}$

$$= \frac{10^5 \times 7.29 \times 10^{-5} \times 0.4}{9.8}$$

$$= \frac{2.92}{9.8}$$

$$= 0.29$$

$$\approx 0.3\,\text{m}$$

The distance $h_B - h_A$ (which is the same as the difference between h_B and $z_0 - z_1$) is therefore about 0.03% of $z_0 - z_1$, so the assumption you made in part (a), that h_B and $z_0 - z_1$ were equal, was a reasonable one.

(c) In the situation described here, as in the example given in the text, $\rho_A > \rho_B$ so the isobar p_1 slopes up towards B and the horizontal pressure gradient force acts from B to A. In the Southern Hemisphere, the Coriolis force acts to the *left* of the direction of current flow, so in conditions of geostrophic equilibrium the horizontal pressure gradient force must be acting to the *right* of the direction of flow. In this case, the current must be flowing out of the page, i.e. towards the south (as station B is due east of station A).

Question 3.8 (a) You should have had no difficulty in identifying the Gulf Stream and the Antarctic Circumpolar Current.

(b) The stronger the geostrophic current, the closer together are the contours of dynamic height. Therefore, according to Figure 3.20 the fastest part of the Antarctic Circumpolar Current is to the east of the southern tip of Africa, at about 60° E.

(c) By inspecting the contour values to the north of the 4.4 dynamic metre closed contour, we can see that it represents a depression of the

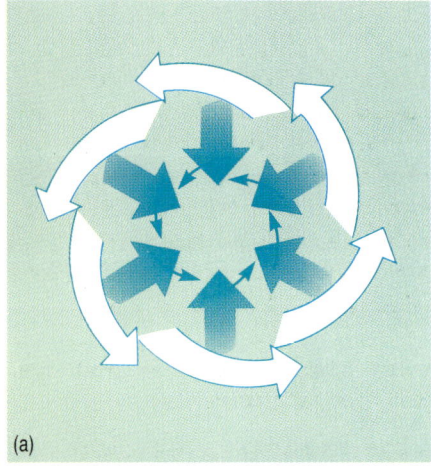

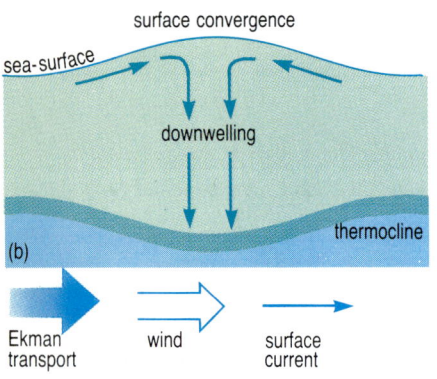

Figure A4 Answer to Question 3.9.

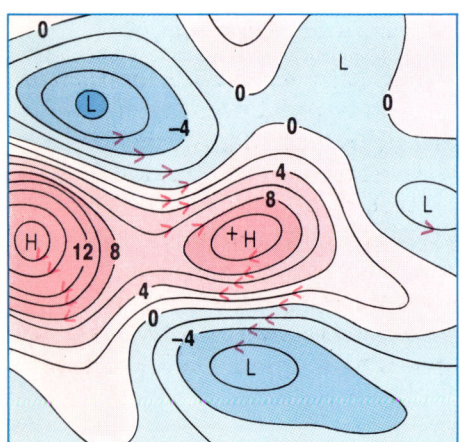

Figure A5 Answer to Question 3.10.

sea-surface, rather than a hill. In the Southern Hemisphere, isobaric surfaces slope up to the *left* of the current direction and so, given that the current is flowing clockwise, we would indeed expect the sea-surface to slope down towards the middle of the region defined by the closed contour.

Question 3.9 The conditions that would lead to convergence and sinking of surface water in the Southern Hemisphere are illustrated in Figure A4(a) and (b).

Question 3.10 In geostrophic flow (to which flow in mesoscale eddies approximates), motion is along contours of dynamic height, in such a direction that (in the Northern Hemisphere) the 'highs' are on the right and the 'lows' are on the left. Thus, the positive numbers correspond to anticylonic (here clockwise) flow, and the negative numbers to cyclonic (anticlockwise flow). As shown in Figure A5, flow in mesoscale eddies resembles that in the atmosphere—cyclonic around 'lows' and anticyclonic around 'highs'.

Question 3.11 (a) Energy is proportional to the *square* of current speed. So, if current speeds associated with eddies are 10 times those associated with the mean flow, the energy must be $10^2 = 100$ times bigger.

(b) The only atmospheric phenomenon shown in Figure 3.31 with an energy close to that of a mesoscale eddy is a severe thunderstorm.

(c) The Antarctic Circumpolar Current has the greatest kinetic energy. This is not surprising as, acted upon by prevailing westerlies, this broad current can flow unimpeded around the globe.

Question 3.12 (a) 1 Wind stress is an external force (i), as it acts on the upper surface of the ocean.

2 Viscous forces are internal frictional forces (ii). However, frictional forces act to oppose motion and so you would also have been justified in choosing (iii).

3 The Coriolis force is a secondary force (iii), as it only acts on water that is already moving.

4 In baroclinic conditions, the horizontal pressure gradient force is an internal force (ii), as it results from the distribution of mass (density) within the body of ocean water; however, horizontal pressure gradient forces that result from sea-surface slopes caused by variations in atmospheric pressure could be argued to be external forces.

5 The tides are produced by the gravitational attraction of the Sun and the Moon, and so the tide-producing forces are external forces (i).

Note that for 4, it is possible to argue that as *all* horizontal pressure gradient forces result from lateral differences in the pressure of overlying water, and these pressures derive ultimately from the gravitational attraction of the water to the mass of the Earth, horizontal pressure forces are themselves external forces.

Note also that the Coriolis force is an *apparent* secondary force. It has been 'invented' to explain movement relative to the Earth that results from the rotation of the Earth itself.

(b) (i) In geostrophic flow, the horizontal pressure gradient force is balancing the Coriolis force.

(ii) The mean flow of the Ekman layer at right angles to the wind is the result of the wind stress balancing the Coriolis force.

Question 3.13 At the end of Section 3.1.2 we stated that the Ekman transport—the total volume transport in the wind-driven layer—may be calculated by multiplying $\bar{u}$, the depth mean current, by the thickness of the wind-driven layer, D. From equation 3.4 $\bar{u}$ is given by $\tau/D\rho f$. Multiplying this by D gives $\tau/\rho f$. A_z does not appear in this expression, and so we can say that the Ekman transport resulting from a given wind stress is independent of the eddy viscosity. This an important result because it means we can calculate Ekman transports without needing an accurate value for the eddy viscosity, which is very hard to estimate.

Question 3.14 (a) Because the water is well mixed, there will be no lateral variations in density, and so conditions will be barotropic.

(b) If the slope of the sea-surface, and all other isobars, is given by tan θ, from the gradient equation (equation 3.11):

$$\tan\theta = \frac{fu}{g}$$

$$= \frac{2\Omega \sin\phi \times 0.2}{9.8}$$

Taking Ω as $7.29 \times 10^{-5} s^{-1}$, and ϕ as 51°N (so that sin ϕ = 0.778),

$$\tan\theta = \frac{2 \times 7.29 \times 10^{-5} \times 0.778 \times 0.2}{9.8}$$

$$= 0.23 \times 10^{-5}$$

As the Straits of Dover are 35 km wide, this means that if the difference in height between the two sides of the Straits is d m,

$$\tan\theta = 0.23 \times 10^{-5} = d/35000$$

i.e. $\quad d = 35000 \times 0.23 \times 10^{-5}$

$$= 0.08 \text{ m}$$

Flow is to the east, so the Coriolis force to the right of the flow will be balanced by the horizontal pressure gradient force acting towards the left, i.e. to the north. In other words, the mean sea-level must be higher on the *southern* side of the Straits, i.e. the French side, by 0.08 m.

Question 3.15 As described in Section 3.3.3, the geostrophic method can of necessity only determine large-scale flows because the oceanographic stations at which profiles of temperature and salinity are taken may be many tens of kilometres apart and the measurements can take an appreciable time to make. This was particularly true before electronic equipment made possible continuous vertical profiling of water properties; until relatively recently, researchers had to depend on individual temperature and salinity measurements made at specific depths. In short, the geostrophic method tends to conceal any small-scale, short-term phenomena and to determine only average conditions which approximate to the mean flow.

Question 3.16 Ekman transport is 90° *cum sole* of the wind direction, i.e. 90° to the right in the Northern Hemisphere and 90° to the left in the

Southern Hemisphere (Section 3.1.2). As a result, cyclonic winds (which are anticlockwise in the Northern Hemisphere and clockwise in the Southern Hemisphere) lead to net motion of the wind-driven layer away from the centre of the low pressure region. This divergent flow of surface waters results in colder water upwelling from below (Section 3.4).

Note: In answering this question, it may have struck you that there appears to be a contradiction between information in Section 2.3.1 and Section 3.4. In the former, we stated that the low atmospheric pressure associated with cyclones leads to a *rise* in sea-level, while in the latter we stated that cyclonic winds lead to a *lowering* of sea-level. There is no conflict here: both mechanisms operate together. Whether the sea-level at a given location is raised or lowered there will depend on a number of factors. In particular, cyclonic winds tend to lead to large rises in sea-level when they blow over shallow, semi-enclosed areas of sea where water can pile up (e.g. the southern North Sea, or the Ganges–Brahmaputra Delta).

Question 3.17 (a) The kinetic energy spectrum in Figure 3.32 was measured in the Drake Passage, i.e. within the flow of the Antarctic Circumpolar Current. We might expect the contributions of mesoscale eddies to the spectrum to be above average because, as stated in the text, this is one of the main regions of the oceans where mesoscale eddies are generated.

(b) Peak A may be attributable to inertia currents set up in various ways (*cf*. the 'inertial period' peak in Figure 3.28). The latitude of the Drake Passage is about 60°S, and the inertial period $T = 2\pi/f$ is therefore $2 \times 3.142/2 \times 7.29 \times 10^{-5} \times \sin 60° = 0.498 \times 10^5$ s. Frequency $= 1/T = 2.01 \times 10^{-5}$ cycles per second ≈ 0.2 cycles per day.

Question 3.18 (a) The regions of the ocean that show the greatest variation in sea-surface height are regions with fast currents: the Gulf Stream, the Kuroshio, the Agulhas Current and the Antarctic Circumpolar Current. Surface currents are associated with slopes in the sea-surface—the faster the current, the greater the slope and the more elevated some areas of the sea-surface become. However, Figure 3.33 shows *variability* in sea-surface height: these fast currents are not large steady 'rivers' of water but are continually shifting and changing, producing meanders and eddies. (Variability also occurs in regions of slow surface currents but in these cases the sea-surface slopes will be small and the variations in the height at a given point, associated with variations in current flow, will also be small.)

(b) The equatorial current systems do not show up clearly on Figure 3.33 mainly because the slope of the sea-surface associated with geostrophic currents is proportional to the Coriolis parameter, f, and hence to the sine of the latitude. The Coriolis force is therefore zero on the Equator and very small close to it, and sea-surface slopes are correspondingly small.

CHAPTER 4

Question 4.1 (a) Both Franklin and Rennell believed that the Gulf Stream was the result of water being piled up in the Gulf of Mexico, so that the Stream was driven by a head of water and flowed in such a way as to tend to even out the horizontal pressure gradient. Maury clearly did not understand that in this case what would be important would be the variation of pressure at a given horizontal level; in other words, he did not understand about horizontal pressure gradients.

(b)   Figure 3.1 shows that the volume of water transported in the Gulf Stream must *increase* between the Straits of Florida and Cape Hatteras because of the addition of water that has recirculated in the subtropical gyre. (The volume transport is also increased as a result of mixing and entrainment of water adjacent to the Gulf Stream.)

Question 4.2   At that time, the idea of eddy viscosity—i.e. of frictional coupling resulting from turbulence—had not been formulated. If the effect of the wind were transmitted by molecular viscosity only, it would indeed take a long time for major ocean currents to be generated (*cf.* Section 3.1.1). Shortly after this, in 1883, Reynolds published his now famous paper on the nature of turbulent flow in pipes; his influential ideas on the generation of frictional forces by turbulence were published in 1894.

Question 4.3   (a)   The component of the Earth's rotation about a vertical axis at latitude ϕ is $\Omega \sin \phi$. Vorticity is $2 \times$ angular velocity, and so the planetary vorticity possessed by a parcel of water on the surface of the Earth at latitude ϕ is $2\Omega \sin \phi$.

(b)   (i) and (ii)   At the North and South Poles, this will be equal to $2\Omega \sin 90°$, i.e. 2Ω, because $\sin 90° = 1$. As illustrated in Figure 4.7(b), the surface of the Earth rotates anticlockwise in the Northern Hemisphere and clockwise in the Southern Hemisphere. If we use the convention, given in the text, that clockwise rotation corresponds to negative vorticity, the planetary vorticity possessed by a parcel of fluid at the South Pole will be written as -2Ω. (iii) At the Equator, $\phi = 0$, and so planetary vorticity $= 2\Omega \sin 0 = 0$.

Question 4.4   (a)   (i)   A body of water carried southwards from the Equator is moving into regions of increasingly negative planetary vorticity (Figure 4.7(b)). (ii) Its absolute vorticity $f + \zeta$ must remain constant, so to compensate for the descrease in f, its relative vorticity ζ must *increase*; so the water must increasingly acquire a tendency to rotate with positive vorticity (i.e. anticlockwise) in relation to the Earth.

(b)   Negative relative vorticity will be acquired by a body of water (i) from winds blowing in a clockwise direction, *and* (ii) from cyclonic winds in the Southern Hemisphere, because these are also clockwise (Figure 4.5).

Question 4.5   (a) and (b)   In both cases, if D is reduced then for f/D to remain constant f must be reduced. While over the shallow bank the current must therefore be diverted *towards* the Equator; this involves it curving southwards in case (a) and northwards in case (b). Note that it does not make any difference what the direction of current flow is initially.

Question 4.6   (a)   The resemblance is fairly close. In both cases the overall motion is clockwise (*cf.* Figure 4.5): the part of the wind field between 15° and 30° of latitude corresponds to the Trade Winds, which have a strong easterly component; the part between 30° and 45° of latitude corresponds to the westerlies forming the higher-latitude limit of the subtropical anticyclones.

(b) Ekman transport is to the right of the wind in the Northern Hemisphere and so will be in the directions shown by the wide arrows in Figure A6.

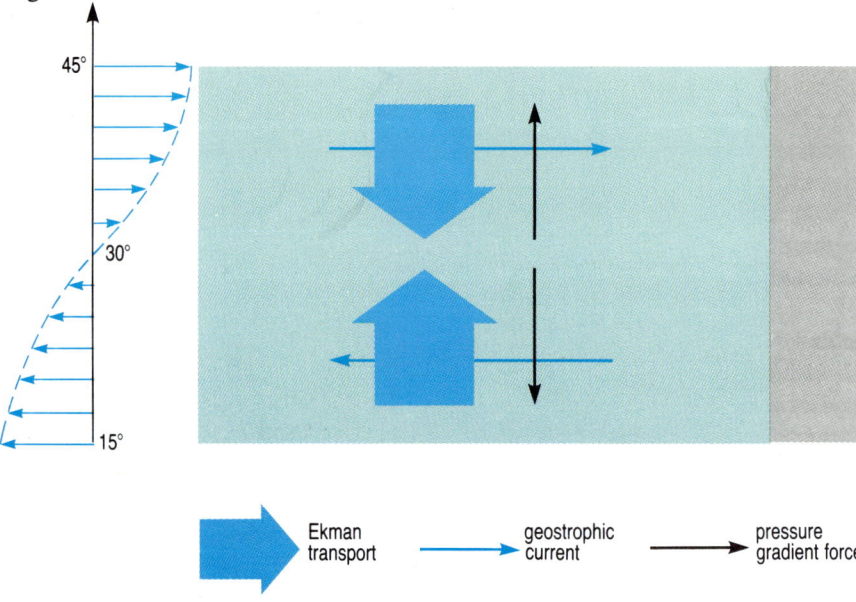

Figure A6 Answer to Question 4.6(b) and (c).

(c) The Ekman transports towards the centre of the ocean will converge and lead to a raised sea-surface (*cf*. Figure 3.23 (c) and (d)). As a result, there will be horizontal pressure gradient forces acting northwards and southwards away from the centre (black arrows on Figure A6), so that geostrophic currents will flow in the directions shown by the thinner of the blue arrows (*cf*. Figure 3.24).

Question 4.7 (a) In (b), the sea-surface has a large topographic high in the middle of the gyre, whereas there is no such feature in (a). The reason for the difference is that in (b), unlike (a), the ocean is on a rotating Earth and a Coriolis force has come into existence. For there to be equilibrium, this must be balanced by horizontal pressure gradient forces, which are brought about by the sea-surface slopes.

(b) The flow pattern in (b) is symmetrical, like that in (a), and so the existence of the Coriolis force *per se* cannot be the reason for the intensification of western boundary currents. However, when the Coriolis force is made to increase with latitude—albeit linearly—the flow pattern *does* show the asymmetry observed in the real oceans. Figure 4.12 therefore strongly suggests that the intensification of western boundary currents is a result of the fact that the Coriolis force increases with latitude.

Question 4.8 (a) If $dw/dt = 0$, and there are no forces acting vertically, other than the weight of the water and the vertical pressure gradient, so that $F_z = 0$, equation 4.3c becomes

$$0 = -\frac{1}{\rho}\frac{dp}{dz} - g$$

$$+\frac{1}{\rho}\frac{dp}{dz} = -g$$

or $dp = -\rho g dz$

This is the hydrostatic equation (3.8), which relates the pressure at a given depth to the weight of the overlying seawater.

Question 4.9 The western boundary currents are fast, narrow and deep, while eastern boundary currents are slow, broad and shallow; this means that the rate at which water is transported polewards in the western boundary currents can be equal to the rate at which water is carried equatorwards in the eastern boundary currents, despite the fact that western boundary currents are narrow, and flow in the eastern boundary currents occupies much of the rest of the ocean. (Here we are ignoring currents flowing into and out of the gyres, e.g. the Labrador Current and the northern branches of the North Atlantic Drift.)

Question 4.10 (a) The sections indicate that flow through the Florida Straits is baroclinic. Figure 4.18(a) and (b) show that there are strong lateral variations in T, S and hence density, and the slopes of the isotherms indicate that the isopycnals slope up to the left so as to intersect the sea-surface (the topmost isobaric surface). Furthermore, the velocity sections (c) and (d) show that the current changed with depth, decreasing to zero above the sea-bed, consistent with baroclinic conditions (*cf.* Figure 3.15(b)).

(b) As this is in the Northern Hemisphere the sea-surface must slope up to the right and the current must flow 'into the page' so that Florida is on the left-hand side of the section and Bimini on the right-hand side.

(c) Isotherms and isohalines (Figure 4.18(a) and (b)) both indicate that, in the upper 100m, the least dense water is in the middle of the section. As a result, although the sea-surface will generally slope up to the right, it will have a 'hump' in the middle (see Figure A7).

Figure A7 Answer to Question 4.10(c).

Question 4.11 Comparison of Figure 4.24(a) and (b) suggests that results obtained by Lagrangian methods may be more difficult to interpret than those obtained by Eulerian methods. On the other hand, they do provide an idea of current flow patterns, and give information about a large area. Eulerian measurements provide information about how a current changes with time at a given place, which Lagrangian methods do not, but a very large number of fixed current meters would be needed to give equivalent spatial coverage. In practice, experiments using drifters do involve a large number of them—in the MODE experiment, 20 Sofar floats were required to monitor adequately a region about 300km wide—but they are very much cheaper than moored current meters.

Question 4.12 (a) The blue area on Figure 4.26(a) must be the cold water of the Labrador Current; and (b) the yellow area must be warm Sargasso Sea and Gulf Stream water, with the reddest region corresponding to the warm core of the Stream (in image (b) these waters are shown blue, as they have low primary (phytoplankton) productivity. Note that the 'edges' of each of these water masses will not exactly correspond to the boundary between one colour and another; the colour changes have to occur at fixed values of temperature which may or may not tie up exactly with the temperature range in a given current, or water mass. Also, as mixing with adjacent water occurs, the temperature of the water in a current, and hence the colour on a CZCS image, progressively changes.

Question 4.13 Eddies form rather in the manner of river meanders cutting off ox-bow lakes, enclosing and pinching off volumes of water so that they end up on the opposite side of the Stream. Eddies on the continental-margin side of the Gulf Stream are therefore *warm*-core eddies. This is borne out by Figure 4.26(a) in which the central region of the eddy is yellow, corresponding to warm Gulf Stream and Sargasso Sea water. (Note that the extent of the eddy is considerably larger than indicated by the area of yellow.)

Question 4.14 Comparison of Figure 2.3(a) and (b) shows that north of Cape Blanc, the North-East Trades blow along the coast all year. The degree to which the Trade Winds penetrate to the south of Cape Blanc depends on the position of the ITCZ, which is furthest north in the northern summer. Coastal upwelling therefore occurs much further south in November (Figure 4.31(b)) than in March–April (Figure 4.31(a)).

Question 4.15 False. Both Franklin and Rennell believed that the Gulf Stream is driven by the 'head' of water in the Gulf of Mexico, i.e. by the horizontal pressure gradient *in the direction of flow*. They were not aware of the idea of geostrophic balance between the Coriolis force and the horizontal pressure gradient at *right angles* to the flow.

Question 4.16 The author is referring to the fact that turbulent eddies provide the internal friction of the ocean—its eddy viscosity.

Question 4.17 As it is a frontal region, the edge of the Gulf Stream is also a *convergence*, where surface debris will tend to accumulate (*cf.* Section 3.4).

Question 4.18 The changes in depth associated with the seamounts mean that D in $(f + \zeta)/D$, the potential vorticity of water in the current, is changing. If $(f + \zeta)/D$ is to remain roughly constant, a reduction in D must be accompanied by decreases in the value of f and/or ζ—i.e. by equatorward meanders (*cf.* Question 4.5) and/or a tendency to clockwise rotation. (Note that as the Gulf Stream is a region of strong velocity shear, we cannot assume that $f \gg \zeta$.)

Question 4.19 Flow is in geostrophic equilibrium when the only forces acting in a horizontal direction are the horizontal pressure gradient force and the Coriolis force, acting in opposite directions and at right angles to the flow. In such circumstances, the flow is steady and non-accelerating. Water circulating in a relatively small-scale meander or eddy experiences a centripetal force towards the centre of the eddy, and its rate of change of speed and direction (i.e. its acceleration) becomes significant.

Question 4.20 (a) This is true. The average velocity will be calculated from the estimates of the distance travelled and time taken. As the actual path taken by the drifter is likely to be complicated, the distance it has travelled may well be underestimated, especially if its position has been fixed infrequently or not at all (as in the case of cheap plastic drifters). Also, if a drifter is washed ashore and not found for a long time, the travel time will be overestimated.

(b) This is not strictly true. Some Lagrangian techniques rely on people voluntarily returning drifters stating when and where they were found. But any oceanographer will tell you that moored current meters are at the mercy of other users of the sea lanes, and may be easily damaged accidentally.

(c) False. The Doppler methods and OSCR are Eulerian methods as they measure the velocity of a patch of water in a given position(s). They provide no information about the path taken by a specific parcel of water, even though OSCR provides information about surface currents over a fairly wide area.

(d) False. The calculation of surface currents using ships' drift is a Lagrangian technique because it relies on the deduction of the path taken by ships under the influence of the current. It is the average current velocity over some distance that is calculated, rather than the current speed at any particular location. A significant proportion of all current measurements obtained by Lagrangian methods have been calculated from ships' drift.

CHAPTER 5

Question 5.1 (i) The sea-surface slopes down from the north towards the divergence at 10° N, leading to a southward horizontal pressure gradient force; if there is to be geostrophic equilibrium, this must be balanced by the Coriolis force acting northwards. The Coriolis force acts to the right of the current in the Northern Hemisphere, and so geostrophic flow must be westward.

(ii) The same argument can be used here.

(iii) In the Southern Hemisphere, on the other hand, the Coriolis force acts to the left of the current. A sea-surface slope down from the south will lead to a northwards horizontal pressure gradient force which may be balanced by the Coriolis force acting to the south, i.e. to the left of a westward-flowing current.

Question 5.2 The position of the ITCZ, the zone in which the Trade Winds meet, does not simply run east–west across the Atlantic. It is distorted by the effect of continental masses, in particular the bulge of the African continent. The wind field that results is reflected in the pattern of surface currents.

Question 5.3 (i) From Figure 5.4(a), the cross-sectional area of the Cromwell Current is about 4° of latitude wide × 200m deep = $(4 \times 110 \times 10^3)$m × 200m = 88×10^6m².

(ii) From Figure 5.4(a) the average velocity in the Cromwell Current is about $0.4\,\mathrm{m\,s^{-1}}$. Therefore, at this locality the volume transport of the Cromwell Current is about $88 \times 0.4 \times 10^6 \approx 35 \times 10^6\,\mathrm{m^3\,s^{-1}}$.

Question 5.4 (a) As discussed in the answer to Question 4.14, to the north of Cape Blanc the Trade Winds blow roughly parallel to the coast so as to cause offshore transport of surface water, and hence upwelling, more or less all the year round. The same is true to the south of Cape Frio. Between Cape Blanc and Cape Frio, wind directions vary markedly in response to the changing position of the ITCZ, and so upwelling occurs only seasonally.

(b) The upwelling in the region between the Equator and about 4° S is at least partly the result of the surface divergence caused by the South-East Trades crossing the Equator (as illustrated in Figure 5.1(a)).

Question 5.5 (a) Winds are cyclonic around regions of low pressure.

(b) When the ITCZ is in its most northerly position, it passes over the Guinea Dome which is then affected by cyclonic winds. As illustrated in Figure 3.23(a) and (b), cyclonic winds lead to a raised thermocline, surface divergence and upwelling, so the effect of the ITCZ is to intensify the 'doming' of the isotherms in the near-surface waters.

Question 5.6 Figure 5.8(a) is the easier of the two maps to use here. It shows that the thermocline is very shallow along the Equator in the eastern Atlantic; more importantly, it shows that the thermocline is also shallow along the coast of the Gulf of Guinea and that the Guinea Dome is well developed. (All these upwelling features may be seen to some extent in Figure 5.8(b).) Figure 5.8 must therefore show the situation in July to September, rather than January to March.

Question 5.7 No, it is not possible. Since the Coriolis force acts to the right of the direction of motion in the Northern Hemisphere and to the left in the Southern Hemisphere, the opposing pressure gradient, resulting from the sea-surface slope, must act to the left in the Northern Hemisphere and to the right in the Southern Hemisphere. This condition is satisfied by Kelvin waves travelling with the coast on the right in the Northern Hemisphere, and on the left in the Southern Hemisphere. (However, this is not the whole story, because the coastal current and the Kelvin wave *can* travel in opposite directions. How this comes about is beyond the scope of this Volume.)

Question 5.8 The Coriolis parameter, f, is equal to $2\Omega \sin \phi$. At 10° N, $\sin \phi = 0.174$, so $f = 2 \times (7.29 \times 10^{-5}) \times 0.174 = 2.5 \times 10^{-5}\,\mathrm{s^{-1}}$. Hence, the Rossby radius of deformation, $L = c/f = 2/(2.5 \times 10^{-5})\,\mathrm{m} = 80\,\mathrm{km}$.

Question 5.9 (i) The South Equatorial Current is the (blue) region of westward-flowing water, extending from the surface to a maximum depth of about 500m, between 4°N and 3°S. (ii) Only the deep flow in the North Equatorial Current may be seen, poleward of about 11° N. (iii) Both a North and a South Equatorial Counter-Current can be seen, one each side of the South Equatorial Current. (iv) The Equatorial Undercurrent may be clearly seen on the Equator, with its core at a depth of about 100m.

Question 5.10 The pressure gradient force driving the Equatorial Undercurrent is opposed by frictional forces caused by turbulence (eddy viscosity) both within the Undercurrent itself and resulting from shear between it and adjacent currents. The Undercurrent is too shallow to be affected by bottom topography.

Question 5.11 Yes, in principle, monsoon winds are the same as land and sea breezes. In both cases, the wind direction depends on differential heating of land and sea, being from land to sea in the winter/night when the atmospheric pressure over the land is greater than that over the sea, and from sea to land in the summer/day when the atmospheric pressure is higher over the sea. Monsoon winds are, of course, on a much greater scale and interact with winds in the upper troposphere.

Question 5.12 It is the Somali Current, which in the northern summer forms a poleward-flowing western boundary current, similar to those found in the other oceans.

Question 5.13 The locations of upwelling in the Indian Ocean differ from those in the Pacific and the Atlantic in that the main regions are along the *western* boundary of the ocean, whereas in the Pacific and Atlantic they are along *eastern* boundaries. Also, there is no significant open ocean upwelling just south of the Equator in the eastern ocean, as there is in the Pacific and Atlantic.

Question 5.14 The Rossby radius of deformation must be somewhat *bigger* than the distance indicated by the separation between the 2.5 contour and the Equator as we really need to know where the displacement becomes negligible. Therefore, L for this baroclinic equatorial Kelvin wave is about 6–7° of latitude, or 660–770 km. (This is unusually large for a baroclinic equatorial Kelvin wave; see Section 5.3.1.)

Question 5.15 In the case of shelf waves, it is the depth D that varies, rather than latitude (i.e. planetary vorticity, f), as is the case in Rossby waves (*cf.* the discussion of topographic steering in Section 4.2.1). If the current flowing parallel to a particular depth contour is displaced either towards or away from the coast, D will decrease or increase. For potential vorticity $(f + \zeta)/D$ to remain constant (and assuming for convenience that latitude remains more or less constant) the relative vorticity ζ must change. The current will oscillate to and fro about the original depth contour, with alternate cyclonic and anticyclonic eddies forming along it (*cf.* Figure 5.15(b)).

Question 5.16 (a) The *Fram* must have travelled with the Transpolar Current.

(b) A distance of 4000 km in three years gives an average current speed of:

$$\frac{4000 \times 10^3}{3 \times 365 \times 24 \times 3600} \approx 0.04 \, \text{m s}^{-1}.$$

This is within the usual range of estimated mean current speeds obtained using ice-drift (0.01–0.04 m s^{-1}).

CHAPTER 6

Question 6.1 The effect of water in the atmosphere is greater in low latitudes than in high latitudes because evaporation occurs at a greater rate from a warmer sea-surface. As discussed in Section 2.3, upward convection of air warmed by the sea carries moisture as well as heat, and warm air can contain more moisture than cold air. This is borne out by Figure 6.1, because the distortion of contours over the oceans is greatest at very low latitudes. Away from tropical regions, the difference in insolation between ocean and land areas does not seem to show any obvious relation to latitude.

Question 6.2 The heat-budget equation is:

energy gained as short-wave radiation = energy lost as long-wave radiation + energy lost as latent heat + energy lost by conduction to atmosphere

i.e. $Q_s = Q_b + Q_e + Q_h$

Question 6.3 (a) The area in question is affected by the Kuroshio, the western boundary current in the Pacific. The water carried from low latitudes in the Kuroshio is significantly warmer than the overlying air, and so the transfer of heat from sea to air by conduction/convection is above average here. In addition, the warming of the air in contact with the sea-surface encourages turbulent convection and leads to high (positive) values of Q_e as well as Q_h. Also, warmer air can contain more water vapour than cooler air before becoming saturated and this will contribute to high evaporation rates.

(b) By contrast, in the eastern equatorial Pacific, sea-surface temperatures are anomalously low because of upwelling of cooler subsurface water (Section 5.1.2). Because the air is warmer than the sea, the sea gains heat from the air by conduction and Q_h is negative. A cool sea-surface results in the overlying air being stable and convection being inhibited. This, along with the cool sea-surface and the relatively high humidity of air cooled by contact with the sea-surface, leads to low values of Q_e.

Question 6.4 The statement is false. The values given for Bowen's ratio (0.1 and 0.45) are both less than 1, indicating that Q_e is greater than Q_h at both high and low latitudes (*cf.* the general statement given at the start of Section 6.1.2). Q_h is generally less than Q_e. In fact, from the values given, Q_h is one-tenth of Q_e in low latitudes, rising to nearly half of Q_e at 70° N.

Question 6.5 (a) In general, surface salinities are higher in the Atlantic than in the Pacific. This is particularly true in the Northern Hemisphere where, at a given latitude, values are about 2 parts per thousand higher in the Atlantic than in the Pacific.

(b) A notable feature of Figure 6.8(b) is the large negative *E–P* value and low salinity just north of the Equator. This is the result of the high precipitation characteristic of the ITCZ.

Question 6.6 (a) From the information given, $F = -7 \times 10^4 \text{m}^3\text{s}^{-1}$. From the estimated values of S_1 and S_2,

$$\frac{S_1}{S_2} = \frac{36.3}{37.8} = 0.9603$$

$$\frac{S_2}{S_1} = \frac{37.8}{36.3} = 1.041$$

Substituting in equations 6.6a and 6.6b, we get

$$V_2 = \frac{-7 \times 10^4}{(1 - 1.041)} = \frac{-7 \times 10^4}{-0.041} = 1.71 \times 10^6 \, m^3 s^{-1}$$

and $V_1 = \dfrac{-7 \times 10^4}{(1 - 0.9603)} = \dfrac{-7 \times 10^4}{0.0397} = -1.76 \times 10^6 \, m^3 s^{-1}$

Thus, the rate of inflow through the Straits of Gibraltar is about $1.76 \times 10^6 \, m^3 s^{-1}$, while the rate of outflow is about $1.71 \times 10^6 \, m^3 s^{-1}$. (The fact that V_1 and V_2 have different signs merely reflects the fact that one is an inflow and the other an outflow.)

An alternative method would have been to use equations 6.3 and 6.4 and hence obtain two equations relating V_1 and V_2, which could then be solved.

(b) It would take: $\dfrac{3.8 \times 10^6 \times 10^9 \, m^3}{1.76 \times 10^6 \, m^3 s^{-1}} = 2.16 \times 10^9 \, s$

or $\dfrac{2.16 \times 10^9}{365 \times 24 \times 3600} = 68$ years

That is, it would take about 70 years for all the water in the Mediterranean to be replaced once.

Question 6.7 Important regions of convergence in mid-latitudes are the centres of the subtropical gyres, where anticyclonic winds lead to Ekman transport towards the centres of the gyres (Figure 3.23(c) and (d)). The high-latitude region of convergence that should have come to mind is, of course, the Antarctic Polar Frontal Zone, or the Antarctic Convergence.

Question 6.8 Yes. In Figure 5.23(a), a 'tongue' of cold water can be seen extending down from the Antarctic Front to a depth of between 500m and 1000m; in the corresponding region in Figure 5.23(b), the near-surface isohalines dip downwards. The flow of Antarctic Intermediate Water down towards the north may be seen from the slope of the isotherms and isohalines from the region of the 'tongue' to a depth of about 1500m at 56° S (the northerly limit of the section).

Question 6.9 A layer of ice 'insulates' the ocean from the atmosphere and significantly reduces heat loss through conduction/convection, Q_h (as well as preventing heat loss through evaporation, Q_e). When there is no ice cover, heat losses to the atmosphere are greatly increased, particularly during winter conditions.

Question 6.10 See Figure A8 for the result. Measurement of segments a and b shows that they are in the proportion of 1 : 2, which means that the mixture contains twice as much of water type II as of water type I (i.e. water type I makes up 33% of the mixture). You could have obtained the same result arithmetically by using values of T (or S). By proportion,

$$\frac{b}{a} = \frac{3-2}{5-3} \left(\text{or } \frac{34.85 - 34.5}{35.5 - 34.85} \right) = \frac{1}{2}$$

However, the graphical method is useful because it helps you to estimate proportions of contributing water masses 'by eye': the closer the position of a mixture to a water mass, the more of that water mass must be in the mixture.

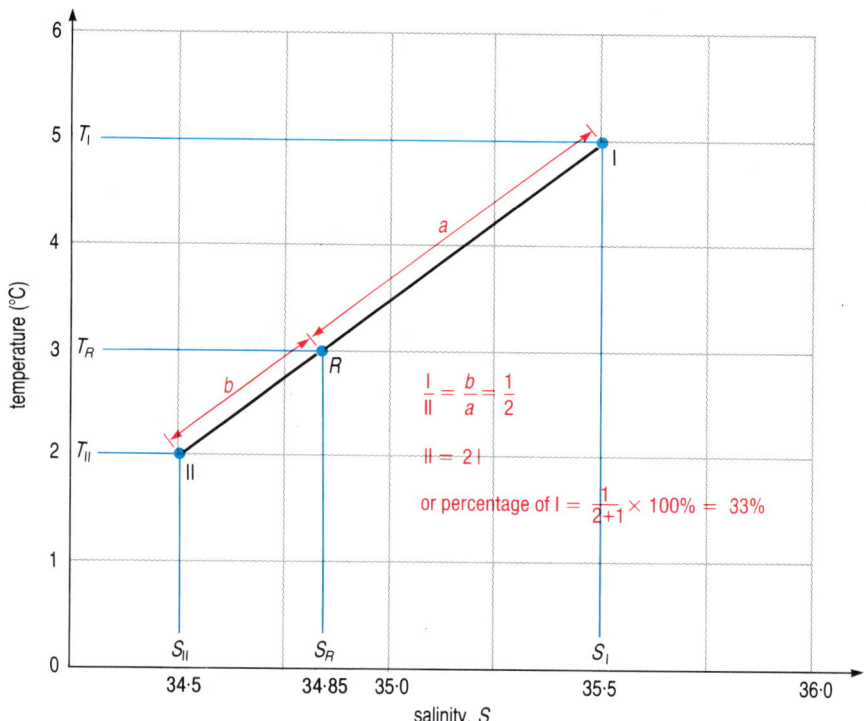

Figure A8 The completed T–S diagram for Question 6.10, showing water types I and II mixed in proportions of 1 : 2, to give a resultant mixture with a temperature of 3 °C and a salinity of 34.85.

Question 6.11 If you find *Meteor* Station 200 on Figure 6.15, you will see that at this location warm saline upper water overlies cooler, fresher AAIW, which in turn is underlain by relatively saline NADW and finally, at depth, AABW. (Remember that the 10 and 0°C isotherms and the 34.8 isohaline on Figure 6.15 are *not* actually boundaries between water masses, although they give fairly good indications of the relative positions of the different water masses.) All the features seen on Figure 6.15 may be identified on the T–S curve in Figure 6.26.

(a) Yes, the water at about 800m is the eroded core of Antarctic Intermediate Water (see Figure A9).

(b) If you draw construction lines as shown in Figure A9 (*cf.* Figure 6.24) and measure the segments a and b, you get:

$$\frac{b}{a+b} \times 100 = 55\%$$

So, the percentage of Antarctic Intermediate Water at 800m depth is about 55%.

For completeness, the proportions of the other water masses that contribute to the mixed water at 800m are about 25% for North Atlantic Deep Water and about 20% for the water at 400m depth.

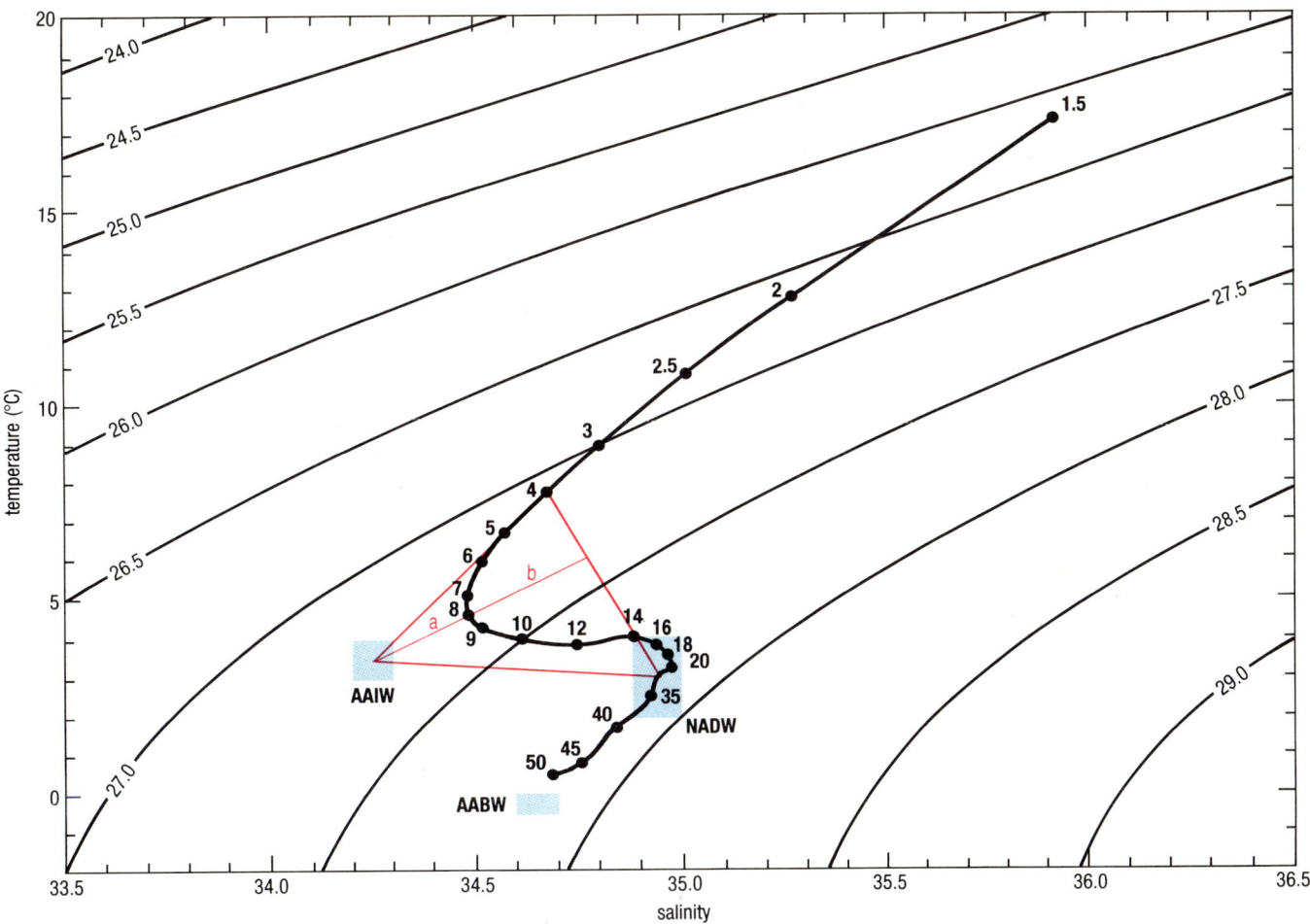

Figure A9 The completed T–S diagram for Question 6.11. The ratio a : b shows that the proportion of Antarctic Intermediate Water in the core is about 55%.

(Note, however, that the graphical method enables you to find out the proportion of one water mass without having to find it for all three.)

Question 6.12 (a)(i) The water mass below North Atlantic Central Water is of high salinity (reaching nearly 35.8) and is warm (9–10°C). This, combined with the location and the depth of the water mass, indicates clearly that the water mass is Mediterranean Water. The trend of the θ–S curve in this region is more or less parallel to the σ_θ contours, reflecting the spread of Mediterranean Water at depths where it is neutrally buoyant. Interestingly, there seems to be a 'double core' of Mediterranean Water. This could be because Mediterranean Water of slightly different densities is spreading out along two different isopycnic surfaces (corresponding to σ_θ ~27.6 and ~27.55). This may indicate a local intrusion of slightly fresher, cooler water between about 900 and 1000 m depth.

(ii) The deepest water mass represented by the θ–S curve is North Atlantic Deep Water, clearly identified by its temperature–salinity combination of ~2–4°C and ~35.0 (as well as by the geographical location of the station). If the θ–S curve had been plotted for sufficiently great depths, some influence of Antarctic Bottom Water would have been detected.

(b) The two water masses are of approximately the same density. However, if you plotted them on Figure 6.27, you would have seen that when they mix together, the resulting water is *denser* than either of the original contributions (with $\sigma_\theta \sim 28.05$ as opposed to 28.00).

Question 6.13 Yes. The lowest silica concentrations ($<40\,\mu\text{mol}\,l^{-1}$), corresponding to North Atlantic Deep Water, are seen between 1500 and 3000m. Below that, concentrations increase again up to $100\,\mu\text{mol}\,l^{-1}$, in Antarctic Bottom Water. The shape of the contours of silica concentration strongly suggests that both North Atlantic Deep Water and Antarctic Bottom Water are flowing along the western boundary of the ocean.

Question 6.14 (a) The three arrows represent the freshwater that leaves the Antarctic in the form of pack ice and icebergs (*cf.* the discussion of the role of coastal polynyas in ice production in Section 6.3.2).

(b) The convergence of freshwater at low latitudes in the North Atlantic must reflect its net *removal* from the ocean, as a result of evaporation of water greatly exceeding precipitation (notwithstanding the negative $E-P$ values in the region of the ITCZ). Both the high surface salinities in the region (Figure 6.8(a)) and the shape of the $E-P$ curve for mid to low latitudes (Figure 6.8(b)) are consistent with this view.

(c) As discussed in Question 6.5, surface salinity is much greater in the Atlantic than it is in the Pacific. Freshwater is removed from the Atlantic by net evaporation and can only be gained in the Pacific by precipitation (or from rivers). We may therefore assume that much of the water removed from the Atlantic by evaporation is supplied to the Pacific as rainfall.

Question 6.15 (a) This is true, because at night Q_s is zero while Q_b will have a value determined by the temperature of the sea-surface and the degree of cloudiness (plus water vapour content of the atmosphere, etc.).

(b) The fogs over the Grand Banks are a result of warm moist air overlying the cold waters of the Labrador Current (not the warm Gulf Stream). However, Figure 2.3 shows that in summer in particular the prevailing winds will pick up heat and moisture from the Stream. So the statement is true, albeit indirectly.

(c) The statement is true: Mediterranean Water forms as a result of high evaporation by dry winds that are also cold, so both evaporation (Q_e) and cooling (Q_h) will lead to heat loss.

(d) This statement is probably true. During the South-West Monsoon, there is upwelling off the coast of Somalia. If the upwelled water at the surface is colder than the overlying air, there will be a gain of heat by the sea by conduction (i.e. Q_h represents a gain). A cold sea-surface tends to promote a stable layer of overlying air and evaporation will be inhibited, and as long as the sea-surface is not warmer than the overlying air by more than 0.3°C, heat will be gained by condensation.

Question 6.16 Meltwater is fresh and therefore of low density. The formation of a layer of low-density water at the surface *increases* the stability of the water column and so tends to 'damp out' vertical convection.

Question 6.17 Features you may have thought of are (i) the convergence of surface waters, (ii) a sloping boundary between the denser and less dense water (*cf.* Figure 6.20), and (iii) a tendency for meanders and eddies to form along the zone (Section 5.5.2).

Question 6.18 (a) The high-salinity 'ridge' evident in diagram 4 corresponds to North Atlantic Deep Water; diagram 4 is therefore for the Atlantic, while diagram 3, with much lower salinities, is for the Pacific.

(b) The low temperature water mass visible on all the diagrams (though clearest in 4, for the Atlantic) is Antarctic Bottom Water.

(c) The peak in diagrams 1, 2 and 3 corresponds to the huge volume of water making up Pacific and Indian Ocean Common Water.

Question 6.19 Our answer is illustrated in Figure A10. The results obtained here differ from those of the researchers who first published estimates of the proportions. They obtained the following values: Antarctic Bottom Water, 60%; North Atlantic Deep Water, 24%; Antarctic Intermediate Water, 16%. The discrepancy probably arises because these researchers allowed for the fact that heat flow through the sea-floor raises the temperature of Antarctic Bottom Water.

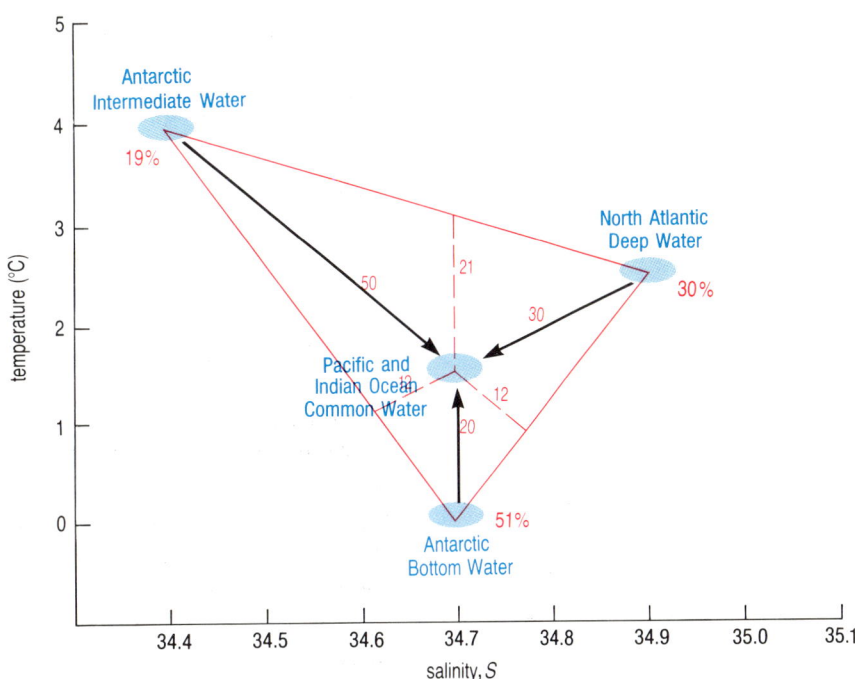

Figure A10 Completed *T–S* diagram for the answer to Question 6.19. The numbers give the lengths of segments a–f in mm.

Question 6.20 The upwelling at the Antarctic Divergence ensures that surface waters are kept well supplied with the nutrient elements (including silicon) that are required by phytoplankton populations. On death, the remains of the phytoplankton sink down to the deep ocean; the dissolution of the siliceous skeletons of diatoms enriches the bottom waters with silica.

Question 6.21 As discussed in Section 6.1.1, the surfaces of the oceans and continents emit long-wave (back-) radiation which is reflected back by clouds and absorbed and re-emitted by water vapour and other gases including carbon dioxide. The *net* amount of long-wave radiation emitted by the surface gives rise to the heat loss, Q_b. Because the concentration of carbon dioxide in the atmosphere is increasing, the amount of heat absorbed in the atmosphere and re-emitted back towards the surface of the Earth is probably also increasing, so that the net *loss* of heat from the sea-surface as Q_b is probably decreasing.

Question 6.22 The main aim of WOCE is to develop a way of predicting climatic change over periods of decades. Climatic change is mediated by the transport of heat and freshwater around the atmosphere–ocean system. We now know that mesoscale eddies are important agents in the transport of heat and freshwater (i.e. salt) both within and between oceans (*cf.* Sections 3.5.2, 4.3.5 and 6.6).

ACKNOWLEDGEMENTS

The Course Team wishes to thank the following: Dr. Martin Angel and Dr. Mike Davey, the external assessors; Dr. John Harvey for a major contribution to Chapter 6; Dr. Peter Killworth, Dr. Denise Smythe-Wright and Dr. Phillip Woodworth for advice on specific Sections; and Mr. Mike Hosken and Mrs. Mary Llewellyn for commenting on the whole Volume. Dr. Malcolm Howe and Dr. Chris Vincent also provided helpful advice on content and level.

The structure and content of this Volume and of the Series as a whole owes much to our experience of producing and presenting the first Open University Course in Oceanography (S334), from 1976 to 1987. We are grateful to the Course Team members who prepared and maintained that Course, to the students and tutors who provided valuable feedback and advice, and to UNESCO for supporting its use overseas.

Grateful acknowledgement is also made to the following for material used in this Volume:

Figures 1.1, 3.21, 3.26(d), 4.4, 4.26, 5.21 and 6.18 NASA; *Figure 1.4(c)* J.W. Hedgpeth (1957) *Treatise on Marine Ecology and Palaeoecology*, Vol. 1, Memoirs of Geological Society of America, **67**; *Figures 1.5, 6.33 and 6.34* J.D. Woods (1985) in *Nature*, **314**, Macmillan; *Figures 2.1 and 2.17* European Space Agency; *Figures 2.2(b) and 2.3* A.H. Perry and J.M. Walker (1977) *The Ocean–Atmosphere System*, Longman; *Figures 2.7 and 2.14(a)* NOAA/NESDIS/NCDC/SDSD; *Figure 2.10(a)* L.R.A. Melton; *Figure 2.10(b)* H.G. Mullett; *Figure 2.12* D.R. Sikka and S. Gadgil (1980) in *Monthly Weather Review*, **108**, American Meteorological Society; *Figure 2.15* J.G. Harvey (1976) *Atmosphere and Ocean*, Artemis Press; *Figure 2.16* J.H. Golden, courtesy of the Meteorological Office; *Figure 3.1* A.N. Strahler (1963) *Earth Sciences*, Harper and Row; *Figures 3.2 and 5.9* J. Crease et al. (eds.) (1985) *Essays on Oceanography, Progress in Oceanography*, **14**, Pergamon; *Figure 3.8* G. Neumann (1968) *Ocean Currents*, Elsevier; *Figures 3.18(a), 4.21, 4.22, 5.20, 6.6 and 6.11* G.L. Pickard and W.J. Emery (1982) *Descriptive Physical Oceanography*, Pergamon; *Figures 3.20, 3.32 and 6.35* B.A. Warren and C. Wunsch (eds.) (1981) *Evolution of Physical Oceanography*, MIT Press; *Figure 3.26(a)* M. Hosken; *Figure 3.26(b)* J.B. Wright; *Figure 3.28* F. Webster (1969) in *Deep-Sea Research*, **16**, Pergamon; *Figure 3.29* P. Rhines (1976) in *Oceanus*, **19**, Woods Hole Oceanographic Institution; *Figure 3.30* J.C. McWilliams (1976) in *Journal of Physical Oceanography* **6**, American Meterorological Society; *Figure 3.31* N. Wells (1986) *The Atmosphere and Ocean*, Taylor and Francis; *Figure 3.33* US Navy; *Figure 4.1(a)* L. de Vorsey Jr (1974) *The Atlantic Pilot* (facsimile reprint of De Brahm (1772)) University of Florida Press; *Figures 4.11, 4.12 and 4.14* H. Stommel (1965) *The Gulf Stream*, University of California Press; *Figures 4.17(b) and 6.16 (base map only)* Marie Tharp, 1 Washington Ave., New York 10960; *Figures 4.18, 6.5 and 6.20* H.U. Sverdrup et al. (1942) *The Oceans*, Prentice-Hall; *Figure 4.19* C.A.M. King (1975) *Introduction to Physical and Biological Oceanography*, Arnold; *Figure 4.20* W.S. von Arx (1962) *An Introduction to Physical Oceanography*, Addison-Wesley; *Figure 4.23(a) and (b)* Wimpol; *Figure 4.24* A.R. Robinson (ed.) *Eddies in*

Marine Science, Springer-Verlag; *Figure 4.27* D. Tolmazin (1985) *Elements of Dynamic Oceanography*, Allen & Unwin; *Figure 4.28* R.E. Cheyney and J.G. Marsh (1981) *Journal of Geophysical Research*, **86**, American Geophysical Union; *Figure 4.31* J.L. Fellous and A. Morel/ESA; *Figures 4.32 and 6.14* A. Defant (1961) *Physical Oceanography*, Vol. 2, Pergamon; *Figure 5.3(a) and (b)* J.A. Knauss (1978) *Introduction to Physical Oceanography*, Prentice-Hall; *Figure 5.3(c)* L. Lemasson and B. Piton, Institut Francaise de Recherche Scientifique pour le Developpement en Corporation; *Figure 5.4(a) and (b)* American Meteorological Society; *Figure 5.4(c) and (d)* P. Pullen and D. Paul/NOAA/PMEL and RD Instruments; *Figures 5.5 and 5.6* B. Voituriez in *Oceanologica Acta*, **4**, Gauthiers Villars; *Figure 5.7* B. Voituriez and A. Herbland, Centre Oceanologique de Bretagne; *Figures 5.8 and 5.11(b)* J. Merle (1980) in *Journal of Physical Oceanography*, **10**, American Meteorological Society; *Figure 5.16* D.B. Enfield (1987) in *Endeavour, New Series*, Pergamon; *Figures 5.22 and 5.23* T. Whitworth III (1988) in *Oceanus*, **31**, Woods Hole Oceanographic Institution; *Figure 5.24* IOC, UNESCO; *Figure 6.7* H.J. McLellan (1965) *Elements of Physical Oceanography*, Pergamon; *Figures 6.10 and 6.23* P. Tchernia (1980) *Descriptive Regional Oceanography*, Pergamon; *Figures 6.12, 6.13 and 6.17* W.J. Emery and J. Meinke (1986) in *Oceanologica Acta*, **9**, Gauthier Villars; *Figure 6.21* M.C. Gregg (1973) in *Scientific American*, **228**, W.H. Freeman; *Figure 6.27* IFREMER, Brest, France; *Figure 6.29* L.D. Talley and M.S. McCartney (1982) in *Journal of Physical Oceanography*, **12**, American Meteorological Society; *Figure 6.30* J.L. Reid *et al.* (1977) in *Journal of Physical Oceanography*, **7**, American Meteorological Society; *Figure 6.31(a) and (c)* D. Smythe-Wright; *Figure 6.32* H. Stommel (1958) in *Deep-Sea Research*, **5**, Pergamon; *Figure 6.36* B. Bolin and H. Stommel (1961) in *Deep-Sea Research*, **8**, Pergamon.

INDEX

Note: page numbers in italic refer to illustrations

AABW *see* Antarctic Bottom Water
AAIW *see* Antarctic Intermediate Water
absolute vorticity 81–3
    see also potential vorticity
adiabatic changes 21
    in atmosphere 21–2
    in ocean 192–3
advection, heat transport by
    in atmosphere 11, 17
    in ocean 31, 163
advection fog 165
'ages' of water masses 198
Agulhas Current *31*, 111
    eddies in 111, 200
    volume transport 111
air, density of 21
air masses 11, *14*, 18–19, *19*, 176
air–sea interface, processes at 6–7, 26, 29, 33–5, 69, 164–9, 172–84 *passim*
Alaska Current *31*, 148
albedo 167
altimetry, satellite 58, *58*, 72, *72*, 112, *113*
amphidromic systems 142
Angola Dome *131*, 132
angular momentum, conservation of 80, *83*
    see also vorticity 80
Antarctic Bottom Water *177*, *180*, 180–4, *184*, 185, 191, *191*, 196, 198
Antarctic Circumpolar Current *31*, 58, 79, 148, 151–4, *152*, 184
    heat transport across 68
    mesoscale eddies in 67–8
    volume transport of 154
Antarctic Circumpolar Water 183–4, *184*
Antarctic Convergence 152, *153*, 156, 206, 230
    see also Antarctic Polar Frontal Zone
Antarctic Divergence 154, *177*, 184, *184*, 208, 231
Antarctic Front 152, *152*, 153, *153*
Antarctic Intermediate Water 176, *177*, 184, *184*, 185
Antarctic Ocean *see* Southern Ocean
Antarctic Polar Current *see* Polar Current
Antarctic Polar Frontal Zone 152–4, *152*, *153*, 176, 184
    mesoscale eddies in 67, 68
anticyclonic flow in ocean 59, 60, *60*, 67, 70, 98, 112, 175
anticyclonic winds *18*, 18–19, 29, 59, *59*, 70, 87, 175
Antillean Current 96, *97*, 98
Arabian Sea 134, *135*, *137*
    seasonal upwelling in 137
Arctic Sea 148–51, *149*
    high salinity outflow from 178
Atlantic
    Equatorial Counter-Current(s) 124, 130, *130*, 131, 132
    equatorial current system 124–5
    Equatorial Undercurrent 128
    meridional heat transport in 200, *201*
    residence time 198
    salinity in 168, 200, 226, 230

    sea-surface slope across 138–9, *138*
    subtropical gyre 79, 96–101, *97*, 106–12, 115–18, 119–21
    upwelling in *131*, 131–2
    water masses in 171, 172–85, *173*, *174*, *175*, *177*, *179*, *180*, *184*, 191, *191*, *193*, 193–4
atmosphere
    circulation of 13–30
    interaction with ocean, 24–9, 146–8
    poleward transport of heat by 9–11, *10*, 17–23, 210
    see also air–sea interface, processes at; convection, in atmosphere
atmosphere–Earth system, radiation balance of 9–11
atmospheric pressure 54
    effect on geostrophic calculations 55
    effect on sea-level 28, 55, 218
    related to global wind system 14–16

Bache, A. D. 78
back-radiation *see* long-wave radiation
Bahama Bank 97
Baltic Sea, inertia currents in 39
baroclinic conditions 44, 46, *47*, 49
    determination of geostrophic velocity in 50–3
baroclinic waves 139–43
barotropic conditions 43–6, *47*, 49, 52
    determination of geostrophic velocity in 50, 72, 217
barotropic waves 139–43
Benguela Current *31*, 79
Blake Plateau *97*, 98
bottom water masses 178, *180*
    see also Antarctic Bottom Water
Bowen's ratio 166, 204
Brazil Current *31*, 79
buoys, surface 103, *103*, *104*, 105

cabelling 194
California Current *31*, 79
Canary Current *31*, 79, 98
    coastal upwelling in 115, *117*
carbon dioxide in atmosphere 160, 161
    see also greenhouse effect
carbon-14 as water-mass tracer 198
central water masses *174*, 174–5, *179*, 188, *194*
Central Waters *see* central water masses
centripetal force 39, 223
chlorofluorocarbons, as water-mass tracers *197*, 197–8
Circumpolar Deep Water *see* Antarctic Circumpolar Water
climatic change 27–8, 163, 202–3, 208, 231–2
clouds 9, *20*, 22–3, 25, 26, *26*, 28
    cumulonimbus, in ITCZ *14*, *23*, *125*
    during an El Niño 146–8, *147*
    cumulus *14*, 22, *23*
    effect on oceanic heat budget *160*, 160–1, 164

cloud streets 61, *62*
coastal polynyas *see* polynyas
Coastal Zone Color Scanner (CZCS) imagery *109*, *117*
cold-core eddies *110*, 110–12, 119
cold front *19*
conduction, heat loss from ocean by 163, 164–8, 176, 180, 182, 204, 226, 227
conservative properties 172, 192, 194
continuity, principle of 95, *96*, 96
convection
    forced 22
    in atmosphere 21–9, 134, 146–8, *147*, 160, 164–6, 204
    in ocean 7, 172, 174, 176, 178, 182–3
convergences 59, 59–62, *61*, *62*
    as regions of water-mass formation 173–5, 176
    see also fronts
coordinate system used in oceanography 93, *93*
core water 190, *190*, *191*, 228
Coriolis force 7, 7–8, 11, 36–7, *38*, 81
    effect on Equatorial Undercurrent 129
    effect on winds 15, 17
    in equations of motion 94
    in Kelvin waves 140
    see also Coriolis parameter; geostrophic currents; planetary vorticity
Coriolis parameter 37, 81, 82
Costa Rica dome 132
Cromwell Current 126–8, *127*, *128*, 129
    volume transport of 128
Cromwell, T. 126
curl, of wind stress 86, 92, *92*, 120
current meters 104, 104–6, *105*
current shear 80–1, *80*
    see also relative vorticity
current velocity, measurement of 101–6, 112–14, *113*
    see also geostrophic currents, determination of
cyclones, tropical 24–8, *26*
    during El Niño 148
    effect on ocean 27–8
    energy of 68
cyclonic flow in the ocean 59, 67, 70, 76, 112
    see also thermal domes
cyclonic winds *18*, 18–19, *20*, 29, 59, 70, 176

dead-reckoning 76, *76*
De Brahm, W. G. 74, *75*
deep circulation 159, 178–91, 196, 198–9, *199*, 203, 205–6
    see also thermohaline circulation
deep water masses 178, *180*, 203–4
    see also Antarctic Circumpolar Water; North Atlantic Deep Water; Pacific and Indian Ocean Common Water
density
    driving ocean circulation 6, 159
    effect on stability of fluid 21 *see also* stability of seawater, related to hydrostatic pressure 54–5

see also geostrophic currents, sigma-*t* (σ_t), sigma-θ (σ_θ), water-mass formation
density distribution in the ocean 35, 43–4, 46, 48, 50–1, 53, *53*, *54*, 124, 192–3
see also geostrophic currents
depression *see* cyclonic winds
depth in oceans, related to pressure 54–5
depth mean current 38
depth of frictional influence 38
divergence of surface water 59, 59–61, *61*, 70, 115
see also Antarctic Divergence; Equatorial Divergence; upwelling
Doldrums *14*, 122–4, *123*
domes, thermal *see* thermal domes
Doppler effect, use in current measurement 114, *128*
double diffusion 186
drag coefficient, for wind 33, 213
Drake Passage
kinetic energy density spectrum for *72*
T and *S* characteristics in 153
possible effect on Antarctic Circumpolar Current 154
'drift bottles' 103
'drift currents' 75
drifters 32, 101–4, *103*, 106, 223
see also Lagrangian methods of current measurement; Sofar floats
dynamic height *see* dynamic topography
dynamic topography 56–8, *57*

Earth, rotation of 6–8, 11, 40, 81
see also Coriolis force; Coriolis parameter; planetary vorticity
Earth–atmosphere system, radiation balance of 9–10, *10*, 11
East Australian Current 79
East Greenland Current 149, 151
easterly waves 24–6, *25*
eastern boundary currents 31, 79, 119, 212, 221
coastal upwelling in 115–18
eddies 6, 34, 79–80, *80*, *81*, 137, 185
mesoscale 65–9, 71, 72, 108–12, 186, 208, 232
see also Meddies; current shear
eddy viscosity 34–5, 37, 69, 89, 92, 125, 154, 219, 225
Ekman, V. W. 35, 36, 69
Ekman layer *36*, 38, 59, 60, 69
Ekman pumping 59
Ekman spiral *36*, 36–8, 69
Ekman transport 38, 59, 69, 71, 217
in association with Antarctic Circumpolar Current 151–2
in equatorial latitudes 122, *123*, 124
in Sverdrup's model of subtropical gyres 85
El Niño 145–8
energy of the ocean
kinetic 56, 63–7, 71, 72
potential 56, 67, 71
energy density spectrum *see* kinetic energy density spectrum
equations of motion 93–6
Equatorial Counter-Currents 124, 130, *130*, 131, 132
equatorial current systems 122-30, *123*
Equatorial Divergence 122, *123*, 131, *131*, 224

Equatorial Undercurrent 125–30, *126*, *127*, *128*, 157, 225
in Atlantic 128
in Indian Ocean 125, 137
see also Cromwell Current
equatorial wave guide 142, 144–5, 156
Eulerian methods of current measurement 104–6, 223
European Basin, residence time of water in 198
evaporation 6
difference between Atlantic and Pacific *169*, 200, 203
global variations *14*, 168, 175
heat loss by 163, 164–6, *165*, *167*, 176
evaporation–precipitation balance (*E* – *P*) 125, 168, *169*, 171, 175, 176

Ferrel, W. 78
fetch 27
floats *see* Sofar floats
flushing time *see* residence time
Folger, T. 74, *75*
Franklin, B. 74, 75, *75*, 77, 78
Freons *see* chlorofluorocarbons
freshwater, global transport of 200–1, *201*
friction
as a generator of relative vorticity 81, 89, 90
internal *see* eddy viscosity
representation in theoretical calculations 92, 94, 95
frictional coupling
between moving fluids and Earth 7, 82
within ocean 33–5, 36, 37, 38
see also eddy viscosity; turbulence
fronts 61
in Antarctic Polar Frontal Zone *152*, 152–3, *153*, 154, 156, 206, 230
see also convergences

geoid 56, 58, *58*, *113*
geostrophic equation 51, 70, 214–15
geostrophic equilibrium 40, 44, *44*, 45, *45*, 46, 222
geostrophic currents 40, 43–58, 70
velocity determination for 48–54
see also baroclinic conditions
geostrophic flow 40, 78, 120
in Antarctic Circumpolar Current 46, 151–2, 153
in atmosphere 44, 78
in equatorial current system 122–4, 130, 155, 223
in Gulf Stream, 46, 98–101, *99*, *100*, 119, 121, 222
in mesoscale eddies 67, 216
see also geostrophic currents
global fluxes of heat and freshwater 199–204
global wind system 13–16, *14*, *16*
gradient equation 45, 48, *49*, 50, 70, 78, 217
Grand Banks 97, 98, 151
advection fogs over 165
greenhouse effect 202, 203, 208, 231
Greenland Sea 148, *149*, 178, *181*
Guinea Basin, residence time of water in 198
Guinea Current 124
Guinea Dome 130, *130*, *131*, 132
Gulf Stream 6, *31*, 73–9, 84–92, 96–102,

97, 106–12, 114, 119, 121
counter-currents 66, 100–1, 199
early charts 74, 75, 76, *77*
as a front 67, 101, 119
geostrophic flow in 46, 98–101, *99*, *100*, 119, 121, 122
kinetic energy of *68*
'rings' (eddies) 65–7, 98, 108–12, *109*, *110*, *111*
theories about 73–5, 77–8, 84–93, 95
transport of heat by 163
volume transport of 98
see also Mid-Ocean Dynamics Experiment
Guyana Current 96, *97*

Hadley cells *14*, 15, 17, 29
Hadley circulation *see* Hadley cells
heat transport
in atmosphere 9–11, 13, 17–23, 27, 29
in ocean 9, 11, 31, 165, 184
see also thermohaline circulation
heat budget of ocean 159–80
heat-budget equation 163–8
Helland-Hansen, B. 51, 78
Helland-Hansen's equation 51, 70
high, atmospheric *see* anticyclonic winds
horizontal pressure gradient force 40, 43
represented in equations of motion 94
role in geostrophic currents 40, 44–53
horizontal pressure gradients
in atmosphere 14–15, 134, 146
in ocean 40, 42–3
see also horizontal pressure gradient force
Humboldt Current *31*, 79, 152
hurricanes *see* cyclones, tropical
hydrostatic equation 41, *41*, 54–5
hydrostatic pressure 41, *41*, 43, 54–5

ITCZ *see* Intertropical Convergence Zone
Ibn Khordazbeh 135
ice cover
meltwater from, effect on stability 180, 206, 230
motion in response to wind and/or currents 36, 151
in northern high latitudes 149–51, *150*, 156
in southern high latitudes *150*, 151, 156
ice formation
as a cause of brine production 178, 181
effect on albedo 167
effect on heat budget 167, 182, 227
see also polynyas
Indian Ocean
Equatorial Undercurrent in 125, 137
current system in 135–7, *136*, 156
deep water in 185
flow to Atlantic from 200
monsoon winds 15, 125, 134–5, 155–6
residence time of water in 198
upwelling in 137
inertia currents 38–40, *39*
insolation 160
see also solar radiation
in situ temperature 192
International Indian Ocean Expedition 137
Intertropical Convergence Zone *14*, 16, *16*, 23, *23*, 24, 29, 122, 212, 223

during El Niño *147*
   effect on density distribution of ocean 125
   effect on upwelling 117, 131, 132, 224
Irminger Current *149*, 151
isobar/isobaric surface 18, *43*, 43–6, *47*
   see also gradient equation
isopycnal/isopycnic surfaces 43, *43*, 44, 46, *47*

Kelvin waves 140–3, *141*, 144, 145, 146
kinetic energy density spectra 63–5, *64*, *72*
kinetic energy of the ocean 56, 63–7, 71, 72
Kuroshio *31*, 79

Labrador Current *31*, 108, 151, 176
Labrador Sea *149*, 151, 178
Labrador Sea Water 176, *177*, *179*, 180, 195
   see also Western Atlantic Sub-Arctic Intermediate Water
Lagrangian methods of current measurement 104, 106, 221
   see also drifters; Sofar floats
Langmuir circulation 61, *62*
latent heat polynyas *see* polynyas
Lomonosov Current 128
long waves *see* baroclinic waves; barotropic waves
long-wave radiation 9, 161
   in heat-budget equation 163, 164, *164*, 166, 167, 204, 206, 208, 231–2
low, atmospheric *see* cyclonic winds

Marianas Trench 55
marine geoid *see* geoid
Maud Rise 181
Maury, M. F. 76, 77, *77*
mean current flow 32, 57, 76
   see also mean motion
mean motion 63, *63*
Meddies 186
Mediterranean Sea 169
   outflow from 171
   residence time of water in 171, 226–7
Mediterranean Water 171, *171*, 172, *173*, *174*, 175, *177*, 180, 184, 186, 194, 229
mesoscale eddies 65–9, 71, 98, 108–12, *109*, *110*, *111*, 119, 137, 185–6, 200, 208, 232
   see also Mid-Ocean Dynamics Experiment
Mid-Ocean Dynamics Experiment 65, *66*, 72, 221
mid-ocean equatorial upwelling *see* upwelling
mixing in the ocean 185–7
   study of, using temperature–salinity diagrams *187*, 187–91, *189*, *190*, *191*, *193*, 194
   see also turbulence
MODE *see* Mid-Ocean Dynamics Experiment
mode water 174–5, 194–5
Mohn, H. 78
molecular viscosity 34, *34*, 219
monsoons 15, 125, 134–5, 155
Munk, W. 90, 92, *92*, 120

NADW *see* North Atlantic Deep Water
Nansen, F. 36

neutrally buoyant floats *see* Sofar floats
Nimbus satellite imagery 150, *181*
   see also Coastal Zone Color Scanner (CZCS) imagery
non-conservative properties 172
   as water-mass tracers 195–8, *196*, *197*
North Atlantic Central Water *174*, 175, 179, 180, 184, 188, 194
North Atlantic Current *see* North Atlantic Drift
North Atlantic Deep Water *177*, 178–80, *179*, *180*, 182, 184, *184*, 185, 191, *191*, *193*, 196–7, 203, 205, *207*, 208, 228, 229, 230, 231
   deep western boundary current of 196, 199
North Atlantic Drift *31*, 77, *77*, *97*, 98, 178
North Atlantic subtropical gyre 32, 60, 73, 79, 96, 119, 120, 121
   observations of 96–109
   theories about 84–93
   water mass formation in 173, 174, 175, 188, 205, 227
   see also eastern boundary currents; Gulf Stream; Sargasso Sea; western boundary currents
North-East Atlantic Deep Water 179, *179*
North-East Monsoon 134
   surface current flow in *136*, 137
North Equatorial Current *31*, 73, 96, *123*, 124, 129, 130, *130*, *131*
   in Indian Ocean 135, *136*
North-West Atlantic Bottom Water 179, *179*
Norwegian Sea 148, *149*, 178, *179*, 180, 183

OSCR *see* Ocean Surface Current Radar
Ocean Surface Current Radar 114
ocean circulation, charts of
   global *31*
   in Indian Ocean *136*
   in northern high latitudes *149*
   in southern high latitudes *152*
oceanic wave guides *see* equatorial wave guide; wave guides
open-ocean polynyas *see* polynyas
oxygen, as a water-mass tracer 195–7, *196*
Oyashio *31*, 148

Pacific
   Equatorial Counter-Current(s) 124
   equatorial current system 124–5
   Equatorial Undercurrent *see* Cromwell Current
   monsoonal circulation 15, 125, 134
   residence time of water in 198
   salinity in 168, 200, 185, 203–4, 230, 231
   sea-surface slope across 126, 127, *127*
   water masses in *174*, 175, *180*, 185, 203–4, 205, *207*, 231
Pacific and Indian Ocean Common Water 185, 205, *207*, 231
Peru Current *31*, 79, *152*
Peter Martyr of Angheira 73
phytoplankton distribution
   as an indicator of flow patterns 108, *109*, 110, 120
   as an indicator of upwelling 60, 115, *116*, 117, 154

Pillsbury, J. E. 78, 98, *99*
planetary vorticity 81–2, *82*, 83, 84, 89, 90, 119, 143
planetary waves *see* Rossby waves
Polar Current 148, *152*, 154, *184*
Polar Front *see* Antarctic Front
polynyas 181
   coastal (latent heat) 181–3, *182*
   open-ocean (sensible heat) 181–2, *181*, *182*, 183, 206, 230
Ponce de Leon 73
potential density *see* sigma-θ (σ_θ)
potential temperature, θ 192–4
potential vorticity 84
   as a water-mass tracer *194*, 194–5, *195*
   conservation of *see* Rossby waves; shelf waves
precipitation–evaporation balance *see* evaporation–precipitation ($E - P$) balance
pressure *see* hydrostatic pressure
pressure gradients *see* horizontal pressure gradients; hydrostatic pressure
principle of continuity *see* continuity, principle of
pycnocline 35, *53*, 124, 125
   see also thermocline
pycnostad 130, 175, 194

radiation *see* insolation; long-wave radiation; solar radiation
radiation balance
   at Earth's surface 161, *161*
   of Earth–atmosphere system 9–11
rainfall *see* evaporation–precipitation ($E - P$) balance
relative currents 50, 52
relative vorticity 81, 82, 83, 84, 85, 89, 90, 119
Rennell, J. 74, 75
residence time
   in deep ocean 198
   in Mediterranean Sea 171, 227
rings *see* Gulf Stream rings
Ross Sea *152*, 180
Rossby radius of deformation *141*, 141–2, 157, 225
Rossby waves 140, 143–5, *144*, 146
rotation of the Earth *see* Earth, rotation of

salinity, distribution of
   in Antarctic Circumpolar Current *153*
   in Gulf Stream 99, 101, *102*
salinity, effect on density of seawater 6, 48, 50, 101, 125, 153, 186
   in water mass formation 172, 175, 176, 178, 180, 182, 183, 184
salt, conservation of 168–72
salt fingering 186
Sandström, J. W. 51, 78
Sargasso Sea 77, *97*, 101, 110, 111, 115, 119, 174
   see also North Atlantic subtropical gyre
Sargassum weed 106
satellite imagery *see* altimetry, satellite; Coastal Zone Color Scanner (CZCS) imagery; *Nimbus* satellite imagery
Savonius rotor 105, *105*
sea-surface slopes 40, 42, 43, 44, 46, 48, 52, 53, 78, 99

across tropical oceans 125–6, *126, 127*, 137, 146, *147*, 155
  see also horizontal pressure gradients
sea-surface temperature
  global distribution *162*
  as an indicator of upwelling 117–18, *118*, 224
  influence on atmospheric convection 24, 26, 27, 148
  in region of Gulf Stream 74, 106–8, *107, 108, 110*
  in tropical Atlantic *133*
  seasonal variations
    in ice cover 149–51, *150*
    in sea-surface slope 137, 138, *138*, 139
    in thermal doming 132, 224
  see also monsoons
sea-ice *see* ice cover; ice formation; polynyas
sensible heat polynyas *see* polynyas
shear *see* current shear; wind shear
shelf waves 158, 225
short-wave radiation *see* solar radiation
sigma-*t* (σ_t) 192
sigma-θ (σ_θ) 192
silica, as a water-mass tracer 195–7, *196*
slippery sea 35
slope currents 50, 52
Sofar floats 65, *66*, 101, 103, 106, 140
  see also drifters
solar radiation 1, 9–10, *10*, 12, *159, 159*–62
  in heat-budget equation 163, 164, 166, 167
Somali Current 135–7, *136*, 139, 156, 157, 225
sound, speed of, use in current measurement 112–14, *113*
sound channel 140
South Atlantic, heat transport from 200
South Equatorial Current *31*, 96, *123*, 124, *130*, 131, 135, *135*, 155, 200, 223, 224
Southern Ocean 148, 151–4, *152*, 156
South-West Monsoon 134, 135, *135*
  surface current flow in 135
stability
  in atmosphere 21–3 *see also* convection in atmosphere
  in deep ocean 192
  in ocean 7, 125, 172, 174, 176, 178, 180, 183, 186
steady state of dissolved constituents in ocean 168
Stefan's law 161
Stommel, H. 87, *88*, 90, 95, 101, 199, *199*
storm surges 28
Straits of Florida 78, *97, 98, 99, 99*, 114
'stream currents' 75
subpolar gyre 148, 176
subtropical gyres 32, 60, 79
  convergence and sinking in 59–60, 173, 174, 175, 227 *see also* central water masses
  eastern boundary currents *31*, 79, 115–18, 119, 212, 221
  theories about 84–93
  western boundary currents 79, 84–92
  see also North Atlantic subtropical gyre; Sargasso Sea
Sub-Antarctic Front   *152, 152*, 153, *153*

surface waves 139 *see* barotropic waves
Sverdrup, H. U. 84, 85, 86, 87, 90, 120
Swallow floats *see* Sofar floats
Swallow, J. 101

temperature
  effect on density of air 15, 21, 22
  effect on density of seawater 6, 35, 48, 50, 52, 101, 125, 153, 186
  in situ 192
  potential 192–3
  vertical distribution in equatorial oceans *126, 127, 128, 129, 130*, 146, 147
  vertical distribution in Gulf Stream 78, *99*, 101, *102, 111*
  vertical distribution in Antarctic Circumpolar Current 153
  see also sea-surface temperature; temperature–salinity (*T–S*) diagrams
temperature inversion, Trade Wind 14, 22–3
temperature–salinity (*T–S*) diagrams *187*, 187–94, *188, 189, 190, 191, 193*, 208, 228–9
  volumetric 206–7, *207*
  oceanic mixing 185–95
  stability 192–5
thermal domes 130, *130, 131*, 132
thermocline 34
  as a barrier to mixing 34
  tropical 125, 126, *126, 127*, 130, 132, *133*, 138, *138*, 139, *139*, 140, 142, *146*, 147
  under convergences and divergences 59, *59*, 216
thermohaline circulation 7, 31, 184
thermostad *see* pycnostad
topographic steering 84, 219
topography, effect on flow patterns 84, 96, 97, 98, 111, 118, 119, 121, 148, *152*, 154, 158, 178, 179, *179*, 183, 219, 222, 225
torque of wind stress 85–6, 92, *92*, 120
tracers of water masses 195–9
Trade Winds 14, 15, 73, 74, 122
  clouds in 23
  collapse of, during El Niño 146, *146, 147*
  driving equatorial current systems 122–7, 142
  inversion 23
transport of heat *see* heat transport
tritium, ³H, as a water-mass tracer 198
tropical cyclones *see* cyclones, tropical
turbulence
  as a braking mechanism 35, 89, 92, 154, 157, 225
  atmospheric, above sea-surface 22, 33, 164, 166, 168, 204
  in ocean 33, 33–5
turbulent mixing 185–6
  see also eddies; eddy viscosity
typhoons *see* cyclones, tropical

upwelling 59, 60, 70
  associated with Kelvin waves 142–3
  in eastern boundary currents 115–18, *116*, 121, 131, *131*
  in high latitudes 154, 184, 231
  in low latitudes 131–3, 137, 155

in mid-ocean 59, *59*, 131
  see also thermal domes
vertical convection *see* convection
volume transports *see under* individual currents
vorticity 79–84
vorticity-balance calculations 89–92, *91*
  see also absolute vorticity; planetary vorticity; potential vorticity; relative vorticity

warm-core eddies *109*, 110–12, 119
water masses 172–95, 205
  analysis *see* temperature–salinity (*T–S*) diagrams
  bottom 178, *180*, 180–3, 185
  deep *177*, 178–83, *179*, 180, *180*, 185, 203–4
  intermediate *175*, 175–6, *177*
  upper *174*, 174–5
  see also under individual water masses
water spouts 28, 28–9
water type 187, 188, 189, 190
water vapour, atmospheric 11, 28, 29
  effect on density of air 21, 22
  effect on long-wave radiation 9, 161, 164
  effect on solar radiation 160, *160*
  effect on stability of air 21, 22, 26
wave guides 140–3, 144–5, 156
waves, long *see* baroclinic waves; barotropic waves
Weddell Sea 151, *152*, 153, 180, 181, *181*, 183, 205
West Wind Drift *see* Antarctic Circumpolar Current
Western Atlantic Sub-Arctic Water 175, *175*, 176, *177*
  see also Labrador Sea Water
Western North Atlantic Central Water 174, *174*, 175
western boundary currents 79, 84–9
  equatorward, in deep ocean 100–1, 196–7, 199, *199*, 230
  theories to explain 84–92
  see also Gulf Stream
winds, global 13, *14*, 15–16, *16*
winds *see* anticyclonic winds; cyclonic winds; monsoons; Trade Winds; wind stress
wind shear 27
wind stress
  curl/torque of 85–6, 92, *92*, 120
  effect on surface waters of 27, 33–8
  in equations of motion 94
  across equatorial ocean *127*
  and geostrophic currents 46
  in theoretical models of ocean circulation 85, 86, 89, 91, 92
  related to upwelling 27, 59, 115–7
windrows *see* Langmuir circulation
World Ocean Circulation Experiment 202, 208